COMPREHENSIVE SERIES IN PHOTOCHEMISTRY AND PHOTOBIOLOGY

Series Editors

Donat P. Häder
Professor of Botany

and

Giulio Jori
Professor of Chemistry

European Society for Photobiology

COMPREHENSIVE SERIES IN PHOTOCHEMISTRY AND PHOTOBIOLOGY

Series Editors: Donat P. Häder and Giulio Jori

Titles in this Series

Volume 1	UV Effects in Aquatic Organisms and Ecosystems Edited by E.W. Helbling and H. Zagarese
Volume 2	Photodynamic Therapy Edited by T. Patrice
Volume 3	Photoreceptors and Light Signalling Edited by A. Batschauer
Volume 4	Lasers and Current Optical Techniques in Biology Edited by G. Palumbo and R. Pratesi
Volume 5	From DNA Photolesions to Mutations, Skin Cancer and Cell Death Edited by É. Sage, R. Drouin and M. Rouabhia
Volume 6	Flavins: Photochemistry and Photobiology Edited by E. Silva and A.M. Edwards

COMPREHENSIVE SERIES IN PHOTOCHEMISTRY AND PHOTOBIOLOGY – VOLUME 6

Flavins

Photochemistry and Photobiology

Editors

Eduardo Silva and Ana M. Edwards
Biological Chemistry Laboratory,
P. Universidad Catolica de Chile,
Macul, Santiago,
Chile

RSC Publishing

The cover image shows a photolayse enzyme bound to a substrate, taken from Chapter 3, by C.W.M. Kay, A. Bacher, M. Fischer, G. Richter, E. Schleicher and S. Weber.

ISBN-10: 0-85404-331-4
ISBN-13: 978-0-85404-331-6

A catalogue record for this book is available from the British Library

Published by The Royal Society of Chemistry,
Thomas Graham House, Science Park, Milton Road,
Cambridge CB4 0WF, UK

Registered Charity Number 207890

For further information see our web site at www.rsc.org

Typeset by Macmillan India Ltd., Bangalore, India
Printed by Henry Ling Ltd, Dorchester, Dorset, UK

Preface for the ESP series in photochemical and photobiological sciences

"Its not the substance, it's the dose which makes something poisonous!" When Paracelsius, a German physician of the 14th century made this statement he probably did not think about light as one of the most obvious environmental factors. But his statement applies as well to light. While we need light for example for vitamin D production too much light might cause skin cancer. The dose makes the difference. These diverse findings of light effects have attracted the attention of scientists for centuries. The photosciences represent a dynamic multidisciplinary field which includes such diverse subjects as behavioral responses of single cells, cures for certain types of cancer and the protective potential of tanning lotions. It includes photobiology and photochemistry, photomedicine as well as the technology for light production, filtering and measurement. Light is a common theme in all these areas. In recent decades a more molecular centered approach changed both the depth and the quality of the theoretical as well as the experimental foundation of photosciences.

An example of the relationship between global environment and the biosphere is the recent discovery of ozone depletion and the resulting increase in high energy ultraviolet radiation. The hazardous effects of high energy ultraviolet radiation on all living systems is now well established. This discovery of the result of ozone depletion put photosciences at the center of public interest with the result that, in an unparalleled effort, scientists and politicians worked closely together to come to international agreements to stop the pollution of the atmosphere.

The changed recreational behavior and the correlation with several diseases in which sunlight or artificial light sources play a major role in the causation of clinical conditions (e.g. porphyrias, polymorphic photodermatoses, Xeroderma pigmentosum and skin cancers) have been well documented. As a result, in some countries (e.g. Australia) public services inform people about the potential risk of extended periods of sun exposure every day. The problems are often aggravated by the phototoxic or photoallergic reactions produced by a variety of environmental pollutants, food additives or therapeutic and cosmetic drugs. On the other hand, if properly used, light-stimulated processes can induce important beneficial effects in biological systems, such as the elucidation of several aspects of cell structure and function. Novel developments are centered around photodiagnostic and phototherapeutic modalities for the treatment of cancer, artherosclerosis, several autoimmune diseases, neonatal jaundice and others. In addition, classic research areas such as vision and photosynthesis are still very active. Some of these developments are unique to photobiology, since the peculiar physico-chemical properties of electronically excited biomolecules often lead to the promotion of reactions which are characterized by high levels of selectivity in space and time. Besides the biologically centered areas, technical developments have paved the way for the harnessing of solar energy to produce warm water and electricity or the development of environmentally friendly techniques for addressing problems of large social impact (e.g. the decontamination of polluted waters). While also in use in Western countries, these techniques are of great interest for developing countries.

The European Society for Photobiology (ESP) is an organization for developing and coordinating the very different fields of photosciences in terms of public knowledge and scientific interests. Due to the ever increasing demand for a comprehensive overview of the photosciences the ESP decided to initiate an encyclopedic series, the "Comprehensive Series in Photochemical and Photobiological Sciences". This series is intended to give an in-depth coverage over all the very different fields related to light effects. It will allow investigators, physicians, students, industry and laypersons to obtain an updated record of the state-of-the-art in specific fields, including a ready access to the recent literature. Most importantly, such reviews give a critical evaluation of the directions that the field is taking, outline hotly debated or innovative topics and even suggest a redirection if appropriate. It is our intention to produce the monographs at a sufficiently high rate to generate a timely coverage of both well established and emerging topics. As a rule, the individual volumes are commissioned; however, comments, suggestions or proposals for new subjects are welcome.

Donat-P. Häder and Giulio Jori
Spring 2002

Volume Preface

The photochemistry and photobiology of flavins is a field which has important implications for chemistry, biology, and medicine. In view of its interdisciplinary diversity a monograph covering the major aspects on this subject must by necessity be a multiauthored book. Therefore, it has been one of our fundamental motivations to secure contributions from experts who would provide the necessary broad coverage. Each one of the chapters provides an overview of historical developments, state-of-the-art, and future perspectives in different areas of interest related to flavins.

The first part of this book is devoted to the properties and applications of flavins in solution, with chapters dealing with the general properties of flavins, their photochemistry in aqueous and organic solvents, the interaction of the excited states of flavins with amines and their application to the initiation of vinyl polymerization. The following chapters address topics which underline the important role performed by flavin-promoted photoreactions in biological systems. such as riboflavin as a visible light sensitizer in the aerobic photodegradation of ophthalmic and sympathomimetic drugs, the antiviral and antibacterial effects observed in solutions exposed to light in the presence of riboflavin and their applications to blood safety and transfusion medicine. The possible role of flavins in light-induced toxicity with special focus on the eye lens is also discussed in depth.

The final part of the book deals with the special class of flavoproteins, for which the properties of the photoexcited states of the different redox states of their flavin cofactor are essential for their biological activities. There is one chapter dealing with the blue-light initiated DNA repair by photolyase, and three chapters devoted to the flavin-based photoreceptors in plants, bacteria and eukaryotic photosynthetic flagellates, respectively. The mechanisms of light activation in flavin-binding photoreceptors are analysed in the last chapter.

The editors remain very grateful to the individual contributors for their effort and enthusiasm in preparing their chapters. We also wish to thank Giulio Jori for his encouragement and for giving us the opportunity to coordinate the publication of this monograph. We also thank Manuela for her patience, help and understanding in the different steps of the preparation of the book.

Eduardo Silva and Ana M. Edwards

Contents

Chapter 1

General Properties of Flavins

ANA M. EDWARDS

Biological Chemistry Laboratory, Faculty of Chemistry, Pontificial Universidad Católica de Chile Casilla 306, Santiago, 6091144, Chile

Abstract

The general properties of flavins in free solution and when bound to flavoproteins as cofactors are analysed. The extremely high chemical versatility of flavins is reflected in the remarkable versatility of flavoproteins, when considered as a whole. However, each flavoprotein is also characterized by a strict specificity, thus implying that one of the most important roles of the protein component is to limit the wide range of possible flavin–protein interactions to those beneficial to the reaction to be catalysed.Many attempts have been made to achieve a rational classification of flavoproteins, depending on their different properties. However, when general classifications are based on the different possible catalytic mechanisms, in most cases there is uncertainty, owing to the wide range of reaction mechanisms potentially available to the flavin cofactor. Despite the significant accumulation of information in recent years on catalysis, biomimetics and structural studies, the factors that determine the specificity of flavoproteins are still poorly understood.

1.1. Introduction

The yellow-coloured compounds with the basic structure of 7,8-dimethyl-10-alkylisoalloxazine are generally termed as flavins. Flavins are ubiquitous in nature, and they take part in many biochemical reactions as coenzymes and photoreceptors. Riboflavin, the precursor of all the biologically important flavins, was first reported as lactochrome, a bright yellow pigment isolated from cow milk in 1879.[1] Later, in the late 1920s and early 1930s, yellow pigments with bright greenish fluorescence were isolated from different sources, and they were named as lactoflavin, ovoflavin, *etc.*, indicating the source from which they had been isolated. Concomitantly, it was recognized that the yellow pigment was a constituent of the vitamin B complex. Two important groups determined the structure and proved it by chemical synthesis.[2,3] The name riboflavin (RF) was given to the compound; it derives from the ribityl side chain and from the yellow-conjugated ring system (see Figure 1).

There is a broad distribution of flavins in tissues, but little is present as free RF. The majority is found in flavocoenzymes, mainly as flavin adenine dinucleotide (FAD), and in lesser amounts as flavin mononucleotide (FMN), the common name of riboflavin-5′-phosphate, despite the known fact that RF is

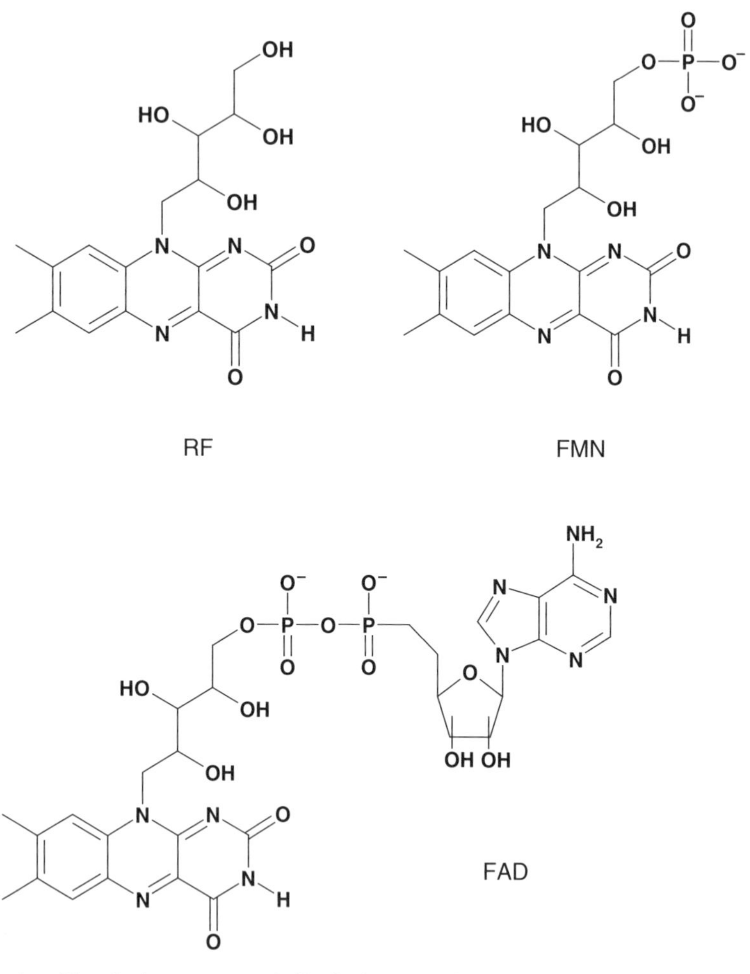

Figure 1. Chemical structures of riboflavin (RF), flavin momonucleotide (FMN) and flavin adenine dinucleotide (FAD)

not a real nucleoside because the linkage between the ribityl chain and the N_{10} of the flavin is not glycosidic; therefore, FMN and FAD are not real nucleotides. The structures are shown in Figure 1.

Since the pioneer study of Theorell,[4] who demonstrated in 1935 that the biochemical basis for the necessity of RF as a vitamin is its role as precursor of the FMN cofactor in enzyme catalysis (coenzyme), and those by Krebs[5] and Warburg[6] who showed its role as precursor of FAD cofactors, hundreds of flavoprotein enzymes have been known, and new ones are reported every year. Most of them contain non-covalent-bound FAD or FMN, and are specific for binding either of the two flavin forms as nature initially provided them with. Nowadays, there are many flavoproteins crystal structures known, which reveals that the majority of the flavin–protein interactions are with the N-10 ribityl side chain of FMN or FAD.[7] A recent study on the sequence-structure relationship in 32 families of FAD-containing proteins, showed that in every case the pyrophosphate moiety binds to the most strongly conserved sequence motif, suggesting that pyrophosphate binding is a significant component of molecular recognition.[8]

1.2. Properties of Flavins

The redox potential for the two electron reduction of the flavin is about −200 mV. However, this value can greatly vary in flavoproteins, due to the crucial role of the protein environment in the properties of flavins, spanning a range from approximately −400 mV to +60 mV. In general, the proximity of a positive charge is believed to increase the redox potential and a negative charge or a hydrophobic environment are expected to lower it.[9] A few flavoenzymes have a covalent-bound FAD molecule, and site-directed mutagenesis studies suggest that the covalent interaction could increase the oxidative power of the flavin.[10]

Since the discovery and characterization of RF and its derivatives FMN and FAD, they have been recognized by their ability of participate in both one- and two-electron transfer processes. This means that flavin molecules can exist in three different redox states: oxidized, one-electron reduced (semiquinone) and two-electron reduced states. Therefore, they can participate in redox reactions as either one- and two-electron mediator making the flavoenzymes very versatile in terms of substrate and type of catalysed reactions. This is a major reason for the ubiquity of flavin-dependent enzymes in biological systems. Flavins have the potential for transfer of single electrons, of hydrogen atoms and of hydride ions. Therefore, they can participate in redox reactions as either one- and two-electron mediator making the flavoenzymes very versatile in terms of substrate and type of reactions, which is a major reason for the ubiquity of flavin-dependent enzymes in biological systems. In addition, the oxidized flavin molecule is susceptible to nucleophilic attack, especially at N-5 and C-4α.[11]

In free solution (when not bound to an enzyme), the equilibrium of the different flavin species is pH dependent, as shown in Scheme 1 (proposed by Heelis, 1982).[12]

This Scheme presents the different redox states: oxidized, one-electron reduced (semiquinone) and two-electron reduced (fully reduced) and also the different protonation states for each of them. From the nine forms in Scheme 1, at least six are physiologically possible on the basis of their pK_a values.

At pH 7, only about a 5% radical is stabilized in an equimolar mixture of oxidized and reduced flavin. The semiquinone can exist in a neutral or anionic form, with a pK_a of ~ 8.5.

On binding to a specific protein, this equilibrium can change dramatically: some enzymes show no stabilization of semiquinone, while others give almost 100% stabilization. In some cases, if the protein can stabilize the neutral radical species over the whole range of pH values at which the enzyme is stable, the pK_a is shifted up significantly from 8.5. In other cases if the semiquinone anion is stabilized, the pK_a is decreased significantly. There are some enzymes, of which glucose oxidase was the first example,[13] which show such a pK_a that the identification of both forms is possible. In addition to these redox/ionic forms (each of them with different canonical forms), there are other electronic states, known as charge-transfer states. They are electronic states that do not belong to any of the three redox states, but are those in which partial charge is transferred to or from one of the three redox states. All these redox, ionic and charge-transfer states are the origin of the different colours of flavins and

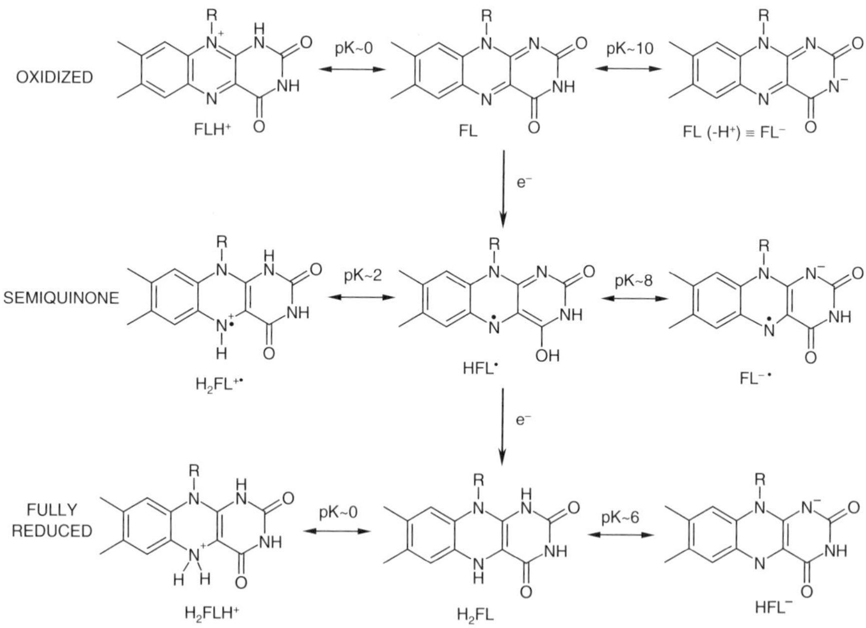

Scheme 1. Redox and acid-base equilibria of flavins

flavoproteins.[14] The large spectral differences between the various flavin redox–ionic–electronic states make it possible to monitor the events occurring in catalysis using the flavin itself as a reporter.

1.3. Classification of Flavoproteins

Flavins show an extremely high chemical versatility, which is reflected in the remarkable versatility of flavoenzymes; however, each of the enzymes is also characterized by a strict specificity. Many attempts have been made to achieve a rational classification of the many different types of flavoproteins, depending on their different properties.

In 1970, Hemmerich *et al.*[15] proposed four classes of flavoproteins according to their behaviour at half reduction in the absence of substrates.

(A) Flavoproteins exhibiting stoichiometric amounts of blue semiquinone
(B) Flavoproteins exhibiting stoichiometric amounts of red semiquinone
(C) Flavoproteins exhibiting no paramagnetic intermediates
(D) Flavoproteins exhibiting non-stoichiometric amounts of semiquinone.

The red radicals of class B have been identified as flavosemiquinone anions (Fl^{-}), and the blue radicals of class A as neutral semiquinones ($FlH^{\bullet}$). According to this classification, glucose oxidase belongs to class A and B, depending on pH. The authors proposed that in class A and B flavopropteins N-5 is protected by forming a strong hydrogen bond in the radical state, so that no "in plane" transfer of electrons can occur. These radicals are stable as long as the hydrogen bond at N-5 is not removed. They identified class D as metalloflavoproteins, where the radical yields at half reduction never exceed 50% of total flavin. Class C includes those flavoproteins that exhibit a cyst(e)ine residue participating in the catalysis. Both internal redox systems (flavin and disulfide) are in covalent contact at 2e-reduction of the enzyme. The point of attachment of the sulfur component to the flavin has been proposed to be C-4α.

Despite the fact that several hundred flavoenzymes have been described, Massey and Hemmerich[16] proposed that a large number of flavoproteins may be classified into two classes, based on distinct spectroscopic properties: regio-specific reactivities of the flavin skeleton and hydrogen bonding patterns with the apoprotein. The results reported by Shinkai *et al.* in 1985[17] support this proposal.

The first group (dehydrogenases/oxidases) is characterized by a "red" semiquinone radical, a bent structure of the reduced form, and hydrogen bonding to the N-1 nitrogen atom, activating the N-5 position.

The second group (electron transferases) is characterized by a blue semiquinone radical, a planar structure of the reduced form, and hydrogen bonding to N-5, activating of the C-4α position.

With the aim of addressing some characteristics of the two classes of flavoenzymes described above, Wouters *et al.*[18] determined the electronic properties of the three different protonated forms HFlox, H_2Flox^{+}(N-5) and

H_2Flox^+(N-1) using lumiflavin as model compound. Lumiflavin is a photoproduct of RF, in which the ribityl side chain is replaced by a methyl group. In order to describe the electrophilic/nucleophilic properties, they investigated the geometry, charge distribution, and HOMO–LUMO topologies of the three forms of lumiflavin adopting the RHF/3-21G level. The electronic spectra were evaluated at the CNDO/CI level with a particular attention to the blue or red shifts respect to the oxidized form. The authors concluded that a larger electron delocalization, demonstrated by a smaller bond length alternation upon protonation, as well as a larger basicity associated with a more negative charge can explain the favourable protonation of N-1 with respect to N-5, and the unusual electrophilic affinity of N-5 when lumiflavin is protonated at N-1. The calculations correctly place the $\pi-\pi^*$ transition as the lowest energy transition for neutral oxidized lumiflavin and predict a blue shift of the low-lying electronic transition upon monoprotonation (formation of the stable N-1 protonated form). They found in their calculation on radical intermediates of lumiflavin, a theoretical rationalization for the experimental classification of flavoenzymes in two distinct classes. Hydrogen bonding at N-5 and the fact that the semiquinone radical is red appear consistent with protonation at N-1 and larger wavelength absorption. Hydrogen bonding at N-1 and the fact that the semiquinone radical is blue are consistent with protonation of N-5 and a smaller wavelength absorption.

Among the known redox coenzymes, flavocoenzymes are unique in that they can participate in both one-electron and two-electron processes. The other redox cofactors usually catalyse exclusively either one- or two-electron processes. Active redox metalloenzymes catalyse only one-electron process, and nicotinamide nucleotides, with wide distribution in biological systems, are involved in only two electron redox reactions, because the radical forms of the pyridine ring are not sufficiently stable as to be involved in enzymatic reactions. For these reasons, flavoenzymes mediate two-electron and one-electron processes, as is the case of the well-known mitochondrial and chloroplast electron transport chains.

The reactions catalysed by a flavoenzyme always involve two separate half-reactions: reductive and oxidative half-reactions, both of which are necessary for the turn over of the enzyme. The former is the process in which a substrate or an electron donor is oxidized with the concomitant flavin reduction. In the latter process, the reduced flavin is oxidized by another substrate or an electron donor. Hemmerich *et al.*[19] proposed in 1977 a scheme consisting of five broad classes of flavoenzymes based on the nature of the substrate involved in the two separate half-reactions.

(i) Transhydrogenase, where two-electron equivalents are transferred, along with the appropriate hydrogen ions, from one organic substrate to another.
(ii) Dehydrogenase–oxidase, where two-electron equivalents are transferred to the flavin from an organic substrate, where molecular oxygen is the oxidizing substrate, being reduced to H_2O_2.

(iii) Dehydrogenase–monooxygenase, where the flavin is reduced generally by a reduced pyridine nucleotide, and where on oxidation with O_2 in the presence of a cosubstrate one atom of oxygen is inserted into the cosubstrate, while the other is reduced to H_2O.
(iv) Dehydrogenase–electron transferase, where the flavin is reduced by two-electron transfer from a reduced substrate and then reoxidized in sequential single electron transfers to acceptors, such as cytochromes and iron–sulphur proteins. This class might be further subdivided to distinguish those enzymes which function in the reverse sense, *i.e.*, those which receive electrons one at a time and then transmit them in a two-electron step in the reduction of a pyridine nucleotide.
(v) Electron transferase, where the flavin is reduced and reoxidized in one-electron steps.
(vi) This scheme was adopted by the International Union of Biochemistry.[20]

In the last decade, several interesting studies about flavins properties have been published. Recently, Miura has proposed an alternative classification of flavoenzymes based on the number of electrons involved in the catalytic cycle.[14] Because these half-reactions can be either one-electron or two-electron processes, there are four possible combinations of the two half-reactions; therefore, the author proposed that flavoenzymes can be classified according to the combination of one- and two-electron processes for reductive and oxidative half-reactions. Although four categories are theoretically possible; 1/1, 1/2, 2/1 and 2/2, the author points that flavoenzymes of the 1/2 category are not known at present. The author proposed that specificity of each flavoenzyme is understood in terms of the regulatory mechanism of the broad reactive potential of flavin. The regulatory mechanisms include hydrogen-bonding networks, electrostatic effects, charge transfer interactions, positioning between substrate and flavin, and modulation of resonance hybridization.[14]

The reaction between the fully reduced flavin H_2Fl and oxygen is essentially irreversible in free solution, according to the redox potentials, but it is also slow because of the spin inversion associated with the reaction of the singlet-reduced flavin and the triplet molecular oxygen. In flavoproteins, the reaction with oxygen may be orders of magnitude faster or slower, depending on the specific flavoprotein. They have been classified in four well-defined groups, according to the rate of their reactions with oxygen and the nature of the products formed, such as electron transferases, dehydrogenases or transhydrogenases, oxidases and monooxygenases. In 1994, Massey published an interesting review about the reactions of flavin and flavoproteins with oxygen.[21] The author discussed the common characteristics of the flavoproteins that are classified according their reactivity with oxygen, and proposed that they have many characteristic properties of a particular group, not shared with other groups of flavoproteins. Considering that the ratios of the rates of reaction with oxygen are at least 10^6 among flavoproteins, and that enormous differences in the decay rates of the hydroperoxides are found, the author also pointed out that there are important questions that remain to be answered What are the flavin–protein interactions that control the rate of reaction with oxygen? What factors determine whether the C-4α hydroperoxide should be formed? What are the factors that determine the stability of this hydroperoxide for different flavoenzymes?

In 1997, Wouters *et al.*[18] published a study of the electronic properties of lumiflavin as model compound. The authors found in their calculation a theoretical rationalization for the behaviour of flavoproteins and its classification in two groups, as described above.[18] They also gave an explanation, in terms of energy level differences, for the red shift observed for the first absorption band of the isoalloxazine ring upon complex formation with appropriate ligands. The favourable stacking observed in the crystal structures of lumiflavin–hydroquinone complexes was also rationalized in terms of the complementarity of the molecular electrostatic potential generated by the two molecules, and/or by the superposition of the HOMO of hydroquinone and the LUMO of the oxidized flavin.[18]

In 2000, Massey[7] published an interesting review, where the catalytic versatility of flavoproteins is explained by the catalytic mechanisms of selected flavoenzymes. The author studied the catalytic mechanisms of the different enzymes by replacing the native flavin with artificial flavins, such as 8-chloroflavins, 8-mercaptoflavins, and the photoreactive 6-azido-FAD. He also replaced the substrate with derivatives such as β-chloro amino acids or β-chloro α-hydroxyacids, and used the information available from crystal structures. He proposed possible reaction mechanisms for each of the groups of flavoenzymes: that catalyse the oxidation of α-hidroxyacids and α-amino acids, disulphide reductases, monooxygenases, and reductases, dehydrogenases and electron transferases (including the old yellow enzyme).

Flavoproteins are now known to have a variety of folding topologies. In 2000, Fraaije and Mattevi[9] examined the three-dimensional structures of flavoenzymes that were available in the protein data bank (PDB),[22] searching for recurrent features in their catalytic apparatus. They focused their analysis to the dehydrogenases, a group of enzymes that catalyse a reaction involving the rupture of a kinetically stable C–H bond, coupled with the transfer of two electrons to the flavin. The authors found that these enzymes share a few invariant features in the hydrogen-bond interactions between the protein and the flavin and that the positioning of the reactive part of the substrate with respect to the flavin is generally conserved. The authors recognized that the mechanistic problems cannot be solved solely on the basis of structural data; however, they pointed out that all the different mechanistic proposals (hydride transfer, radical mechanism and carbanion mechanisms) require, at least to some extent, juxtaposition between the flavin N-5-C4α locus and the reactive C–H group of the substrate. Although many of the reviewed enzymes are proposed to function via hydride transfer, there is no general consensus about the exact mechanisms for some of them. Nonetheless, the stereochemical principles underlying the mutual interactions between the substrate C–H group and the flavin are surprisingly well conserved, and they represent a validation test for the plausibility of any proposed mechanism, which must be compatible with the observed stereochemistry of substrate binding. The authors proposed that in the future the stereochemical requirements of each of the proposed mechanisms would be defined, to evaluate their compatibility with the three-dimensional structures.[9]

Dym and Eisenberg published in 2001[8] an interesting review, where they analyzed structure–sequences relationships in 32 families of FAD-binding proteins. Their work illustrates how similar cofactors are utilized by nature in a wide variety of proteins families. The availability of the three-dimensional structures of a large collection of FAD-proteins (150 flavoprotein X-ray structures solved in the presence of FAD were retrieved from PDB) allow to identify four different FAD-family folds, each with distinctive conserved sequence motifs. Furthermore, the FAD pocket shape can be distinguished from one fold to another.

In contrast to the diversity of the FAD fold and its sequence motif, the pyrophosphate moiety makes hydrogen bonds with residues from the most conserved sequence motifs of each fold family. The authors proposed that the pyrophosphate moiety is crucial for molecular recognition, whereas the isoalloxazine ring is involved in catalytic function and the adenine ring stabilizes cofactor binding, both rings interacting with protein residues in partially conserved sequence motifs. As a result, the isoalloxazine and adenine rings interact with different residues in the different proteins belonging to the same FAD-family fold. This is consistent with the variability in bond type (covalent or non-covalent) and cofactor conformation (elongated or bent) observed in the FAD-family folds. However, in each FAD-family fold there is a clear correlation between the fold, the shape and position of the pocket and the cofactor conformation. This is in agreement with the finding that the cofactor exhibits directionality, which is highly conserved within a family fold.[8]

The same year, Fitzpatrick[23] reported a review about recent progress on the elucidation of the mechanisms of oxidation of organic substrates of flavoenzymes that catalyse dehydrogenation of bonds between carbon and either nitrogen or oxygen. The work is focused on the oxidation of alcohols, amino and hydroxy acids, amines and nitroalkanes. The author combined information obtained from site-directed mutations, effect of inhibitors and substrate analogues, kinetic isotope effects and structural information, with an initial goal to develop unified mechanisms for each of the different families of flavoprotein oxidases. In most, if not all the cases, multiple mechanisms still remain viable despite intensive study. This uncertainty can be attributed to the wide range of reaction mechanisms potentially available to the flavin cofactor. The author concluded that radical and nucleophilic mechanisms must be considered in addition to the more accepted hydride transfer mechanism. The possibility that the reaction observed with substrate analogues may diverge from the normal catalytic mechanisms must also be considered.[23]

Our understanding of flavin chemistry has increased in the last years by the wealth of information obtained by catalysis and biomimetics as well as from structural studies of flavoproteins. However, as discussed in this work, versatility and specificity are essential concepts to describe flavoproteins/flavoenzymes.

The wide range of possible reactions give such a high versatility to these proteins as to make very complex any attempt to describe their behaviour by general patterns. However, each flavoprotein has a strict specificity, thus implying that one of the most critical roles of the protein component is to limit the whole range of possible flavin–protein interactions to those beneficial to the reaction to be catalysed. Despite the significant progress that has been made in recent years, the factors that determine the specificity of flavoproteins are still poorly understood, and many research groups are contributing to this area.

In this context, advances in the research on the photochemical and photobiological properties of flavins and flavoproteins are necessary for a complete understanding of these important compounds.

Acknowledgements

The financial support received from Fondecyt (Chile) Grant N°1040667 is acknowledged.

References

1. A. Wynter Blyth, The composition of cows' milk in health and disease, *J. Chem. Soc. Trans.*, 1879, **35**, 530–538.
2. R. Kuhn and F. Weygand, Synthetic vitamin B2, *Ber.*, 1934, **67B**, 2084–2085.
3. P. Karrer, K. Schopp and F. Benz, Synthesis of flavins, IV, *Helvetica Chim. Acta*, 1935, **18**, 426–429.
4. H. Theorell, Purification of the active group of the yellow enzyme, *Biochem. Z.*, 1935, **275**, 344–346.
5. H.A. Krebs, Metabolism of amino acids. III. Deamination of amino acids, *Biochem. J.*, 1935, **29**, 1620–1644.
6. O. Warburg and W. Christian, The yellow enzyme and its functions, *Biochem. Z.*, 1933, **266**, 377–411.
7. V. Massey, The chemical and biological versatility of riboflavin, *Biochem. Soc. Trans.*, 2000, **28**, 283–296.
8. O. Dym and D. Eisenberg, Sequence–structure analysis of FAD-containing proteins, *Prot. Sci.*, 2001, **10**, 1712–1728.
9. M.W. Fraaije and A. Mattevi, Flavoenzymes: diverse catalysts with recurrent features, *TIBS*, 2000, **25**, 126–132.
10. M.W. Fraaije, R.H. van den Heuvel, W.J. van Berkel and A. Mattevi, Covalent flavinylation is essential for efficient redox catalysis in vanillyl–alcohol oxidase, *J. Biol. Chem.*, 1999, **274**, 35514–35520.
11. F. Muller, *Free flavins: synthesis, chemical and physical properties*, in *Chemistry and Biochemistry of Flavoenzymes*, F. Muller (ed), **vol I**, CRC Press, Boca Raton, FL, 1991, 1–71.
12. P.F. Heelis, The photophysical and photochemical properties of flavins (isoalloxazines), *Chem. Soc. Rev.*, 1982, **11**, 15–39.

13. V. Massey and G. Palmer, On the existence of spectrally distinct classes of flavoprotein semiquinones. A new method for the quantitative production of flavoprotein semiquinones, *Biochemistry*, 1966, **5**, 3181–3189.
14. R. Miura, Versatility and specificity in flavoenzymes: control mechanisms of flavin reactivity, *Chem. Rec.*, 2001, **1**, 183–194.
15. P. Hemmerich, G. Nalgelschneider and C. Veerger, Chemistry and molecular biology of flavins and flavoproteins, *FEBS Lett.*, 1970, **8**, 69–83.
16. V. Massey and P. Hemmerich, Active-site probes of flavoproteins, *Biochem. Soc. Trans.*, 1980, **3**, 246–257.
17. S. Shinkai, N. Honda, Y. Ishikawa and O. Manabe, Coenzyme models. 41. On the unusual reactivities of *N*(5)-hydrogen-bonded flavin. An approach to regiospecific flavin activation through hydrogen bonding, *J. Am. Chem. Soc.*, 1985, **107**, 6286–6292.
18. J. Wouters, F. Durant, B. Champagne and J.M. André, Electronic properties of flavins: implications on the reactivity and absorption properties of flavoproteins, *Int. J. Quantum Chem.*, 1997, **64**, 721–733.
19. P. Hemmerich, V. Massey and H. Fenner, Flavin and 5-deazaflavin: a chemical evaluation of 'modified' flavoproteins with respect to the mechanisms of redox biocatalysis, *FEBS Lett.*, 1977, **84**, 5–21.
20. International Union of Biochemistry, *Nomenclature of Electron-Transfer Proteins in Enzyme Nomenclature, Recommendations 1984*, Academic Press, Orlando, FL, 1984.
21. V. Massey, Activation of molecular oxygen by flavins and flavoproteins, *J. Biol. Chem.*, 1994, **269**, 22459–22462.
22. F.C. Bernstein, T.F. Koetzle, G.J. Williams, E.F. Meyer Jr., M.D. Brice, J.R. Rodgers, O. Kennard, T. Shimanouchi and M. Tasumi, The protein data bank: a computer-based archival file for macromolecular structures, *Arch. Biochem. Biophys.*, 1978, **185**, 584–591.
23. P.F. Fitzpatrick, Substrate dehydrogenation by flavoproteins, *Acc. Chem. Res.*, 2001, **34**, 299–307.

Chapter 2

Photochemistry of Flavins in Aqueous and Organic Solvents

IQBAL AHMAD[a,b] AND FAIYAZ H.M. VAID[b]

[a] Dubai Pharmacy College, Dubai, P.O. Box 19099, United Arab Emirates
[b] Department of Pharmaceutical Chemistry, Faculty of Pharmacy, University of Karachi, Karachi 75270, Pakistan

Abstract

The photochemical aspects of flavins have been reviewed with reference to the analysis of flavins and photoproducts in degraded solutions, spectral and photophysical properties, photodegradation reactions, photosensitization reactions, and photostability of riboflavin in pharmaceutical preparations. The photochemistry of flavins in aqueous solution involves two major photodegradation reactions, viz intramolecular photoreduction and photoaddition, occurring separately or concomitantly in the presence of phosphate buffer, leading to the formation of lumichrome and cyclodehydroriboflavin, respectively. The photoreduction is catalyzed by H_2PO_4- ions and the photoaddition by HPO_4^{2-} ions. Intermolecular photoreduction also takes place in a large number of flavin-substrate interactions involving the flavin triplet state. Knowledge of the spectral and photophysical properties, and photochemical behaviour of flavins is necessary to understand the nature of flavin–protein interactions, flavin photosensitization reactions, and their role in biological processes. The study of the photochemical behaviour of riboflavin as a function of pH is a prerequisite to achieve optimum stability of the vitamin in pharmaceutical preparations. Technological processes are being developed to utilize riboflavin as a photosensitizer in the pathogenic inactivation of blood components. Riboflavin-sensitized photooxidation of amino acids in protein may have important biological implications to control photoinduced damage in biological systems. The application of femtosecond-resolved fluorescence spectroscopy and transient absorption techniques may provide useful information on the dynamics of flavoproteins.

2.1. Introduction

Flavins are among the most widely investigated class of compounds in view of their biological function,[1–3] photosensitivity,[4,5] and nutritional importance.[6,7] The photochemistry of flavins has been the subject of intense research over the past 50 years and has facilitated the understanding of their degradation mode, interactions, and biochemical role. Several reviews have been published in this field [8–15] and the previous work is well documented. A number of international symposia have been held on various aspects of flavins and flavoproteins, including the photochemistry, to take account of the progress made in specialized areas of research[16–23] and to understand the mechanisms involved in flavoprotein catalysis.

The development of laser-flash photolysis methodologies and their applications to flavin photochemistry has played a significant role in the investigation of flavin-mediated electron transfer reactions and in the understanding of redox interprotein and intraprotein mechanisms.[15,24] The applications of new spectroscopic techniques such as Stark spectroscopy, as well as those of the significantly refined resonance Raman and ultra-fast spectroscopy have facilitated the study of electron transfer reactions of flavins and flavoproteins.[25] The widely used spectroscopic techniques in the study of flavins and flavoproteins have been reviewed.[26] Subnanosecond-resolved fluorescence spectroscopy[27] and femtosecond-resolved fluorescence up-conversion and transient absorption techniques have been used to study the dynamics of flavoproteins.[28]

Riboflavin (RF) and its analogues exhibit strong absorption in the ultraviolet and visible region and act as sensitizers in the photodegradation of a wide range of substrates by free radical or singlet oxygen-mediated mechanisms.[29] The dye-sensitized reactions, such as the flavin-sensitized photooxidation reactions, are commonly termed as photodynamic action. They may cause damage to the biological system and have extensively been studied.[30] The use of nanomaterials as singlet oxygen photosensitizers and their potential applications in photodynamic therapy has recently been reviewed.[31] Quintero and Miranda[32] have discussed the mechanisms of photosensitization induced by drugs in detail.

There has not been much progress in the basic photochemistry of flavins during the last 20 years. Excellent reviews on the photochemical reactions of flavins, and related aspects have been published by Heelis.[13,14] The present review would mainly focus on the analytical aspects, spectroscopic and photophysical properties of flavins, developments in flavin photochemical reactions, some factors influencing photochemical reactions, flavin photosensitized reactions, and photostability of RF in pharmaceutical preparations.

2.2. Analysis of Riboflavin and Photodegradation Products

One of the major problems in the study of the stationary photolysis of RF and its analogues, such as 7,8-dimethyl-10-formyl-methylisoalloxazine (formylmethylflavin, FMF), an intermediate product in the photolysis of RF in aqueous solution,[33] has been the non-availability of suitable assay methods for the determination of these compounds and their photodegradation products. The structural similarity and spectral characteristics of flavins make it difficult to determine a particular compound or a mixture of these compounds simultaneously during photolysis unless a specific method is available. The various studies involving the determination of RF in photolyzed solutions at 445 nm did not take into consideration the interference caused by degradation products absorbing at this wavelength during photolysis. Thus, the analytical information or kinetic data obtained from these studies[33–43] may not be accurate.

The development of specific multicomponent spectrophotometric methods for the determination of RF and its major photodegradation products, FMF, lumichrome (LC), and lumiflavin (LF), or FMF and its photodegradation/hydrolysis products, LC and LF,[44] and their applications to the study of photolysis[45] and hydrolysis[46] of FMF and of photolysis of RF[47] lead to the evaluation of the kinetics of these reactions, and the determination of their rate-pH profiles. Similarly, a modification of the previously reported analytical procedure[48] enabled the determination of the photolysis (*i.e.*, FMF, LC, LF) as well as photoaddition (cyclodehydroriboflavin, CDRF) products of RF and to study the kinetics of simultaneous photolysis and photoaddition reactions of RF in aqueous solution.[48–50] The application of the method[47] to the determination of RF and photodegradation products, FMF, LC, and LF, during the photolysis at pH 9.0 using a low intensity 4.5 W Mazda lamp (Associated Electronic Industries, U.K.) is shown in Table 1, indicating the role of FMF as an intermediate product in this reaction. The assay method gives an almost constant molar balance, with time. The slightly higher values of molar balance (about 2%), compared to that of the initial concentration of RF, are due to the presence of some minor photodegradation products detected by TLC, and absorbing in the region of the analytical wavelengths (unpublished data).

Recently, the photochemical conversion of RF to LC in aqueous solution at pH 7.3 has been studied,[51] with the assumption that the absorption at 374 and

Table 1. Photolysis of 10^{-4} M riboflavin at pH 9.0 (Na borate/HCl buffer). Concentrations of riboflavin and photoproducts

Time (h)	RF(M × 10^5)	FMF(M × 10^5)	LC(M × 10^5)	LF(M × 10^5)	Total (M × 10^5)
0	10.00				10.00
1	9.11	0.78	0.27	0.08	10.24
2	8.19	1.36	0.58	0.12	10.25
3	7.30	1.63	1.12	0.17	10.22
4	6.54	1.87	1.58	0.21	10.20
5	5.70	2.11	2.15	0.25	10.21
6	5.22	2.10	2.65	0.28	10.25
7	4.74	1.90	3.27	0.31	10.22
8	4.38	1.85	3.64	0.33	10.20
9	3.91	1.78	4.18	0.34	10.21
10	3.57	1.68	4.60	0.35	10.20

446 nm during photolysis is only because of RF and LC and the concentrations of these compounds can be evaluated using the equations:

$$A_{374\,\text{nm}} = [\text{RF}]\varepsilon_{374\,\text{nm (RF)}} + [\text{LC}]\varepsilon_{374\,\text{nm (LC)}}$$
$$A_{446\,\text{nm}} = [\text{RF}]\varepsilon_{446\,\text{nm (RF)}} + [\text{LC}]\varepsilon_{446\,\text{nm (LC)}}$$

It may be pointed out that at pH 7.3 LF is also produced in the photolysis of RF and exhibits strong absorption at 374 and 446 nm. These wavelengths are in fact the absorption maxima of both RF and LF, whereas LC exhibits an absorption maximum at 356 nm[44,45,52] and has weak absorption at both 374 and 446 nm. Therefore, there may be considerable interference from LF absorption at the analytical wavelengths and the assay values of RF and LC during photolysis may be influenced.

2.3. Spectral and Photophysical Properties of Flavins

The electronic absorption spectrum of a flavin such as RF in aqueous solution exhibits four peaks at 445, 375, 265, and 220 nm. All the absorption maxima possess high-molar absorptivities ($> 10^4$ M^{-1} cm^{-1}), indicative of $\pi \rightarrow \pi^*$ transitions. The precise positions of the absorption maxima and the values of molar absorptivities depend on the environment of the flavin chromophore. Flavins exhibit an intense yellow-green fluorescence at 520 nm in aqueous solution. Only the neutral forms of flavins are fluorescent and their anions or cations are non-fluorescent.[13,53,54] The early studies on the spectral and photophysical properties of flavins have been reviewed previously.[4,11–14,55–57]

During the last decade, Sikorski and co-workers have carried out extensive studies on the spectroscopic and photophysical properties of flavins including RF and iso-(6,7)-riboflavin,[58] lumiflavin and 3-methyllumiflavin,[59,60] 3-ethyllumiflavin,[61] 3-benzyllumiflavin,[62] dimethyl- and tri-methylisoalloxazines,[60] LC and analogues,[59,63,64] 9-methylalloxazine,[65] methyl-, dimethyl-, trimethyl-, and tetramethyl-alloxazines,[60,66] isoalloxazines,[67,68] alloxazines,[68–70] using aqueous and organic solvents. These workers also studied the spectroscopic and photophysical properties of 1-methyllumichrome and mono- and dimethyl-alloxazines on cellulose.[71,72] The various spectroscopic and photophysical data for the singlet-excited and triplet-excited states of flavins are reported in Tables 2 and 3, respectively. The solvent effects on the absorption and emission bands of flavins may be explained on the basis of hydrogen-bonding interactions. Fluorescence quantum yields for LFs are an order of magnitude larger than those of LCs, due to their lower non-radiative rate constants. Substitution patterns of flavins have profound effects on both spectral and photophysical properties with varying fluorescence quantum yields.[59,60,64]

The electronic and structural properties of protonated flavins suggest the favourable protonation of N-1 with respect to N-5.[73] The absorption and fluorescence behaviour of RF shows that the molecule exists in three different forms depending on the pH of the solution. There is an equilibrium between the cationic form ($RFl_{ox}H_2^+$) and the neutral form ($RFl_{ox}H$) at low pH ($pK_c = 0.4$), and between the neutral form and the anionic form (RFl_{ox}^-) at high pH ($pK_a = 9.75$).[74] The photon-induced degradation and fluorescence in micro-particles and aqueous droplets of RF lead to the formation of a blue-fluorescence band around 420 nm.[75] The photoinduced excited-state double proton transfer reaction of substituted alloxazines shows that the fluorescence decay times of the alloxazinic forms in 1,2-dichloroethane are of the order of hundreds of picoseconds.[76] The solvent effect on the excited-state proton transfer of LC suggests the formation of a 1:1 lumichrome–acetic acid hydrogen-bonded complex in the ground state.[77]

Time-resolved fluorescence and fluorescence correlation spectroscopy have been used to study the dynamics and excited-state properties of flavins and flavoproteins.[78–82] The quantum yields of triplet formation of RF and FMN in aqueous solution have been determined as 0.375 and 0.225, respectively.[82] Resonance Raman spectroscopic studies have been conducted to investigate FMN photoreduction,[83] flavin–protein interactions,[84–86] and the structure of flavin radical complex of DNA photolysate.[87] NMR spectroscopy has been applied to study flavoproteins in their oxidized and reduced states,[88,89] flavo-enzyme-cofactor interactions,[90,91] and the magnetic field dependence of chemically induced dynamic nuclear polarization (CIDNP) of proteins in the presence of a flavin photosensitizer.[92]

2.4. Photochemical Reactions of Flavins

The main photochemical reactions of flavins involve intramolecular and intermolecular photoreduction, intramolecular and intermolecular photoaddition, and intramolecular photodealkylation. These reactions have been reviewed in detail by Hemmerich[10] and Heelis.[13,14] The nature and magnitude of the photochemical reactions of flavins depend on factors, such as solvent

Table 2. Spectroscopic and photophysical data for the singlet-excited state of selected flavins[a]

Compound	Solvent	λ_2 (nm)	λ_1 (nm)	λ_F (nm)	ϕ_F	τ_F (ns)	k_r (10^8 s^{-1})	Σk_{nr} 10^8 s^{-1})	Reference
Riboflavin	Water, pH 7				0.28	5.1	0.55	1.41	58
	Methanol	360	444	532	0.39	6.3	0.62	0.97	58,97,195
Iso-(6,7)-riboflavin	Methanol	343	447	552	0.20	4.2	0.48	1.9	58
Lumiflavin	Water, pH 6	367	445 (10930)	530	0.28	2.5	11.2	2.9	59,63
	Acetonitrile	342	443 (13400)	533	0.16	7.7	0.21	1.1	59
	Methanol	351	442 (12200)	531	0.13	6.8	0.19	1.3	59
	1,4-Dioxane	332	441 (15900)	531	0.19	9.1	0.21	0.89	59
3-Methyl-lumiflavin	Water, pH 6	364	444	520	0.15	4.5	0.33	1.8	59
	Acetonitrile	342	444 (10100)	531	0.17	7.1	0.24	1.2	59,60
	Methanol	351	444 (9900)	533	0.15	6.3	0.24	1.3	59
	1,4-Dioxane	334	442 (12000)	531	0.29	8.4	0.24	0.95	59
3-Benzyl-lumiflavin	Methanol	353	448 (13000)	534	0.10	5.8	0.17	1.6	62
Lumichrome	Water, pH 6	353 (10400)	385 (7600)	479	0.088	2.7	0.32	3.4	63
	Acetonitrile	334	380 (8300)	437	0.028	0.64	0.43	15.2	64
	Methanol	339	384 (7700)	453	0.032	1.04	0.30	9.3	64,98
	1, 4-Dioxane	327	379 (8910)	445	0.027	0.45	0.60	21	59
1-Methyl-lumichrome	Water, pH 6	354	386 (7200)	475	0.079	2.2	0.32	4.1	63
	Acetonitrile	334	379 (7600)	437	0.027	0.63	0.43	15	59
	Methanol	340	385 (7510)	453	0.037	0.94	0.35	10	59
	1,4-Dioxane	328	381 (8100)	445	0.028	0.51	0.55	19	59
3-Methyl-lumichrome	Water, pH 6	353	384 (7400)	470					59
	Acetonitrite	353	379 (8100)	436	0.026	0.64	0.41	15	59
	Methanol	340	383 (8000)	460	0.032	1.0	0.32	9.7	59
	1,4-Dioxane	328	379 (8700)	443	0.026	0.47	0.55	21	59

[a]λ_1 and λ_2 are the positions of the two lowest energy bands in the absorption spectra, molar absorptivities in parenthesis; λ_F is the fluorescence emission maximum; ϕ_F is the fluorescence quantum yield; τ_F is the fluorescence life time; k_r is the radiative rate constant; and Σk_{nr} is the sum of non-radiative rate constants.

Table 3. Spectroscopic and photophysical data for the triplet-excited state of lumiflavin and lumichromes[a]

Compound	Solvent	λ_3 (nm)	λ_2 (nm)	λ_1 (nm)	ϕ_T	τ_T (μs)	k_{ic} (10^8 s^{-1})	k_{isc} (10^8 s^{-1})	Reference
Lumiflavin	Water, pH 6	370	510	640	0.67	67	0.20	2.6	63,99,100
	Acetonitrile	370	510	640					68
Lumichrome	Water, pH 6					17	0.82	2.6	63
	Acetonitrile	360	450	540	0.68	11	0.46	1.0	64
1-Methyl-lumichrome	Water, pH 6				0.50	18	1.9	2.3	63
	Acetonitrile	370	460	560	0.66	6.9	0.62	1.3	64
3-Methyl-lumichrome	Acetonitrite	360	430	520		15			64

[a] λ_1, λ_2, and λ_3 are the positions of the maxima in the transient absorption spectra, ϕ_T is the quantum yield of intersystem crossing, the rate constant for internal conversion is k_{ic} and for intersystem crossing is k_{isc}.

polarity,[11] pH of the medium,[47,93] buffer kind and composition,[48,49,93] ionic strength,[41] oxygen content,[93,94] and light intensities and wavelengths,[39,50] and may proceed through the participation of both singlet-excited and triplet-excited states.[93,95] Since the appearance of the previous review on the photochemistry of flavins,[14] most of the work in this field has been carried out on intermolecular photoreduction and its relevance to flavin–protein and other interactions to understand the mechanism of the electron transfer reactions involved.[15,24] Some recent work on photodegradation reactions of RF includes the development of specific analytical methods for the determination of RF and its photodegradation products in complex mixtures (Section 2.2), determination of the rate-pH profiles for the photolysis of RF, study of the simultaneous photoreduction (normal photolysis) and photoaddition (intramolecular photoaddition) reactions of RF, and evaluation of the buffer and irradiation wavelength effects on photodegradation reactions.[47–50] These aspects would be presented in this section.

2.4.1. Intramolecular Photoreduction and Photoaddition Reactions

Intramolecular photoreduction is the basis of the normal photolysis of RF in aqueous solutions.[13,14,95] A recent study of the reaction has led to the understanding of the photochemical behaviour of RF over a wide range of pH.[47] However, in the presence of phosphate buffer and at a pH value >6, both intramolecular photoreduction and intramolecular photoaddition reactions occur simultaneously giving rise to a number of products, depending upon the reaction conditions. These reactions have recently been studied and the second-order rate constants for the mono-valent and di-valent phosphate ion-catalyzed photodegradation of RF have been determined.[48,49] Since the formation of CDRF by photoaddition is catalyzed by HPO_4^{2-} ions,[93] it has been suggested that $H_2PO_4^-$ ions may be involved in the formation of LC by normal photolysis.[49] Thus, the two phosphate species appear to catalyze the major reaction pathways (*i.e.*, photoaddition and photoreduction/photodealkylation) of RF photodegradation, respectively. In the light of the previous studies,[13,14,93] and the recent work,[47–50] the photodegradation of RF in neutral solution in the presence of phosphate buffer may be represented by the following major intramolecular processes.

Pathway A: Photoreduction/photodealkylation

$$\mathrm{RF} \xrightarrow{h\nu} [{}^1\mathrm{Fl}^*_{\mathrm{ox}}] \underset{k_1,\mathrm{H_2PO_4^-}}{\xrightarrow{ko}} \mathrm{LC} \tag{1}$$

$$[{}^1\mathrm{Fl}^*_{\mathrm{ox}}] \xrightarrow{isc} [{}^3\mathrm{Fl}^*_{\mathrm{ox}}] \underset{\text{phosphate}}{\longrightarrow} \underset{\text{leucodeuteroflavin}}{\mathrm{DFH_2}} \tag{2}$$

$$\mathrm{DFH_2} \xrightarrow{O_2} \mathrm{FMF} + \text{side-chain products} \tag{3}$$

$$\text{FMF} \xrightarrow{h\nu} \text{LC} + \text{side chain products} \tag{4}$$

$$\text{FMF} \xrightarrow{\text{H}^+,\,\text{OH}^-} \text{LC} + \text{LF} + \text{side-chain products} \tag{5}$$

The excited singlet state ($^1\text{Fl}^*_{\text{ox}}$) is considered to be the major species involved in photodealkylation of RF Equation (1).[95] The driving force for this reaction appears to be the polarity of $^1\text{Fl}^*_{\text{ox}}$. The participation of excited triplet state ($^3\text{Fl}^*_{\text{ox}}$) its reduction to leucodeuteroflavin Equation (2) leading to FMF Equation (3) and its conversion to LC Equation (4)[96] has also been suggested. FMF may also produce LC and LF by hydrolytic degradation Equation (5).[37,46]

Pathway B: photoaddition

$$\text{RF} \underset{\text{HPO}_4^{2-}}{\longrightarrow} \underset{\text{complex}}{\left[\text{RH} - \text{PO}_4^{2-}\right]} \xrightarrow{h\nu} \left[^1\text{Fl}^*_{\text{ox}}\right] \tag{6}$$

$$\left[^1\text{Fl}^*_{\text{ox}}\right] \xrightarrow{k_2} \underset{\text{Intermediate}}{[\text{Dihydroflavin}]} \xrightarrow{\text{autoxidation}} \text{CDRF} \tag{7}$$

The photoaddition of RF has been proposed to occur via the $\text{RF} - \text{HPO}_4^{2-}$ complex, which creates sterically favourable conditions for C (9)/O (2′α)- interaction.[93] The involvement of excited singlet state $^1\text{Fl}^*_{\text{ox}}$ in this reaction Equation (6) has been suggested on the basis of quenching experiments. The presence of HPO_4^{2-} ions may facilitate the re-orientation of C-2′ hydroxyl group to effect photoaddition. The autoxidation of dihydroflavin intermediate leads to the formation of CDRF Equation (7).

At constant pH and light intensity both LC and CDRF are produced in the presence of phosphate ions by photodegradation of RF through pathways A and B, respectively, the relative rates of the two reactions being dependent upon the phosphate concentration. The involvement of flavin $-\text{HPO}_4^{2-}$ complex in pathway B causes fluorescence quenching, and hence the suppression of pathway A involving the formation of LC directly from the excited singlet state Equation (1) or through the triplet state Equation (2).

2.4.2. Intermolecular Photoreduction Reactions

Flavins play a vital role in biological oxidation–reduction processes. The redox properties of flavins and flavoproteins determine the rates of electron-transfer reactions.[101,102] The mechanisms of flavoprotein-catalyzed reactions are complex[103–105] and initially involve the reduction of the isoalloxazine ring system in two steps via a semiquinone free radical intermediate.[106] The photoinduced electron-transfer reactions of flavins and deazaflavins have been dealt in detail.[107] The use of flavin photochemistry and the applications of laser-flash photolysis time-resolved spectrophotometry have facilitated the investigations of intraprotein and interprotein electron transfer mechanisms.[15,24] A “parallel plate" electrostatic model has been developed to investigate the ionic strength dependence of bimolecular rate constants as applied to electron-transfer proteins. Examples of mono- and di-phasic ionic strength dependencies of several electron protein systems are known.[108] Tollin and co-workers have done pioneering work on intermolecular photoreduction reactions of flavins and flavoproteins using laser-flash photolysis and time-resolved spectroscopic techniques. The details of these reactions are given below.

Laser-flash photolysis has been used to determine the second-order rate constants for the reduction of horse heart cytochrome *c* and cytochrome *c*–cytochrome oxidase complex by the semiquinone and fully reduced forms of various flavin analogues (FH· and FH^-, respectively). Substitution in the dimethylbenzene ring of the flavin causes appreciable changes in the rate constants, whereas substitutions at the N-10 position do not. A charged group attached to the dimethylbenzene ring produces a large ionic strength effect on the rate constants.[109] Both FH· and FH^- are effective in reducing the cytochrome *c*–cytochrome oxidase complex. The interaction site for electron transfer to cytochrome *c* is the same in the complex as with the free protein, and small reactants like the flavins are not sterically hindered in their access to the bound cytochrome *c*. The presence of cytochrome *c* greatly enhances electron transfer from reduced flavins to cytochrome oxidase.[110] A correlation has been found between the rate constants and the difference in redox potentials of the reactants for electron-transfer reactions between oxidized cytochromes and photoproduced semiquinones. The RF-cytochrome rate constants are about 70% of those for reduction by LF, probably because of steric hindrance by the ribityl side chain.[111]

The above work has been extended by studying the kinetics of reduction by free flavin semiquinone of the individual components of 1:1 complexes of yeast ferric and ferryl cytochrome *c* peroxidase and the cytochrome *c* of horse, tuna, and yeast. Complex formation decreases the rate constant for reduction of ferric peroxidase by 44%, suggesting a decrease in the dynamic motions of the peroxidase at the peroxidase access channel caused by complexation. Further, that the orientations of the three cytochromes within the complex are not equivalent.[112] The complex formed by covalent cross-linking between the peroxidase and cytochrome *c* at low ionic strength has kinetic properties both similar to and different from those of the electrostatic complex. In the electrostatic complex, a net positive charge is present, whereas in the covalent complex, an essentially neutral charge is present. For the covalent complex, intracomplex electron transfer rate is invariant to changes in the ionic strength, whereas for the electrostatic complex, the rate is highly dependent on ionic strength.[113] The second-order rate constants for the reduction of cytochrome *f* by various flavin semiquinones at extreme values of ionic strength and pH 7.0 show wide variation and are governed by the local positive charge of the interaction domain (the exposed heme edge).[114] The kinetics of reduction of the flavoprotein domain of P450BM-3 (BMR) with NADPH, suggests the occurrence of the consecutive first-order reaction: (a) complex formation between BMR and NADPH; (b) reduction of FAD with formation of the

$NADP^+$-FADH-charge-transfer complex; and (c) transfer of the first electron from FADH to FMN to form an ionic, red FMN semiquinone leaving the FAD as the neutral, blue semiquinone. The low potential of the ionic FMN semiquinone can be utilized to reduce the P450 heme[115] and is pH dependent.[116] The influence of an aromatic side chain of yeast iso-1-cytochrome *c* on the kinetics of its electron transfer reactions has been investigated to compare a series of site-specific mutant cytochromes in their reduction by free flavin semiquinone. Significant differences are observed between some of the mutants and wild-type cytochromes in the rate constants for reduction by LF semiquinone, which do not correlate with side chain aromaticity.[117] The rate constant for the oxidation of LF semiquinone increases dramatically with increasing pH for the redox proteins. The pH-rate profiles for the redox proteins closely follow the ionization of the proton at the N-5 position of the neutral LF semiquinone, suggesting a higher intrinsic reactivity of the anionic semiquinone.[118] The ionization potentials of the anionic semiquinone and hydroquinone states of flavins serve as accurate predictors of their one- and two-electron reduction potentials.[119]

Molecular orbital calculations (MINDO/3) on semiquinone and fully reduced LF suggest that the planar-reduced flavin has a smaller ionization potential than the bent-reduced flavin. This could account for the low redox potential of protein-bound reduced flavins.[120] The oxidized flavins are most stable in the planar configuration but also highly flexible about the N (5)–N (10) axis, requiring only 1 kcal mol^{-1} for a 10^0 bend. Reduced flavin geometry depends on methyl substitution pattern. The N (10) substituted forms are bent with typical fold angles around 155°, whereas the unsubstituted reduced form is planar.[121] The NMR shielding and electron correlation studies reveal remarkable behaviour of flavin N-5 redox centre. It shows a large decrease in shielding upon protonation of the oxidized flavin at N-1, and subsequent activation of N5 for electrophilic attack.[122] The flavin reactivity is controlled by hydrogen bonding, electrostatic effect, charge-transfer interaction, positioning between a substrate/ligand and flavin, and modulation of resonance hybridization.[3] The photoreaction between riboflavin tetra-acetate (RFTA) and nucleosides has been investigated using time-resolved infrared spectroscopy (TRIR), laser-flash photolysis, and density function theory calculations. The TRIR data for RFTA are in excellent agreement with the calculated spectra for the triplet flavin, and in the presence of silylated guanosine, with the formation of the most stable hydroflavin radical (RFTH·), by an electron transfer–proton transfer mechanism.[123] The photogenerated triplet states of RF, examined by time-resolved EPR spectroscopy at 80 K, exist in the non-equilibrated electron spin-polarized state. The protonated states of flavins can be distinguished by the kinetics of the transient EPR signals, the fastest decay being observed for the protonated RF triplets, followed by the deprotonated triplets.[124] The excited triplet state of LF is predicted to be 42 kcal mol^{-1} higher in energy than the singlet ground state, which compares to a solution- phase triplet energy of 49.8 kcal mol^{-1} of RF[125]. The dynamics of ultra-fast fluorescence quenching reactions of flavin chromophore in protein nanospace has been reported.[126] The methodology for global and target analysis of time-resolved spectra used in biological research has been reviewed.[127]

The electron transfer reactions of flavins show an increase in electrophilicity of flavin-excited states upon hydrogen bonding.[128] The excited singlet state of a deprotonated reduced flavin exhibits at pH 8 a broad absorption band with a maximum at $\sim$490 nm and a lifetime of 100 $\pm$ 15 ps.[129] In contrast to flavins, the triplet state of flavin enzyme mimic benzodipteridine is not more basic than the ground state, in agreement with the calculated shift in electron density upon excitation.[130] The reactivities of the triplet-excited states of flavins with different N (10)-phenyl substituents have been found to be very similar suggesting that the substituents are nearly perpendicular to the plain of the main ring due to their steric interactions with the main isoalloxazine ring system.[131] A time-resolved total internal reflection fluorescence study of the photoinduced redox cycle of RF in the presence of *N*,*N*-dioctadecyl-[1,3,5]triazine-2,4,6-triamine at a water/CCl_4 interface revealed a TIR fluorescence spectrum with a maximum at 480 nm. This maximum has been ascribed to 1,5-dihydroflavin (RFH_2).[132] The electron transfer reactions of diflavin reductases have recently been reviewed.[133]

The effects of solvent on the rates of flavin redox reactions have been studied by laser-flash photolysis. These include electron transfer to the flavin triplet state and the oxidation of flavin semiquinone by oxidized flavin radical, oxidized phenol radical, and quinone. The rate constant for triplet quenching by 2,6-dimethylphenol was proportional to the inverse of solvent viscosity. The flavin semiquinone yield was linearly dependent on the solvent dielectric implying the existence of a polar intermediate along the reaction pathway. The rate constants for semiquinone oxidation reactions exhibit a biphasic dependence on solvent dielectric, being virtually independent of dielectric at low values and sharply increasing at high values, indicating a change in mechanism with solvent polarity.[134]

A study of the aerobic photolysis of formylmethylflavin in aqueous and organic solvents has been conducted by the authors. The second-order rate constants for the bimolecular redox reaction are dependent upon the solvent dielectric constant (unpublished observation). This implies the existence of a polar intermediate along the reaction pathway as suggested earlier.[134]

2.5. Flavin-Sensitized Photoreactions

Flavins play an important role in photosensitized reactions involving a wide range of substrates, which act as electron donors. The bimolecular quenching rate constants for the reactions of flavin singlet-excited and triplet-excited states with various substrates are reported in Table 4. Silva *et al.* have carried out a large number of studies on flavin-sensitized photooxidation reactions of proteins and amino acids. The details of these studies along with other studies on flavin-sensitized reactions are given in the following sections.

Table 4. Bimolecular quenching rate constants (k_q 10^9 M^{-1} s^{-1}) for the reactions of flavin singlet-excited (^1RF) and triplet-excited (^3RF) states

Substrate	Solvent	^1RF	^3RF	Reference
Proteins				
Bovine serum albumin	Water		0.22	147
β-lactoglobulin			0.36	147
Cytochrome *c*	Water, pH 7.5		1.5	137
Cytochrome c_{552}			2.8	137
Cytochrome c_{552} Spinach plastocyanin			1.1	136
α-crystallin	Water, pH 7.4		3.31×10^{-5a}	135
β_H-crystallin			4.16×10^{-5a}	135
β_L-crystallin			7.33×10^{-5a}	135
Amino acids				
Cysteine	Water		12	147
Tryptophan			1.4	147
			3.0	135
			1.7	149
			3.7	196
Tyrosine			1.75	147
			1.4	149
N-acetyl tyrosine			1.0	150
Aliphatic acids				
Glycolate	Water, pH 5.0		1.6×10^{-4}	170
Glyoxalate			1.3×10^{-3}	170
Lactate			5.5×10^{-4}	170
Proprionate			0.9×10^{-4}	170
Pyruvate			1.4×10^{-2}	170
Sulfanilic acid and sulfonamides				
Sulfanilic acid	MeOH–water	4.17	2.13	43
Dapsone		6.52	1.40	43
Sulfadiazine		5.57	1.17	43
Sulfisoxazole		4.76	1.27	43
Aliphatic and aromatic amines				
Tri-*n*-butylamine	MeOH	4.7	1.5	195
Tripropylamine		3.0	0.85	195
Triethanolamine		4.5	0.34	195
Triethylamine		2.6	0.15	195
Di-*n*-butylamine		1.4	0.005	195
n-Butylamine		0.44	6.4	195
N, *N'*- Dimethylaniline		11.7	7.8	195
N-Methylaniline		12.1	4.3	195
Aniline		11.7	7.8	195
Flavones				
3-Hydroxyflavone	MeOH	3.5	2.3	42
7-Hydroxyflavone		3.0	0.53	42

Table 4 (*continued*)

Substrate	Solvent	^{1}RF	^{3}RF	Reference
Atropine derivatives				
Homatropine	Water	10^{-2}		157
Scopolamine		10^{-2}		157
Miscellaneous substances				
2-Amino-4-hydroxy-6-methylpyrimidine	Water, pH 6	2.4	0.027	181
Ascorbate	Water , pH 7		1.92	145
Isoproterenol	Water	2.54	1.5	158
Norflurazon	Water	0.9	0.059	182
Urocanic acid	Water, pH 7.4	6.7	0.2	156
O_2	Water		1.2	196
I^-	Water, pH 4.4		6.0	197
$K_3Fe(CN)_6$	Water, pH 7.4		1.74	156

[a]units: ml mg s^{-1}.

2.5.1. *Photooxidation*

The most widely studied reactions include the RF-sensitized photooxidation of proteins, such as eye lens crystallins,[135] plastocyanins,[136] cytochromes,[137] and amino acids in proteins,[30] amino acid content of parenteral nutrition infusion,[138] and individual amino acids such as tryptophan,[139–147] and tyrosine.[143,147–150] Other photooxidation reactions involve ascorbic acid,[145,146,151–154] glucose,[155] glucose 6-phosphate,[156] antimuscarinic drugs, homatropine and scopolamine,[157] sympathomimetic drug, isoproterenol,[158] anticancer drug, doxorubicin,[159] phenothiazines,[160] folic acid,[161] substituted phenols,[162] furocoumarin (psoralens),[163] bitter principles of beer, isohumulones,[164–166] constituents of Boldo and tea infusion,[167] bilirubin,[168] and lipids.[169]

2.5.2. *Photodecarboxylation*

Flavins have been found to sensitize the photodecarboxylation of α-substituted acetic acids by a radical mechanism.[170,171]

2.5.3. *Photoisomerization*

Bilirubin undergoes photoisomerization in the presence of RF. This is followed by photooxidation to degradation products of the tetrapyrrole skeleton via biliverdin as intermediate.[168]

2.5.4. *Photomonomerization*

The flavin-sensitized photomonomerization of the *cis*, *syn*-cyclobutane dimer of 1,3-dimethylthymine using RFTA and a 5-deazaflavin derivative has been studied.[172] A photochemically induced dynamic nuclear polarization (photo-CIDNP) study of the carboxymethyllumiflavin-sensitized splitting of pyrimidine dimers has been carried out.[173]

2.5.5. *Photodegradation*

Flavins are involved in photosensitized degradation of a number of compounds including hydroxyflavones,[42] sulpha drugs,[43,174] amines,[175] cyanocobalamin,[176–178] atrazine,[179,180] pyrimidine,[181] norflurazon,[182] retinoids,[183] DNA,[184] RNA,[185] and lactoglobulin.[186]

2.5.6. *Photoinactivation*

The action spectrum of RF-photosensitized inactivation of lambda phage has been reported.[51] RF is involved in the inactivation of blood components,[187] lysozyme,[188] dihydroorotate dehydrogenase,[189] and stress kinases.[190]

2.5.7. *Photoinduction*

RF forms photoinduced adducts with lens proteins, tryptophan, and indole-3-acetic acid.[191,192] The photoproducts of indole-3-acetic acid–riboflavin have been found to induce tumour cell death by an apoptotic mechanism. The effect is greater than those of the tryptophan–riboflavin photoproducts.[191] The development of monoclonal antibodies against riboflavin–tryptophan adduct has been reported.[192]

2.5.8. *Photomodification*

RF is involved in photosensitized modification of lens proteins. The irradiation of these proteins in the presence of RF leads to modification of the proteins with an increase in the proportion of the high molecular weight fraction.[193] RF-sensitized photodynamic modification of high molecular weight kininogen isolated from sheep plasma has been observed.[194]

2.6. Photostability of Riboflavin in Pharmaceutical Preparations

Another area of considerable interest is the evaluation of the photostability of RF in pharmaceutical products,[198] parenteral nutrition admixtures,[199–206] and food products,[207] and its photostabilization.[208–212] A knowledge of the photochemical behaviour of RF and its analogues in these products has important pharmaceutical implications for the formulator and is required to predict the shelf life of the vitamin under specified storage conditions. The photostability studies of light-sensitive compounds are an integral part of drug-development process in assuring the quality of both the active pharmaceutical ingredient and the drug product. These studies are described in International Conference on

Harmonization (ICH), Photostability Guidance and FDA Draft Stability Guidance,[213] and other literature.[214,215]

Acknowledgements

The authors are very grateful to Professor Eduardo Silva of the Departamento de Quimica Fisica, Facultad de Quimica, Pontificia Universidad Catolica de Chile, Santiago, Chile for providing the copies of his research publications and for his kind help in the preparation of the manuscript. They are also grateful to Professor Marek Sikorski of the Faculty of Chemistry, A. Mickiewicz University, Grunwaldzka, Poznan, Poland for sending his research publications and related material.

References

1. D.E. Edmondson and S. Ghisla, Flavoenzyme structure and function: approaches using flavin analogues, *Methods Mol. Biol.*, 1999, **131**, 157–179.
2. V. Massey, The chemical and biological versatility of riboflavin, *Biochem. Soc. Trans.*, 2000, **28**, 283–296.
3. R. Miura, Versatility and specificity in flavoenzymes: control mechanisms of flavin reactivity, *Chem. Rec.*, 2001, **1**, 183–194.
4. G.R. Penzer and G.K. Radda, Photochemistry of flavins, *Methods Enzymol.*, 1971, **18B**, 479–506.
5. K. Parfitt (ed), *Martindale The Complete Drug Reference*, Pharmaceutical Press, London, 1999, 1362.
6. H.J. Powers, Riboflavin and health, *Am. J. Clin. Nutr.*, 2003, **77**, 1352–1360.
7. D.B. McCormick, On becoming a nutritional biochemist, *Ann. Rev. Nutr.*, 2004, **24**, 1–11.
8. B. Holmstrom, Mechanism of the photoreduction of riboflavin, *Arkiv. Kemi*, 1964, **22**, 329–346.
9. G.R. Penzer and G.K. Radda, The chemistry and biological function of isoalloxazine (flavines), *Q. Rev.*, 1967, **21**, 43–65.
10. P. Hemmerich, The present status of flavin and flavocoenzyme chemistry, *Fortsch. Chem. Org. Naturst.*, 1976, **33**, 451–527.
11. P.S. Song, *Chemistry of flavins in their excited states*, in *Flavins and Flavoproteins*, H. Kamin (ed), University Park Press, Baltimore, 1971, 37–61.
12. F. Muller, Spectroscopy and photochemistry of flavins and flavoproteins, *Photochem. Photobiol.*, 1981, **34**, 753–759.
13. P.F. Heelis, The photophysical and photochemical properties of flavins (isoalloxazines), *Chem. Soc. Rev.*, 1982, **11**, 15–39.
14. P.F. Heelis, *The photochemistry of flavins*, in *Chemistry and Biochemistry of Flavoenzymes*, F. Muller (ed.), Vol. 1, CRC Press, Boca Raton FL, 1991, 171–193.
15. G. Tollin, Use of flavin photochemistry to probe intraprotein and interprotein electron transfer mechanisms, *J. Bioenerg. Biomembr.*, 1995, **27**, 303–309.
16. R.C. Bray, P.C. Angel and S.G. Mayhew (eds), *Flavins and Flavoproteins, 1984*, Proceedings of the 8th International Symposium, Brighton, UK, July 9–13, 1984, Walter de Gruyter, Berlin, 1984.

17. D.E. Edmondson and D.B. McCormick (eds), *Flavins and Flavoproteins, 1987*, Proceedings of the 9th International Symposium, Atlanta, GA, USA, June 7–12, 1987, Walter de Gruyter, Berlin, 1987.
18. B. Curti, S. Ronchi and G. Zanetti (eds), *Flavins and Flavoproteins, 1990*, Proceedings of the 10th International Symposium, Como, Italy, July 15–20, 1990, Walter de Gruyter, Berlin, 1991.
19. K. Yagi (ed), *Flavins and Flavoproteins, 1993*, Proceedings of the 10th International Symposium, Nagoya, Japan, July 27–31, 1993, Walter de Gruyter, Berlin, 1994.
20. K.J. Stevenson, V. Massey and C.H. Williams (eds), *Flavins and Flavoproteins 1996*, Proceedings of the 11th Symposium, Calgary, Alberta, Canada, June 30–July 6, 1996, University of Calgary Press, Calgary, 1997.
21. S. Ghisla, P. Kroneck, P. Macheroux and H. Sund (eds), *Flavins and Flavoproteins 1999*, Proceedings of the 13th International Symposium, Konstanz, Germany, August 29–September 4, 1999, Rudolph Weber Agency for Scientific Publcations, Berlin, 1999.
22. S. Chapman, R. Perham and N. Scrutton (eds), *Flavins and Flavoproteins 2002*, Proceedings of the 14th International Symposium, Cambridge, UK, July 14–18, 2002, Rudolph Weber Agency for Scientific Publication, Berlin, 2002.
23. T. Nishinno, R. Mura, M. Tanokura and K. Fukui (eds), *Flavins and Flavoproteins 2005*, Proceedings of the 15th International Symposium, Kanagawa, Japan, April 17–22, 2005, Architect Publishing, Tokyo, 2005.
24. G. Tollin, J.K. Hurley, J.T. Hazzard and T.E. Meyer, Use of laser flash photolysis time-resolved spectrophotometry to investigate interprotein and intraprotein electron transfer mechanisms, *Biophys. Chem.*, 1993, **48**, 259–279.
25. R.J. Stanley, Advances in flavin and flavoprotein optical spectroscopy, *Antiox. Redox Signal.*, 2001, **3**, 847–866.
26. *Methods in Molecular Biology, vol. 131: Flavoprotein Protocols*, S.K. Chapman and G.A. Reid (eds), Humana Press, Totowa, NJ, 1999.
27. P.A.W. van den Berg, K.A. Feenstra, A.E. Mark, H.J.C. Berendsen and A.J.W. G. Visser, Dynamic confirmation of flavin adenine dinucleotide: simulated molecular dynamics of the flavin cofactor related to the time-resolved characteristics, *J. Phys. Chem. B*, 2002, **106**, 8858–8869.
28. D. Zhong and A. Zewail, Femtosecond dynamics of flavoproteins: charge separation and recombination in riboflavin (vitamin B_2)-binding protein and in glucose oxidase enzyme, *Proc. Natl. Acad. Sci. USA*, 2001, **98**, 11867–11872.
29. M.B. Tayler and G.K. Radda, Flavins as photosensitizers, *Methods Enzymol.*, 1971, **18**, 496–506.
30. E. Silva, Sensitized photooxidation of amino acids in proteins: some important biological implications, *Biol. Res.*, 1996, **29**, 57–67.
31. S. Wang, R. Gao, F. Zhou and M. Silke, Nano materials and singlet oxygen photosensitizers: potential applications in photodynamic therapy, *J. Mater. Chem.*, 2004, **14**, 487–493.
32. B. Quintero and M.A. Miranda, Mechanisms of photosensitization induced by drugs: a general survey, *Ars Pharmaceutica*, 2000, **41**, 27–46.
33. E.C. Smith and D.E. Metzler, The photochemical degradation of riboflavin, *J. Am. Chem. Soc.*, 1963, **85**, 3285–3288.
34. M. Halwer, The photochemistry of riboflavin and related compounds, *J. Am. Chem. Soc.*, 1951, **73**, 4870–4874.
35. B. Holmstrom, Spectral studies of the bleaching of riboflavin phosphate, *Arkiv. Kemi.*, 1964, **22**, 281–301.

36. C.K. Radda and M. Calvin, Chemical and photochemical reduction of flavin nucleotides and analogues, *Biochemistry*, 1964, **3**, 384–393.
37. P.S. Song, E.C. Smith and D.E. Metzler, Photochemical degradation of flavins. IV. The mechanism of alkaline hydrolysis of 6,7-dimethyl-9-formylmethylisoalloxazine, *J. Am. Chem. Soc.*, 1965, **87**, 4181–4184.
38. M.M. McBride and M.W. Moore, The photochemistry of riboflavin. II. Polarographic studies on the hydroxyethyl and formylmethyl analogs of riboflavin, *Photochem. Photobiol.*, 1967, **6**, 103–113.
39. Y. Sato, M. Yokoo, S. Takahashi and T. Takahashi, Biphasic photolysis of riboflavin with a low-intensity light source, *Chem. Pharm. Bull. (Jpn)*, 1982, **30**, 1803–1810.
40. Y. Sato, S. Yamasato and Y. Suzuki, Biphasic photolysis of riboflavin. II. Effect of gelatin on the photolysis, *Chem. Pharm. Bull. (Jpn)*, 1983, **31**, 4167–4171.
41. Y. Sato, H. Chaki and Y. Suzuki, Biphasic photolysis of riboflavin. III. Effect of ionic strength on the photolysis, *Chem. Pharm. Bull. (Jpn)*, 1984, **32**, 1232–1235.
42. P. Montana, N. Pappano, N. Debattista, V. Avila, A. Posadaz, S.G. Bertolotti and N.A. Garcia, The activity of 3- and 7-hydroxyflavones as scavengers of superoxide radical anion generated from photo-excited riboflavin, *Can. J. Chem.*, 2003, **81**, 909–914.
43. M. Diaz, M. Luiz, S.G. Bertolotti, S. Miskoshi and N.A. Garcia, Scavenging of photogenerated singlet molecular oxygen and superoxide radical anion by sulpha drugs – kinetics and mechanism, *Can. J. Chem.*, 2004, **82**, 1752–1759.
44. I. Ahmad and H.D.C. Rapson, Multicomponent spectrophotometric assay of riboflavin and photoproducts, *J. Pharm. Biomed. Anal.*, 1990, **8**, 217–223.
45. P.F. Heelis, G.O. Phillips, I. Ahmad and H.D.C. Rapson, The photodegradation of formylmethylflavin – a steady state and laser flash photolysis study, *Photobiochem. Photobiophys.*, 1980, **1**, 125–130.
46. I. Ahmad, H.D.C. Rapson, P.F. Heelis and G.O. Phillips, Alkaline hydolysis of 7,8-dimethyl-10-(formylmethyl) isoalloxazine. A kinetic study, *J. Org. Chem.*, 1980, **45**, 731–733.
47. I. Ahmad, Q. Fasihullah, A. Noor, I.A. Ansari and Q.N.M. Ali, Photolysis of riboflavin in aqueous solution: a kinetic study, *Int. J. Pharm.*, 2004, **280**, 199–208.
48. I. Ahmad, Q. Fasihullah and F.H.M. Vaid, A study of simultaneous photolysis and photoaddition reactions of riboflavin in aqueous solution, *J. Photochem. Photobiol. B: Biol.*, 2004, **75**, 13–20.
49. I. Ahmad, Q. Fasihullah and F.H.M. Vaid, Effect of phosphate buffer on photodegradation reactions of riboflavin in aqueous solution, *J. Photochem. Photobiol. B: Biol.*, 2005, **78**, 229–234.
50. I. Ahmad, Q. Fasihullah and F.H.M. Vaid, Effect of light intensity and wavelengths on photodegradation reactions of riboflavin in aqueous solution, *J. Photochem. Photobiol. B: Biol.*, 2006, **82**, 21–27.
51. C.B. Martin, E. Wilfong, P. Ruane, R. Goodrich and M. Platz, An action spectrum of the riboflavin-photosensitized inactivation of lambda phage, *Photochm. Photobiol.*, 2005, **81**, 474–480.
52. R.M.C. Dawson, D.C. Elliott, W.H. Elliott and K.M. Jones, *Data for Biochemical Research*, Clarendon Press, Oxford, 1986, 137.
53. G. Weber, Fluorescence of riboflavin and flavin adenine dinucleotide, *Biochem. J.*, 1950, **47**, 114–121.

54. A.J.W.G. Visser and F. Muller, Absorption and fluorescence studies on neutral and cationic isoalloxazine, *Helv. Chim. Acta*, 1979, **62**, 593–608.
55. M. Sun, T.A. Moore and P.S. Song, Molecular luminescence studies of flavins. I. The excited states of flavins, *J. Am. Chem. Soc.*, 1972, **94**, 1730–1740.
56. W. Ostrowski (ed), *Flavins and Flavoproteins – Physicochemical Properties and Function, 1976*, Proceedings of the 6th International Symposium, Cracow, Poland, September 28–29, 1976, Polish Scientific Publishers, Warsaw, 1977.
57. J. Koziol, A. Koziolowa, J. Konarski, D. Panek-Janc and J. Dawidowski, in *Flavins and Flavoproteins*, K. Yagi and T. Yamano (eds), Japan Scientific Societies Press, Tokyo, 1980, 475–484.
58. E. Sikorska, I. Khmelinskii, A. Komasa, J. Koput, L.F.V. Ferreira, J.R. Herance, J.L. Bourdelande, S.L. Williams, D.R. Worrall, M. Insinska-Rak and M. Sikorski, Spectroscopy and photophysics of flavin related compounds: riboflavin and iso-6,7)-riboflavin, *Chem. Phys.*, 2005, **314**, 239–247.
59. E. Sikorska, I.V. Khmelinskii, W. Prukala, S.L. Williams, M. Patel, D.R. Worrall, J.L. Bourdelande, J. Koput and M. Sikorski, Spectroscopy and photophysics of lumiflavins and lumichromes, *J. Phys. Chem. A.*, 2004, **108**, 1501–1508.
60. E. Sikorska, I.V. Khmelinskii, D.R. Worrall, J. Koput and M. Sikorski, Spectroscopy and photophysics of iso- and alloxazines: experimental and theoretical study, *J. Fluorescence*, 2004, **14**, 57–64.
61. E. Sikorska, J.R. Herance, J.L. Bourdelande, I.V. Khmelinskii, S.L. Williams, D.R. Worrall, G. Nowacka, A. Komasa and M. Sikorski, Spectroscopy and photophysics of flavin-related compounds: 3-ethyllumiflavin, *J. Photochem. Photobiol. A: Chem.*, 2005, **170**, 267–272.
62. M. Insinska-Rak, E. Sikorska, J.R. Herance, J.I. Bourdelande, I.V. Khmelinskii, M. Kubicki, W. Prukala, I.F. Machado, A. Komasa, L.F.V. Ferreira and M. Sikorski, Spectroscopy and photophysics of flavin-related compounds: 3-benzyl-lumiflavin, *Photochem. Photobiol. Sci.*, 2005, **4**, 463–468.
63. M. Sikorski, E. Sikorska, A. Koziolowa, R.G. Moreno, J.L. Bourdelande, R.P. Steer and F. Wilkinson, Photophysical properties of lumichrome in water, *J. Photochem. Photobiol. B: Biol.*, 2001, **60**, 114–119.
64. E. Sikorska, D.R. Worrall, J.L. Bourdelande and M. Sikorski, Photophysics of lumichrome and its analogs, *Polish J. Chem.*, 2003, **77**, 65–73.
65. E. Sikorska, I.V. Khmelinskii, W. Prukala, S.L. Williams, D.R. Worrall, J.L. Bourdelande, A. Bednarek, J. Koput and M. Sikorski, Spectroscopy and photophysics of 9-methylalloxazine. Experimental and theoretical study, *J. Mol. Struct.*, 2004, **689**, 121–126.
66. E. Sikorska, I.V. Khmelinskii, S.L. Williams, D.R. Worrell, J.R. Herance, J.L. Bourdelande, J. Koput and M. Sikorski, Spectroscopy and photophysics of 6,7-dimethyl-alloxazine: experimental and theoretical study, *J. Mol. Struct.*, 2004, **697**, 199–205.
67. E. Sikorska, I.V. Khmelinskii, J. Koput, J.L. Bourdelande and M. Sikorski, Electronic structure of isoalloxazines in their ground and excited states, *J. Mol. Struct.*, 2004, **697**, 137–141.
68. E. Sikorska, M. Sikorski, R. P. Steer, F. Wilkinson and D.R. Worrall, Efficiency of singlet oxygen generation by alloxazines and isoalloxazines, *J. Chem. Soc., Faraday Trans.*, 1998, **94**, 2347–2353.
69. M. Mir, E. Sikorska, M. Sikorski and F. Wilkinson, Study of the effect of β-cyclodextrin on the photophysics of alloxazines in the solid state, *J. Chem. Soc., Perkin Trans.*, 1997, **2**, 1095–1098.

70. M. Sikorski, E. Sikorska, F. Wilkinson and R.P. Steer, Studies of the photophysics and spectroscopy of alloxazine and related compounds in solution and in the solid state, *Can. J. Chem.*, 1999, **77**, 472–480.
71. M. Sikorski, Photophysics of dimethyl-alloxazines and 1-methyllumichrome on cellulose, *Phys. Chem. Chem. Phys.*, 2002, **4**, 211–215.
72. M. Sikorski, E. Sikorska, I.V. Khmelinskii, R. Gonzalez-Moreno, J.L. Bourdelande and A. Siemiarczuk, Photophysics of alloxazines on cellulose, *Photochem. Photobiol. Sci.*, 2002, **1**, 715–720.
73. J. Wouters, F. Durant, B. Champagne and J.-M. Andre, Molecular structure, dynamics, and function of biological systems. Electronic properties of flavins: implications on the reactivity and absorption properties of flavoproteins, *Int. J. Quant. Chem.*, 1997, **64**, 721–733.
74. P. Drossler, W. Holzer, A. Penzkofer and P. Hegemann, pH dependence of the absorption and emission behaviour of riboflavin in aqueous solution, *Chem. Phys.*, 2002, **282**, 429–439.
75. Y.-L. Pan, R.G. Pinnick, S.C. Hill, S. Niles, S. Holler, J.R. Bottiger, J.-P. Wolf and R.S. Chang, Dynamics of photon-induced degradation and fluorescence in riboflavin microparticles, *Appl. Phys. B: Lasers Optics*, 2001, **72**, 449–454.
76. E. Sikorska and A. Koziolowa, Excited state proton transfer of methyl- and cyano-substituted alloxazines in the presence of acetic acid, *J. Photochem. Photobiol. A: Chem.*, 1996, **95**, 215–221.
77. E. Sikorska, A. Koziolowa, M. Sikorski and A. Siemiarczuk, The solvent effect on the excited-state proton transfer of lumichrome, *J. Photochem. Photobiol. A: Chem.*, 2003, **157**, 5–14.
78. R. Leenders, M. Kooijman, A. van Hoek, C. Veeger and A.J.W.G. Visser, Flavin dynamics in reduced flavodoxins. A time-resolved polarized fluorescence study, *Eur. J. Biochem.*, 1993, **211**, 37–45.
79. P.A.W. van den Berg, A. van Hoek, C.D. Walentas, R.N. Perham and A.J.W.G. Visser, Flavin fluorescence dynamics and photoinduced electron transfer in *Escherichia coli* glutathione reductase, *Biophys. J.*, 1998, **74**, 2046–2058.
80. L.J. Kricka and P.E. Stanely, Time-resolved fluorescence: 1996–1998, *Luminescence*, 1999, **14**, 47–61.
81. P.A.W. van den Berg, J. Widengren, M.A. Hink, R. Rigler and A.J.W.G. Visser, Fluorescence correlation spectroscopy of flavins and flavoenzymes: photochemical and photophysical aspects, *Spectrochim. Acta Part A*, 2001, **57**, 2135–2144.
82. S.D.M. Islam, A. Penzkofer and P. Hegemann, Quantum yield of triplet formation of riboflavin in aqueous solution of flavin mononucleotide bound to the LOV1 domain of Phot1 from *Chlamydomonas reinhardtii*, *Chem. Phys.*, 2003, **291**, 97–114.
83. M. Sakai and H. Takahashi, One-electron photoreduction of flavin mononucleotide: time-resolved resonance Raman and absorption study, *J. Mol. Struct.*, 1996, **379**, 9–18.
84. A.J. Visser, J. Vervoot, D.J. O'Kane, J. Lee and L.A. Carreira, Raman spectra of flavin bound in flavodoxins and in other flavoproteins. Evidence for structural variations in the flavin-binding region, *Eur. J. Biochem.*, 1983, **131**, 639–645.
85. T. Picaud and A. Desbois, Electrostatic control of the isoalloxazine environment in the two-electron reduced states of yeast glutathione reductase, *J. Biol. Chem.*, 2002, **277**, 31715–31721.
86. Y. Zheng and P.R. Carey, Raman spectrum of fully reduced flavin, *J. Raman Spectrosc.*, 2004, **35**, 521–524.

87. D.H. Murgida, E. Schieicher, A. Bacher, G. Richter and P. Hildebrandt, Resonance Raman spectroscopic study of the neutral flavin radical complex of DNA photolyase from *Escherichia coli*, *J. Raman Spectrosc.*, 2001, **32**, 551–556.
88. K.J. Griffin, G.D. Degala, W. Eisenreich, F. Muller, A. Bacher and F. Freeman, ^{31}P-NMR spectroscopy of human and *Paracoccus denitrificans* electron transfer flavoproteins, and ^{13}C- and ^{15}N-NMR spectroscopy of human electron transfer flavoprotein in the oxidized and reduced states, *Eur. J. Biochem.*, 1998, **255**, 125–132.
89. C.G. Van Schagen and F. Mullar, A comparative ^{13}C-NMR study on various reduced flavins, *Helv. Chim. Acta*, 2004, **63**, 2187–2201.
90. A. Niemz and V.M. Rotello, Model systems for flavoenzyme activity. The effects of specific hydrogen bonds on the ^{13}C- and ^{1}H-NMR of flavins, *J. Mol. Recognit.*, 1999, **9**, 158–162.
91. G. Fleischmann, F. Lederer, F. Muller, A. Bacher and H. Ruterjans, Flavin–protein interactions in flavocytochrome b_2 as studied by NMR after reconstitution of the enzyme with ^{13}C- and ^{15}N-labelled flavin, *Eur. J. Biochem.*, 2000, **267**, 5156–5167.
92. C.E. Lyon, J.J. Lopez, B.-M. Cho and P.J. Hore, Low field CIDNP of amino acids and proteins: characterization of transient radicals and NMR sensitivity enhancement, *Mol. Phys.*, 2002, **100**, 1261–1269.
93. M. Schuman Jorns, G. Schollnhammer and P. Hemmerich, Intramolecular addition of the riboflavin side chain. Anion-catalysed neutral photochemistry, *Eur. J. Biochem.*, 1975, **57**, 35–48.
94. G.E. Treadwell, W.L. Cairns and D.E. Metzler, Photochemical degradation of flavins. V. Chromatographic studies of the products of photolysis of riboflavin, *J. Chromatogr.*, 1968, **35**, 376–388.
95. W.L. Cairns and D.E. Metzler, Photochemical degradation of flavins. VI. A new photoproduct and its use in studying the photolytic mechanism, *J. Am. Chem. Soc.*, 1971, **93**, 2772–2777.
96. P.S. Song and D.E. Metzler, Photochemical degradation of flavins. IV. Studies of the anaerobic photolysis of riboflavin, *Photochem. Photobiol.*, 1967, **6**, 691–709.
97. P. Drossler, W. Holzer, A. Penzkofer and P. Hegemann, Fluorescence quenching of riboflavin in aqueous solution by methionin and cystein, *Chem. Phys.*, 2003, **286**, 409–420.
98. S.G. Bertolotti, C.M. Previtali, A.M. Rufs and M.V. Encinas, Riboflavin/triethanolamine as photoinitiator system of vinyl polymerization. A mechanistic study by laser flash photolysis, *Macromolecules*, 1999, **32**, 2920–2924.
99. P.F. Heelis and G.O. Phillips, A laser flash photolysis study of the triplet states of lumichrome, *J. Phys. Chem.*, 1985, **89**, 770–774.
100. M.S. Grodowski, B. Veyret and K. Weiss, Photochemistry of flavins. II. Photophysical properties of alloxazines and isoalloxazines, *Photochem. Photobiol.*, 1977, **26**, 341–352.
101. M.T. Stankovich, *Redox properties of flavins and flavoproteins*, in *Chemistry and Biochemistry of Flavoenzymes*, F. Muller (ed.), Vol. 1, CRC Press, Boca Raton FL, 1991, 401–425.
102. D.B. McCormick, *Oxidation–reduction reactions*, in *Encyclopedia of Life Sciences*, Nature Publishing Group, New York, 2001/www.els.net.
103. S. Ghisla and V. Massey, Mechanisms of flavoprotein-catalysed reactions, *Eur. J. Biochem.*, 1989, **181**, 1–17.
104. S. Bornemann, Flavoenzymes that catalyse reactions with no net change, *Nat. Prod. Rep.*, 2002, **19**, 761–772.

105. N. Carrillo and E.A. Ceccarelli, Open questions in ferredoxin-NADP$^+$ reductase catalytic mechanism, *Eur. J. Biochem.*, 2003, **270**, 1900–1915.
106. P.A. Mayes, *Biological oxidation*, in *Harper's Biochemistry*, R.K. Murray, D.K. Granner, P.A. Mayes and V.W. Rodwell, 24th edn, Appleton & Lange, Stamford, CT, 1996, 116–117.
107. S. Fukuzumi and T. Tanaka, *Flavins and Deazaflavins*, in *Photoinduced Electron Transfer Part*, C.M.A. Fox and M. Chanon (eds.), Elsevier, Amsterdam, 1988, 636–688.
108. J.A. Watkins, M.A. Cusanovich, T.E. Meyer and G. Tollin, A "parallel plate" electrostatic model for bimolecular rate constants applied to electron transfer proteins, *Protein Sci.*, 1994, **3**, 2104–2114.
109. I. Ahmad, M.A. Cusanovich and G. Tollin, Laser-flash photolysis studies of electron transfer between semiquinone and fully reduced free flavins and horse heart cytochrome *c*, *Proc. Natl. Acad. Sci. USA*, 1981, **78**, 6724–6728.
110. I. Ahmad, M.A. Cusanovich and G. Tollin, Laser flash photolysis studies of electron transfer between semiquinone and fully reduced free flavins and the cytochrome *c*–cytochrome oxidase complex, *Biochemistry*, 1982, **21**, 3122–3128.
111. T.E. Meyer, J.A. Watkins, C.T. Przysiecki, G. Tollin and M.A. Cusanovich, Electron-transfer reactions of photoreduced flavin analogues with *c*-type cytochromes: quantitation of steric and electrostatic factors, *Biochemistry*, 1984, **23**, 4761–4767.
112. J.T. Hazzard, T.A. Poulas and G. Tollin, Kinetics of reduction by free flavin semiquinones of the components of the cytochrome *c*–cytochrome *c*-peroxidase complex and intracomplex electron transfer, *Biochemistry*, 1987, **26**, 2836–2848.
113. J.T. Hazzard, S.J. Moench, J.E. Erman, J.D. Satterlee and G. Tollin, Kinetics of intracomplex electron transfer and the reduction of the components of covalent and noncovalent complexes of cytochrome *c* and cytochrome *c* peroxidase by free flavin semiquinones, *Biochemistry*, 1988, **27**, 2002–2008.
114. L. Quin and N.M. Kostic, Electron transfer reactions of cytochrome *f* with flavin semiquinone and with plasocyanin. Importance of protein–protein electrostatic interactions and of donar–acceptor coupling, *Biochemistry*, 1992, **31**, 5145–5150.
115. I. Sevrioukova, C. Shaffer, D.P. Ballou and J.A. Paterson, Equlibrium and transient state spectrophotometric studies of the mechanism of reduction of the flavoprotein domain of P450BM-3, *Biochemistry*, 1996, **35**, 7058–7068.
116. S.G. Mayhew, The effect of pH and semiquinone formation on the oxidation–reduction potentials of flavin mononucleotides. A reappraisal, *Eur. J. Biochem.*, 1999, **265**, 698–702.
117. J.T. Hazzard, A.G. Mauk and G. Tollin, Laser flash photolysis studies of electron transfer mechanisms in cytochromes: an aromatic residue at position 82 is not required for cytochrome *c* semiquinones or electron transfer from cytochrome *c* to cytochrome oxidase, *Arch. Biochem. Biophys.*, 1992, **298**, 91–95.
118. C.T. Przysiecki, G. Tollin, T.E. Meyer, J.E. Staggers and M.A. Cusanovich, Effect of pH and exocyclic substitution on flavosemiquinone reactivity with redox proteins and inorganic oxidants, *Arch. Biochem. Biophys.*, 1985, **238**, 334–343.
119. J.D. Walsh and A.-F. Miller, Flavin reduction potential tuning by substitution and bending, *J. Mol. Struct. (Theochem)*, 2003, **623**, 185–195.
120. M.F. Teitell and J.L. Fox, MO study of flavin reduction, *Int. J. Quant. Chem.*, 1982, **22**, 583–594.

121. L.H. Hall, B.J. Orchard and S.K. Tripathy, The structure and properties of flavins: molecular orbital study based on totally optimized geometries. 1. Molecular geometry investigations, *Int. J. Quant. Chem.*, 1987, **31**, 195–216.
122. J.D. Walsh and A.-F. Miller, NMR shielding and electron correlation reveal remarkable behaviour on the part of the flavin N_5 reaction center, *J. Phys. Chem. B*, 2003, **107**, 854–863.
123. C.B. Martin, X. Shi, M.-L. Tsao, D. Karweik, J. Brooke, C.M. Hadad and M.S. Platz, The photochemistry of riboflavin tetraacetate and nucleosides. A study using density functional theory, laser flash photolysis, fluorescence, UV-VIS, and time-resolved infrared spectroscopy, *J. Phys. Chem. B*, 2002, **106**, 10263–10271.
124. R.M. Kowalczyk, E. Schleicher, R. Bittl and S. Weber, The photoinduced triplet of flavins and its protonation states, *J. Am. Chem. Soc.*, 2004, **126**, 11393–11399.
125. C.B. Martin, M.-L. Tsao, C.M. Hadad and M.S. Platz, The reaction of triplet flavin with indole. A study of the cascade of reactive intermediates using density functional theory and time-resolved infrared spectroscopy, *J. Am. Chem. Soc.*, 2002, **124**, 7226–7234.
126. N. Mataga, H. Chorsowjan and Y. Shibata, Dynamics and mechanisms of ultrafast fluorescence quenching reactions of flavin chromophores in protein nanospace, *J. Phys. Chem. B*, 2000, **104**, 10667–10677.
127. I.H.M. van Stokkum, D.S. Larsen and R. van Grondelle, Global and target analysis of time-resolved spectra, *Biochim. Biophys. Acta*, 2004, **1657**, 82–104.
128. P.F. Heelis and A. Koziolowa, The effect of hydrogen bonding on the electron transfer reactions of excited singlet and triplet states of flavins, *J. Photochem. Photobiol. B: Biol.*, 1991, **11**, 365–370.
129. P.F. Heelis, R.F. Hartman and S.D. Rose, Detection of the excited singlet of a deprotonated reduced, flavin, *Photochem. Photobiol.*, 1993, **57**, 1053–1055.
130. P.F. Heelis, B.J. Parsons and Y. Yano, Spectral and redox properties of benzodipteridine. A pulse radiolysis, laser flash photolysis and semi-emperical orbital study, *J. Chem. Soc., Perkin Trans.*, 1997, **2**, 795–798.
131. F. Bosca, L. Fernandez, P.F. Heelis and Y. Yano, Substituent effects on electrophilicity of flavins: an experimental and semi-empirical molecular orbital study, *J. Photochem. Photobiol. B: Biol.*, 2000, **55**, 183–187.
132. S. Ishizaka and N. Kitamura, Photoinduced redox cycle of riboflavin at a water/oil interface, *Anal. Sci.*, 2004, **20**, 1587–1592.
133. M.B. Murataliev, R. Feyereisen and F.A. Walker, Electron transfer by diflavin reductases, *Biochim. Biophys. Acta*, 2004, **1698**, 1–26.
134. I. Ahmad and G. Tollin, Solvent effect on flavin electron transfer reactions, *Biochemistry*, 1981, **20**, 5925–5928.
135. G. Viteri, A.M. Edwards, J. De la Fuente and E. Silva, Study of the interaction between triplet riboflavin and the α-, β_H- and β_L-crystallins of the eye lens, *Photochem. Photobiol.*, 2003, **77**, 535–540.
136. J.A. Navarro, M.A. De la Rosa and G. Tollin, Transient kinetics of flavin-photosensitized oxidation of reduced redox proteins, *Eur. J. Biochem.*, 1991, **199**, 239–243.
137. M. Roncel, M. Hervas, J.A. Navarro, M.A. De la Rosa and G. Tollin, Flavin-photosensitized oxidation of reduced *c*-type cytochromes. Reaction mechanism and comparasion with photoreduction of oxidized cytochromes by flavin semiquinones, *Eur. J. Biochem.*, 1990, **191**, 531–536.
138. J. Garcia and E. Silva, Flavin-sensitized photooxidation of amino acids present in a parenteral nutrition infusate protection: by ascorbic acid, *Nutr. Biochem.*, 1997, **8**, 341–345.

139. E. Silva, V. Ruckert, E. Lissi and E. Abuin, Effects of pH and ionic micelles on the riboflavin-sensitized photoprocesses of tryptophan in aqueous solution, *J. Photochem. Photobiol. B: Biol.*, 1991, **11**, 57–68.
140. E. Silva, M. Salim-Hanna, A.M. Edwards, M.I. Becker and A.E. De Ioannes, *A light-induced tryptophan-riboflavin binding: biological implications*, in: *Series: Nutritional and Toxicological Consequences of Food Processing*, M. Friedman (ed), Plenum Press, New York, 1991, 33–48.
141. E. Silva, C. Almarza, D. Berndt and L. Larrondo, Photoreactions of riboflavin with spermine and their role in tryptophan photoconsumption induced by riboflavin, *J. Photochem. Photobiol. B: Biol.*, 1993, **21**, 197–201.
142. E. Silva, P. Ugarte, A. Andrade and A.M. Edwards, Riboflavin-sensitized photoprocesses of tryptophan, *J. Photochem. Photobiol. B: Biol.*, 1994, **23**, 43–48.
143. E. Silva and A.M. Edwards, Generation of radical species by irradiation of tryptophan and tyrosine solutions sensitized by riboflavin. Biological implications, *Cienc. Cult. (Sao Paulo)*, 1996, **48**, 47–50.
144. A.M. Edwards, C. Bueno, A. Saldano, E. Silva, K. Kassab, L. Polo and G. Jori, Photochemical and pharmacokinetic properties of selected flavins, *J. Photochem. Photobiol. B: Biol.*, 1999, **48**, 36–41.
145. A. de La Rochette, E. Silva, I. Biriouez-Aragon, M. Mancini, A.M. Edwards and P. Morliere, Riboflavin photodegradation and photosensitizing effects are highly dependent on oxygen and ascorbate concentration, *Photochem. Photobiol.*, 2000, **72**, 815–820.
146. A. de La Rochette, I. Birlouez-Aragon, E. Silva and P. Morliere, Advanced glycation endproducts as UVA photosensitizers of tryptophan and ascorbic acid: consequences for the lens, *Biochim. Biophys. Acta*, 2003, **1621**, 235–241.
147. D.R. Cardoso, D.W. Franco, K. Olsen, M.L. Andersen and L.F. Skibsted, Reactivity of bovine whey proteins, peptides, and amino acids toward triplet riboflavin as studied by laser flash photolysis, *J. Agric. Food Chem.*, 2004, **52**, 6602–6606.
148. E. Silva and J. Godoy, Riboflavin-sensitized photooxidation of tyrosine, *Int. J. Vit. Nutr. Res.*, 1994, **64**, 253–256.
149. C.-Y. Lu and Y.-Y. Liu, Electron transfer oxidation of tryptophan and tyrosine by triplet states and oxidized radicals of flavin sensitizers: a laser flash photolysis study, *Biochim. Biophys. Acta*, 2002, **1571**, 71–76.
150. J.J. Lopez, M.A.G. Garter, Yu.P. Tsentalovich, O.B. Morozova, A.V. Yurkovskaya and P.J. Hore, Effects of surfactants on the photosensitized production of tyrosine radicals studied by photo-CIDNP, *Photochem. Photobiol.*, 2002, **75**, 6–10.
151. P.F. Heelis, B.J. Parsons, G.O. Phillips and J.F. McKeller, The flavin sensitized photooxidation of ascorbic acid. A continuous and flash photolysis study, *Photochem. Photobiol.*, 1981, **33**, 7–13.
152. M.Y. Jung, S.M. Kim and S.Y. Kim, Riboflavin-sensitized photooxidation of ascorbic acid: kinetics and amino acid effects, *Food Chem.*, 1995, **53**, 397–403.
153. H. Kim, L.J. Kirschenbaum, I. Rosenthal and P. Riesz, Photosensitized formation of ascorbate radicals by riboflavin: an ESR study, *Photochem. Photobiol.*, 1993, **57**, 777–784.
154. F.H.M. Vaid, R.H. Shaikh, I.A. Ansari and I. Ahmad, Spectral study of the photolysis of aqueous thiamine hydrochloride and ascorbic acid solutions in the presence and absence of riboflavin, *J. Chem. Soc. Pak.*, 2005, **27**, 227–232.
155. E. Silva, A.M. Edwards and D. Pacheco, Visible light-induced photooxidation of glucose sensitized by riboflavin, *J. Nutr. Biochem.*, 1999, **10**, 181–185.

156. E. Silva, L. Herrera, A.M. Edwards, J.R. De la Fuente and E. Lissi, Enhancement of riboflavin-mediated photooxidation of glucose 6-phosphate dehydrogenase by urocanic acid, *Photochem. Photobiol.*, 2005, **81**, 206–211.
157. S. Criado, C. Guardianelli, J. Tuninetti, P. Molina and N.A. Garcia, Scavenging of photogenerated oxidative species by antimuscarinic drugs: atropine and derivatives, *Redox Rep.*, 2002, **7**, 385–394.
158. W.A. Massad, S. Bertolotti and N. Garcia, Kinetics and mechanism of the vitamin B_2-sensitized photooxidation of isoproterenol, *Photochem. Photobiol.*, 2004, **79**, 428–433.
159. A. Ramu and M.M. Mehta, The riboflavin-mediated photooxidation of doxorubicin, *Cancer Chemoth. Pharm.*, 2000, **46**, 449–458.
160. K. Uekama, T. Irie, F. Hirayama and F. Yoneda, Riboflavin-sensitized photooxidation of phenothiazines in aqueous solution by laser-irradiation, *Chem. Pharm. Bull.*, 1979, **27**, 1039–1042.
161. M.J. Akhtar, M. Ataullah and I. Ahmad, Effect of riboflavin on the photolysis of folic acid in aqueous solution, *J. Pharm. Biomed. Anal.*, 2000, **23**, 1039–1044.
162. K. Tatsumi, H. Ichikawa and S. Wada, Flavin-sensitized photooxidation of substituted phenols in natural water, *J. Contam. Hydrol.*, 1992, **9**, 207–219.
163. S. Caffieri, Furocoumarin photolysis: chemical and biological aspects, *Photochem. Photobiol. Sci.*, 2002, **1**, 149–157.
164. K. Huvaere, K. Olsen, M.L. Andernon and L.H. Skibsted, Riboflavin-sensitized photooxidation of isohumulones, *Photochem. Photobiol. Sci.*, 2004, **3**, 337–340.
165. K. Huvaere, B. Sinnaeve, J.V. Boexlaer and D. De Keukeleire, Photooxidative degradation of beer bittering principles: product analysis with respect to light-struck flavour formation, *Photochem. Photobiol. Sci.*, 2004, **3**, 854–858.
166. A. Heyerick, K. Huvaere, D. De Keukeleire and M.D.E. Forbes, Fate of flavins in sensitized photodegradation of isohumulones reduced derivatives: studies on formation of radicals via EPR combined with detailed product analyses, *Photochem. Photobiol. Sci.*, 2005, **4**, 412–419.
167. E. Silva, M. Jopai, A.M. Edwards, E. Lemp, J.R. De la Fuente and F. Lissi, Protective effect of Boldo and tea infusions on the visible light-mediated prooxidant effects of vitamin B_2, riboflavin, *Photochem. Photobiol.*, 2002, **75**, 585–590.
168. E. Knobloch, F. Mandys, R. Hodr, R. Hujer and R. Mader, Study of the mechanism of the photoisomerization and photooxidation of bilirubin using a model for the phototherapy of hyperbilirubinemia, *J. Chromatogr.*, 1991, **566**, 89–99.
169. E. Silva, T. Gonzalez, A.M. Edwards and F. Zuloaga, Visible light induced lipoperoxidation of a parenteral nutrition fat emulsion sensitized by flavins, *J. Nutr. Biochem.*, 1998, **9**, 149–154.
170. I. Ahmad and G. Tollin, Flavin triplet quenching and semiquinone formation by aliphatic α-substituted acetic acids: intermediates in flavin sensitized photodecarboxylation, *Photochem. Photobiol.*, 1981, **34**, 441–445.
171. M. Novak, A. Miller, T.C. Bruce and G. Tollin, The mechanism of flavin 4a substitution which accompanies photolytic decarboxylation of α-substituted acetic acids. Carbanion vs. radical intermediates, *J. Am. Chem. Soc.*, 1980, **102**, 1465–1467.
172. K. Miyake, Y. Masaki, I. Miyamoto, S. Yanagida, T. Ohno, A. Yoshimura and C. Pac, Flavin-photosensitized monomerization of dimethylthymine cyclobutane dimer in the presence of magnesium perchlorate, *Photochem. Photobiol.*, 1993, **58**, 631–636.

173. R.F. Hartman, S.D. Rose, P.J. Pouwels and R. Kaptein, Flavin-sensitized photochemically induced dynamic nuclear polarization detection of pyrimidine dimer radicals, *Photochem. Photobiol.*, 1992, **56**, 305–310.
174. O.W. Parks, Photodegradation of sulpha drugs by fluorescent light, *J. Assoc. Off. Anal. Chem.*, 1985, **68**, 1232–1234.
175. M.V. Encinas, S.G. Bertolotti and C.M. Previtali, The interaction of ground and excited states of lumichrome with aliphatic and aromatic amines in methanol, *Helv. Chim. Acta*, 2002, **85**, 1427–1438.
176. R.P. Patel and F.K. Soni, Photolysis of cyanocobalamin in the presence of riboflavin, *Indian J. Pharm.*, 1964, **26**, 35–37.
177. I. Ahmad and W. Hussain, Multicomponent spectrophotometric assay of cyanocobalamin, hydroxocobalamin and riboflavin, *Pak. J. Pharm. Sci.*, 1992, **5**(2), 121–127.
178. I.A. Ansari, F.H.M. Vaid and I. Ahmad, Spectral study of photolysis of aqueous cyanocobalamin solutions in presence of vitamins B and C, *Pak. J. Pharm. Sci.*, 2004, **17**(2), 83–93.
179. H. Cui, H.M. Hwang, K. Zeng, H. Glover, H. Yu and Y. Liu, Riboflavin-photosensitized degradation of atrazine in a freshwater environment, *Chemosphere*, 2002, **47**, 991–999.
180. H. Glover, H.M. Hwang and K. Zeng, Effect of riboflavin photoproducts on microbial activity during photosensitization of atrazine transformation, *Environ. Toxicol.*, 2003, **18**, 361–367.
181. E. Haggi, S. Bertolotti, S. Miskoski, F. Amat-Guerri and N.A. Garcia, Environmental photodegradation of pyrimidine fungicides – Kinetics of the visible-light-promoted interactions between riboflavin and 2-amino-4-hydroxy-6-methylpyrimidine, *Can. J. Chem.*, 2002, **80**, 62–67.
182. W. Massad, S. Criado, S. Bertolotti, A. Pajares, J. Gianotti, J.P. Escalada, F. Amat-Guerri and N.A. Garcia, Photodegradation of the herbicide norflurazon sensitized by riboflavin. A kinetic and mechanistic study, *Chemosphere*, 2004, **57**, 455–461.
183. P.P. Fu, S.-H. Cheng, L. Coop, Q. Xia, S.J. Culp, W.H. Tollesen, W.G. Wamer and P.C. Howard, Photoreaction, phototoxicity, and photocarcinogenicity of retinoids, *J. Environ. Sci. Health Part C: Environ.*, 2003, **21**, 165–197.
184. M. Korycka-Dahl and T. Richardson, Photodegradation of DNA with fluorescent light in the presence of riboflavin and photoprotection by flavin-triplet state quenchers, *Biochim. Biophys. Acta*, 1980, **610**, 229–234.
185. P. Burgstaller, T. Hermann, C. Huber, E. Westhof and M. Famulok, Isoalloxazine derivatives promote photocleavage of natural RNAs at G U base pairs embedded within helicles, *Nucleic Acids Res.*, 1997, **25**, 4018–4027.
186. M.Y. Jung, K.H. Lee and S.Y. Kim, Riboflavin-sensitized photochemical changes in beta-lactoglobulin in an aqueous buffer solution as affected by ascorbic acid, *J. Agric. Food Chem.*, 2000, **48**, 3847–3850.
187. F. Corbin III, Pathogenic inactivation of blood components. Current status and introduction of an approach using riboflavin as a photosensitizer, *Int. J. Hematol.*, 2002, **76**, 253–257.
188. A.M. Edwards and E. Silva, Effect of visible light on selected enzymes: vitamins and amino acids, *J. Photochem. Photobiol. B: Biol.*, 2001, **63**, 126–131.
189. A. Saha, Photo-induced inactivation of dihydroorotate dehydrogenase in dilute aqueous solution, *Int. J. Radiat. Biol.*, 1997, **72**, 55–61.

190. A. Mahns, I. Melchheier, C.V. Suschek, H. Sies and L.O. Klotz, Irradiation of cells with ultraviolet-A (320–400) in the presence of cell culture medium elicits biological effects due to extracellular generation of hydrogen peroxide, *Free Radical Res.*, 2003, **37**, 391–397.
191. A.M. Edwards, F. Barredo, E. Silva, A.E. De Ioannes and M.I. Becker, Apoptosis induction in nonirradiated human HL-60 and murine NSO/2 tumor cells by photoproducts of indole-3-acetic acid and riboflavin, *Photochem. Photobiol.*, 1999, **70**, 645–649.
192. M. Diaz, M.I. Becker, A.E. De Ioannes and E. Silva, Development of photoinduced adduct: reactivity to eye lens proteins, *Photochem. Photobiol.*, 1996, **63**, 762–767.
193. R. Ugarte, A.M. Edwards, M.S. Diez, A. Valenzuela and E. Silva, Riboflavin-photosensitized anaerobic modification of rat lens proteins. A correlation with age-related changes, *J. Photochem. Photobiol. B Biol.*, 1992, **13**, 161–168.
194. S.P. Baba, D.K. Patel and B. Bano, Modification of sheep plasma kininogen by free radicals, *Free Radical Res.*, 2004, **38**, 393–403.
195. G. Porcal, S.G. Bertolotti, C.M. Previtali and M.V. Encinas, Electron transfer quenching of singlet and triplet excited states of flavins and lumichrome by aromatic and aliphatic electron donors, *Phys. Chem. Chem. Phys.*, 2003, **5**, 4123–4128.
196. A. Yoshimura and T. Ohno, Lumiflavin-sensitized photooxygenation of indole, *Photochem. Photobiol.*, 1988, **48**, 561–565.
197. P.F. Heelis, C.A. Rowley-Williams, R.F. Hartman and S.D. Rose, Detection of a reduced-flavin triplet state in free flavins and DNA photolyase, *J. Photochem. Photobiol. B Biol.*, 1994, **23**, 155–159.
198. E. DeRitter, Vitamins in pharmaceutical formulations, *J. Pharm. Sci.*, 1982, **71**, 1073–1096.
199. M.C. Allwood, Compatibility and stability of TPN mixtures in big bags, *J. Clin. Pharm. Ther.*, 1984, **9**, 181–198.
200. M.C. Allwood and M.C.J. Kearney, Compatibility and stability of additives in parenteral nutrition admixtures, *Nutrition*, 1998, **14**, 697–706.
201. P.C. Buxton, S.M. Conduit and J. Hathaway, Stability of parentrovit in infusion fluids, *Br. J. Intravenous Ther.*, 1983, **4**, 5–12.
202. M.F. Chen, H.W. Boyce and L. Triplett, Stability of the B vitamins in mixed parenteral nutrition solution, *JPEN*, 1983, **7**, 462–464.
203. J.L. Smith, J.E. Canham and P.A. Wells, Effect of phototherapy light, sodium bisulfite and pH on vitamin stability in total parenteral nutrition admixtures, *JPEN*, 1988, **12**, 394–402.
204. H.J. Martens, Stability of water soluble vitamins in various infusion bags, *Krankenhauspharmazie*, 1989, **10**, 359–361.
205. X.C. Zhan and G.K. Yin, Obtaining natural light photostability of drugs from lamp light exposure experiments, *Acta Pharm. Sin.*, 1992, **27**, 544–548.
206. K. Yamaoka, Y. Nakajima and T. Terashima, Stability of vitamins in TPN mixture in a double-chambered bag, *Jpn. J. Hosp. Pharm.*, 1995, **21**, 357–364.
207. D.B. Min and J.M. Boff, Chemistry and reactions of singlet oxygen in foods, *CRFSFS*, 2002, **1**, 58–72.
208. G. Casini, N. De Laurentis, N. Maggi and S. Ottolino, Protective effect of hydroxybenzoic acid esters on the photodegradation of riboflavin, *Farm. Ed. Prat.*, 1981, **36**, 553–558.

209. A.F. Asker and M.J. Habib, Effect of certain Stabilizers on photobleaching of riboflavin solutions, *Drug. Dev. Ind. Pharm.*, 1990, **16**, 149–156.
210. M.J. Habib and A.F. Asker, Photostabilization of riboflavin by incorporation into liposomes, *J. Parent. Sci. Technol.*, 1991, **45**, 124–127.
211. Y.L. Loukas, P. Jayasekera and G. Gregoriadis, Novel liposome-based multicomponent systems for the protection of photolabile agents, *Int. J. Pharm.*, 1995, **117**, 85–94.
212. Y.L. Loukas, V. Vraka and G. Gregoriadis, Use of nonlinear least-square model for the kinetic determination of the stability constant of cyclodextrin inclusion complexes, *Int. J. Pharm.*, 1996, **144**, 225–231.
213. M.A. Geigel, *The stability testing program*, in *Compliance Handbook for Pharmaceuticals, Medical Devices, and Biologics*, C. Medina (ed), Marcel Dekker, New York, 2004, 195.
214. H.H. Tonnesen (ed), *The Photostability of Drugs and Drug Formulations*, Culinary and Hospatility Industry Publications Services, Weimer, TX, 2004.
215. H.H. Tonnesen and J. Karlsen, A comment on photostability testing according to ICH guidelines: calibration of light sources, *PharmEuropa*, 1997, **9**, 735–736.

Chapter 3

Excited States Interaction of Flavins with Amines: Application to the Initiation of Vinyl Polymerization

MARÍA V. ENCINAS[a] AND CARLOS M. PREVITALI[b]

[a] Facultad de Química y Biología, Universidad de Santiago de Chile, Santiago, Chile

[b] Departamento de Química, Facultad de Ciencias Exactas, Físico-Químicas y Naturales, Universidad Nacional de Río Cuarto, Argentina

3.1. Introduction

The polymerization of vinyl monomers induced by light has been the subject of current interest in the last decades. This is mainly due to its easy handling as well as its numerous applications. The kinetics of free-radical polymerization can be represented by the classical scheme that comprises the photochemical radical production, the initiation chain, the propagation, and the termination steps. Then, a key in these processes is the photoinitiator, which produces radicals through a photochemical process. From here the increased interest of understanding the mechanism involved as well as to introduce new and efficient compounds that can act as photoinitiators.

Among the principal characteristics that a photoinitiator has to meet are ;to present a high absorption in the visible region. The use of this spectral region is now of great interest. Besides the practical advantages of using visible light, they are found applications that range from technological processes to biological materials. Thus, photoinitiators that absorb visible light allow, for example: (i) to find spectral windows when pigmented formulations have to be polymerized, (ii) the use of the sunlight for the curing of outdoor coatings, (iii) the employ of several laser beams as light source. Also, photoinitiators that can be used in polymerizations in aqueous media have received interest because of the ecological advantages that they provide. Riboflavin (Rf) (vitamin B_2), a natural pigment that absorbs in the 400–500 nm region, fulfils with these two requirements.

The use of synthetic dyes as photosensitizers in photoinitiating systems operating in the visible region has been known from many years.[1–3] In most cases the systems operate by an electron transfer reaction from a donor, generally an aliphatic amine, to the excited dye. The use of natural pigments in the place of synthetic dyes has been much less explored. Flavins are natural occurring pigments with absorption bands in the near UV and visible region that may act as a sensitizer for the photoinitiating system. In particular, the photoreduction of flavin derivatives by electron donor is a well-known process that eventually could lead to the production of active radicals suitable for vinyl polymerization. This led us to a detailed investigation of the potentialities of flavins as photoinitiators of radical polymerization. The results are collected in this review. In order to understand the action mechanism of Rf/donor initiating system, a more detailed knowledge of the interaction of excited states of flavins with electron donors is necessary, and this is discussed in the first section of this review.

3.2. Excited States Quenching by Electron Donors

The ground state interactions of flavins with a number of organic molecules, especially aromatics, are known from a long time ago.[4] Also the photophysics and photochemistry of flavins have been reviewed in several opportunities.[5–7] The decay kinetics of the triplet state of lumiflavin (Lf) was studied by Naman and Tegner.[8] Rate constants for the first order decay to ground state and for the second order triplet–triplet annihilation and ground state self-quenching were determined in phosphate buffer at pH 6.9. The photoreduction of flavins by amino acid and EDTA was reported by Heelis *et al.*[9] A mechanistic investigation of the electron transfer quenching of the excited states of Lf by methyl and methoxy substituted benzenes was carried out by Traber *et al.*[10,11] A picosecond laser photolysis study of the quenching of Lf and Rf tetrabutyrate by indole and *N*-methyl indole in various organic solvents has been reported.[12] The photoreduction of Lf by several electron donors, among them EDTA, hydroxylamine and triethanolamine, was studied by Heelis *et al.*[13] in water in the pH range 3–14 using laser flash photolysis. More recently, the electron transfer from 2′-deoxyguanosine-5′-monophosphate to the triplet state of Rf was investigated in aqueous solution.[14] The transient intermediates were identified and the kinetics of several electron transfer processes involving Rf triplet state were determined. The transient absorption spectra of Rf semireduced form and the cation radical of the deoxyguanosine were reported.[14] All these studies provided information on the single-electron transfer from different donors to the excited states of the flavins. However, a detailed and systematic analysis of the electron transfer kinetics of the photoreduction process by a series of electron donors was not found in the literature.

Moreover, most of these studies were carried out in aqueous media, due to the biological importance of flavins. However, in biological systems in many cases the environment where photochemical processes take place is of low aqueous contents. Therefore, it was of interest to analyse the electron transfer processes of the excited states in a non-aqueous solvent. These studies are also of interest in relation to the use of flavins as photoinitiating system of polymerization in non-aqueous media. We have recently reported on the quenching of the excited states of lumichrome (LC) by aromatics amines and triethanolamine (TEOHA) in methanol.[15] It is well known that LC and others alloxazines unsubstituted at *N*(1) in the presence of compounds able to act as proton acceptor or donor are subjected to a proton transfer reaction in the excited singlet to form the corresponding isoalloxazine. This reaction takes place in the presence of aliphatic amines of intermediate basicity. However, with amines of lower basicity such as TEOHA only the electron transfer deactivation of the excited states of LC was observed.

In a more recent paper we reported a detailed investigation in methanol of the quenching of the excited states of Rf, Lf, and LC by aromatic electron donors and for the first two by aliphatic amines.[16] The deactivation of the excited singlet was investigated by steady state and time-resolved (single photon counting) fluorescence measurements. The triplet quenching and the identification of the products of the quenching reaction were carried out by nanosecond laser flash photolysis. In this way the bimolecular rate constants were obtained for the reaction,

$$\mathrm{Fl}^* + \mathrm{D} \rightarrow \mathrm{Fl}^- + \mathrm{D}^+ \qquad (1)$$

where Fl* stands for the flavin (or LC) in its excited singlet or triplet state.

The rate constants for the quenching of Lf, Rf, and LC are collected in Table 1 for aromatic electron donors, while in Table 2 are the corresponding values for the quenching of Lf and Rf by aliphatic amines.

A correlation of the rate constants with the Gibbs free energy change ΔG^0 for the electron transfer process, reaction 1, could be established. The ΔG^0 was calculated from the redox potentials of the donor $E_{(\mathrm{D/D}^+)}$ and acceptor $E_{(\mathrm{A/A}^-)}$, and the energy E^* of the excited state involved, using the Rhem–Weller equation[17]

$$\Delta G^0 = E_{\mathrm{D/D}^+} - E_{\mathrm{A/A}^-} E * + C \qquad (2)$$

where C is the coulombic energy term, for which −0.06 eV has been calculated in methanol for unitary charges. The redox potentials were measured as 0.80, 0.70, and 1.3 V for Rf, Lf, and LC, respectively, using a SCE reference electrode in acetonitrile (measured in our laboratory by cyclic voltammetry). The quencher oxidation potentials in acetonitrile *vs.* SCE were taken from different sources as quoted in ref. 16, and are also given in Tables 1 and 2. The excited singlet energies were taken as 2.48 eV for Rf and Lf, and 2.92 eV for LC.[18] Similarly, for the triplet energy we used 2.17 eV for Rf and Lf, and 2.40 eV for LC.[18] ΔG^0 was assumed to be the same in methanol and acetonitrile based on

Table 1. Quenching rate constants (in units of 10^9 M^{-1} s^{-1}) by aromatic electron donors in methanol

		Lumiflavin		Riboflavin		Lumichrome	
Quencher	$E_{(D/D^+)}$[a]	**Singlet**	**Triplet**	**Singlet**	**Triplet**	**Singlet**	**Triplet**
N,N,N′N′-tetramethyl-*p*-phenylenediamine	0.12	13.7	5.3	12.3	6.9	13	4.6
p-phenylendiamine	0.26	10.9	4.3	11.6	–	10.8	
N,N,N′N′-tetramethyl-benzidine	0.43		10.3		8.8		6.4
p-anisidine	0.66	11.1	10.4			9.6	4.9
o-anisidine	0.74	11.3				9.7	4.9
N,N′-dimethylaniline	0.76	11.5	4.7	11.5	4.7	13.1	6.4
p-toluidine	0.78	9.4	5.8			13	2.8
N-methylaniline	0.81			12.1		12	4.0
o-toluidine	0.84			12.2	7.8	12.5	4.3
Aniline	0.93	10.5	4.4	11.7	4.3	14	4.0
Diphenylamine	0.94					10.6	3.7
2-methylindole	1.07	10.5	5.1			11	2.3
1,2-dimethylindole	1.09					12.7	4.3
1,2,4-trimethoxy-benzene	1.12	9.8	4.7			10.4	3.9
1-methylindole	1.20	10.5	5.6			10	3.8
Indole	1.21	13.0	5.3		5.1	8.5	0.98
1,4-dimethoxybenzene	1.30	9.2	4.4		3.2	6.9	0.17
1,2-dimethoxybenzene	1.47	7.3	0.96			6.8	0.034
1,3-dimethoxybenzene	1.50	7.8	1.46				

[a]Redox potential of the donor, in Volts *vs.* SCE from ref. 16.

Table 2. Quenching rates constants (in units of g 10^9 M^{-1} s^{-1}) of excited states of Lf and Rf in methanol by aliphatic amines

		Lumiflavin		Riboflavin	
Amine	$E_{(D/D^+)}$[a]	**Singlet**	**Triplet**	**Singlet**	**Triplet**
DABCO	0.56	5.9	1.6		1.5
Tributylamine	0.78	3.9	1.2	3.17	
Tripropylamine	0.78	5.2	1.6	3.03	
Triethanolamine	0.90	6.2 (5.0)[b]	1.4 (0.51)[b]	4.5	0.85
Triethylamine	0.95	3.0	0.56	2.6	0.34
Dipropylamine	1.02	1.6			
Dibutylamine	1.07	1.8	0.72	1.4	0.15
Di-isopropylamine	1.31	1.3		1.01	
Di-isobutylamine	1.16	1.6	0.39	1.30	
Pyperidine	1.30	1.1	0.19	1.6	–
Butylamine	1.39	0.85	0.0066	0.60	0.005

[a]Redox potential of the amine, in Volts vs. SCE from ref. 16.
[b]From ref. 13.

the similar polarity of these solvents. The plots of the quenching rate constants *vs.* ΔG^0 are shown in Figures 1–3. It can be seen that the rate constants steeply decrease when ΔG^0 becomes positive, and that they reach a limiting value at high exergonicity. This behaviour is typical of an electron transfer quenching. It is interesting to note that this correlation is hold for compounds of very

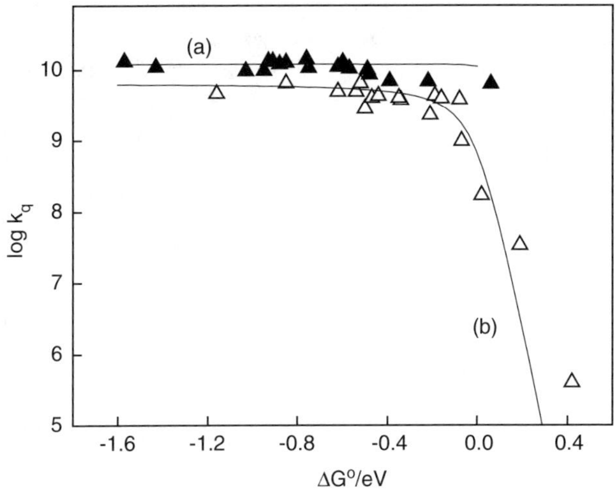

Figure 1. Dependence of the quenching rate constants for excited states of LC on the ΔG^0, in methanol. (▲) Singlet quenching by aromatic donors; (Δ) triplet quenching by aromatic donors. The line (a) was calculated according to Equation (7). The curve (b) was calculated with Equations (5) and (6)

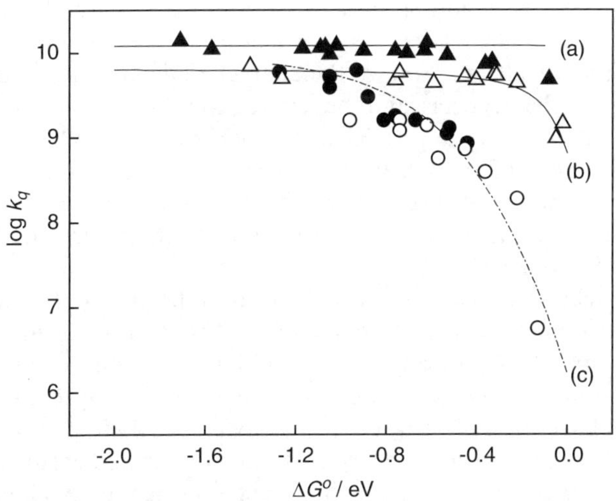

Figure 2. Dependence of the quenching rate constants for excited states of Lf on the ΔG^0, in methanol. (▲) Singlet quenching by aromatic donors; (Δ) triplet quenching by aromatic donors; (●) singlet quenching by aliphatic amines; (○) triplet quenching by aliphatic amines. The line (a) was calculated according to Equation (7). Curves (b) and (c) were calculated with Equations (5) and (6)

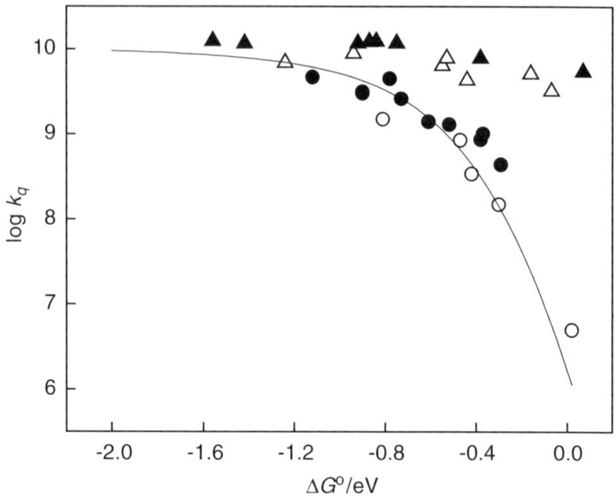

Figure 3. Dependence of the quenching rate constants for excited states of Rf on the ΔG^0, in methanol. (▲) Singlet quenching by aromatic donors; (Δ) triplet quenching by aromatic donors; (●) singlet quenching by aliphatic amines; (○) triplet quenching by aliphatic amines. Solid curve was calculated with Equations (5) and (6)

different chemical structure. Thus, indole derivatives follow the same relation than aromatic amines and methoxy benzenes.

The electron transfer nature of the quenching process was further confirmed by the transient absorption spectra as determined by laser flash photolysis. These spectra reveal that laser flash irradiation of the flavins and LC in the presence of the quenchers produce the characteristic absorption bands of the radical cation of the quencher and the semireduced form of the flavins or LC.[13] A typical transient spectrum of Rf in the presence of TEOHA in MeOH is shown in Figure 4. At 3 μs after the laser flash the broad band with a maximum around 600 nm characteristic of the Rf triplet state can be seen. At longer times the triplet decays to the spectrum corresponding to the semireduced form of Rf with $\lambda_{max} = 570$ nm.

Another conclusion that can be extracted from the plots in Figures 2 and 3 is that aliphatic amines are less efficient quenchers than aromatic amines or other aromatic donors of similar redox potential. Similar results have been reported for electron transfer quenching of excited states for other systems.[19–22] Several explanations have been offered for this behaviour, including differences in the Coulomb term, reorganization energies, steric factors, and exciplex formation. However, to our knowledge, singlet and triplet quenching by aliphatic amines has not been compared with the quenching of both states by aromatics for the same system. It can be seen in Figures 2 and 3 that the rate constants for aliphatic amines increase in a regular trend from triplet to singlet quenching, falling on a single correlation with ΔG^0. In this case the rate constants do not reach the diffusional limit.

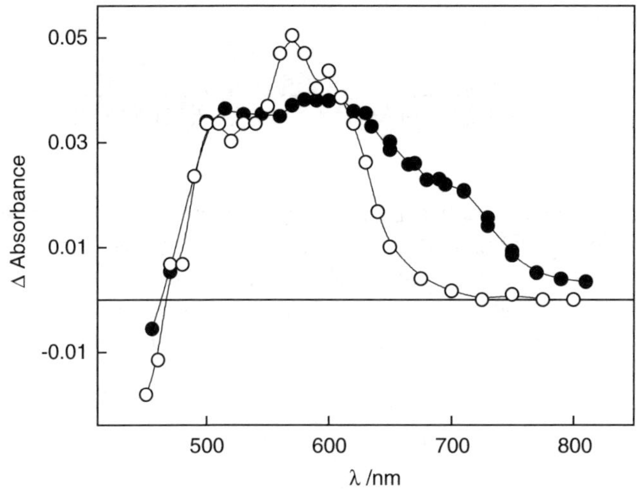

Figure 4. Transient absorption spectra of Rf in methanol in the absence (●) and in the presence of TEOHA 0.6 mM (○)

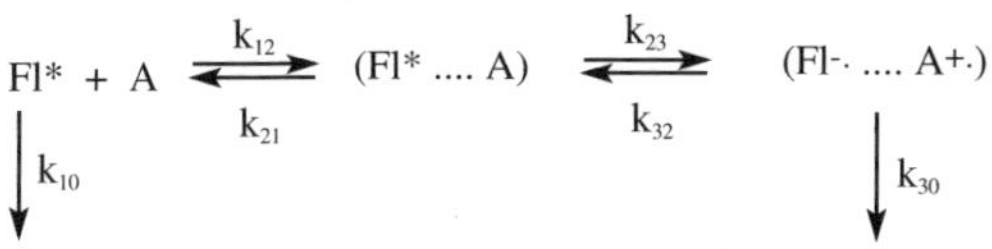

Scheme 1.

For those systems that present a plateau at high exergonicity, the data in Tables 1 and 2 and the plots in Figures 1–3 show that the limiting value of the rate constants for the singlet quenching by aromatic donors is higher than those for the triplet quenching.

These results can be rationalized in terms of the Rehm–Weller mechanism for electron transfer quenching, Scheme 1.

where Fl* stands for the flavin or LC in its excited state and A for the amine and other donor quenchers.

According to this mechanism the stationary state rate constant is

$$k = \frac{k_{12}}{1 + \frac{k_{21}}{k_{23}}\left(\frac{k_{32}}{k_{30}} + 1\right)} \tag{3}$$

Introducing the Gibbs energy change for the electron transfer process

$$\frac{k_{23}}{k_{32}} = \exp\left(-\Delta G^0/RT\right) \tag{4}$$

and $\quad k_{23} = k_{23}^0 \exp\,(-\Delta G^*/RT)$

Equation (3) becomes

$$k = \frac{k_{12}}{1 + \frac{k_{21}}{k_{23}^{0}} \exp(\Delta G^{*}/RT) + \frac{k_{21}}{k_{30}} \exp(\Delta G^{0}/RT)} \tag{5}$$

Several relationships between ΔG^* and ΔG^0 have been proposed, including the classical parabolic Marcus equation.[23] However, in order to fit the experimental results for bimolecular charge separation reactions, an expression of ΔG^* that tends asymptotically towards zero for highly negative values of ΔG^0 is necessary. The Rehm–Weller relationship[17] (Equation (6)) meets this requirement.

$$\Delta G^{*} = \frac{\Delta G^{0}}{2} + \left[\left(\frac{\Delta G^{0}}{2} \right)^{2} + \Delta G^{*}(0)^{2} \right]^{1/2} \tag{6}$$

The limit of Equation (5) for very exergonic processes is

$$k = \frac{k_{12}}{1 + \frac{k_{21}}{k_{23}^{0}}} \tag{7}$$

The plots of log k *vs.* ΔG^0 in Figures 1–3 were fitted to the Rehm–Weller model with Equations (5) and (6). In Table 3 the fitting parameters are collected. According to the Equation (7) a limiting value of k_{12} is expected for the quenching rate constant if k_{23}^{0} is much higher than k_{21}, this is commonly observed in singlet quenching by an electron transfer process. However, a value lower than k_{12} results if k_{23}^{0} is of the order of k_{21}. It can be seen in Table 3 that the value of $k_{23}{}^{0}$ necessary for fitting the triplet-quenching curve by aromatic donors is two orders of magnitude lower than that for singlet quenching. This was explained by considering some degree of non-adiabaticity in the triplet quenching.[16]

Table 3. Kinetics parameters from the curve fitting of the free energy correlation of electron transfer quenching rate constants of LC, Lf, and Rf by aromatic electron donors and aliphatic amines in methanol

	Quenching of LC and Lf by aromatic electron donors		**Quenching of Lf and Rf by aliphatic amines**
	Singlet	**Triplet**	**Singlet and triplet**
k_{12} M^{-1} s^{-1}	1.2×10^{10}	1.2×10^{10}	1.2×10^{10}
k_{21} s^{-1}	2.0×10^{10}	2.0×10^{10}	2.0×10^{10}
k_{23} s^{-1}	5.0×10^{12}	2.5×10^{10}	5.0×10^{11}
k_{30} s^{-1}	1×10^{11}	1.0×10^{11}	1.0×10^{11}
$\Delta G^*(0)$/eV	0.08	0.08	0.32

It can also be observed in the Table 2 that the singlet and triplet quenching by aliphatic amines could be fitted by the same kinetic parameters. However, while for the singlet and triplet quenching by aromatic donors the intrinsic barrier $\Delta G^*(0)$ in Equation (5) was found to be of the order 0.08 eV, similar to the solvent reorganization energy, the corresponding value for aliphatic amines was of 0.3 eV.[16] The higher value for aliphatic quenchers is most likely due to the presence of an internal contribution to the intrinsic barrier, related to changes in bond lengths and angles in the quenchers, necessary to attain the transition state. The observed dispersion of the data points could be related, at least in part, to differences in the internal reorganization energy for each amine. Some contribution of the uncertainty of the irreversible oxidation potentials of the aliphatic amines cannot be disregarded.

In summary, photoinduced electron transfer from aromatic donors to singlet- and triplet-excited states of flavins and LC is very efficient process. Preliminary results on the quantum yields for the charge separation process, indicate that in spite of being less effective as quenchers, aliphatic amines produce free radicals in a substantial yield in the quenching reaction.

3.3. Flavin/Amine Systems as Photoinitiators of Vinyl Polymerization

The photopolymerization of acrylamide initiated by Rf was reported long time ago by Oster *et al.*[24] More recently the polymerization photoinitiated by Rf has received more attention due to the widespread use of polyacrylamide gel electrophoresis for the separation of proteins and nucleic acids. Polyacrylamide gels are formed by copolymerization of acrylamide and *bis*-acrylamide. Polymerization is initiated by a free radical-generating system containing a photoinitiator, ammonium persulfate and tetramethylenediamine (TEMED).[25,26] Even, other dyes are used as photoinitiation systems, Rf takes the advantage of obtaining highly reproducible gels with high resolution. For the polymerization photoinitiated by Rf it has been reported that the persulfate radicals add to the monomer double bond and TEMED accelerates the polymerization.[26] However, the photoinitiation mechanism is not clearly understood.

We have recently studied the systems Rf/triethanolamine (TEOHA) and LC/triethanolamine as photoinitiators of the polymerization of 2-hydroxyethyl methacrylate (HEMA) in a medium monomer/methanol 1:1.[27] The same system was employed later in aqueous medium for the polymerization of HEMA.[28] In a third paper we investigated the effect of the amine structure on the efficiency of the photoinitiating system.[29]

The irradiation of Rf in HEMA/methanol (1/1) mixture does not lead to polymerization. However, when the irradiation was carried out in the presence of amines the polymerization of HEMA was observed.[27] The polymerization rate (R_p) increases with the TEOHA concentration reaching a constant value (Figure 5). Similar results were obtained when LC was used as photoinitiator.

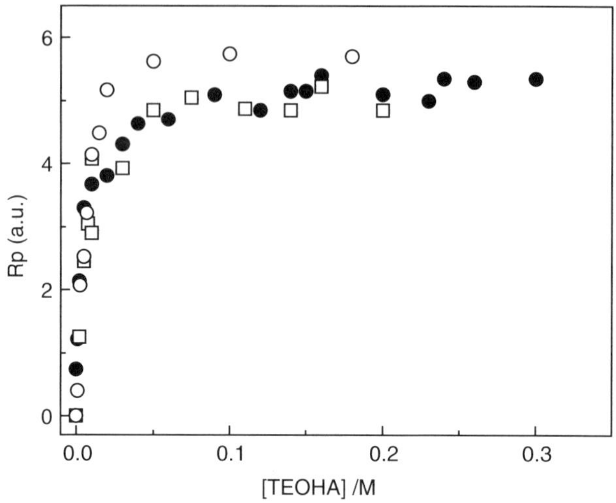

Figure 5. Polymerization rates vs. TEOHA concentration: (●) Rf, λ = 366 nm; (□) Rf, λ = 450 nm; (○) LC, λ = 366 nm. HEMA/methanol (1/1)

We have determined[27,28] that the kinetics of the HEMA polymerization photoinitiated by flavin in the presence of amines follows the classical rate law,

$$R_p = \left(\frac{k_p}{k_t^{1/2}}\right)\Phi_{\text{inic}}^{1/2} I_a^{1/2}[M] \tag{8}$$

where k_p and k_t are the propagation and termination rate constants, Φ_{inic} is the initiation quantum yield, I_a is the light absorbed intensity and $[M]$ is the monomer concentration.

Assuming that Φ_{inic} is proportional to the quantum yield of active radicals Φ_{rad} from Equation (8), *i.e* radicals that add to the monomer, at constant monomer concentration and light intensity,

$$Rp^2 = cte.\ \Phi_{\text{rad}} \tag{9}$$

The radical quantum yield produced in the deactivation of Rf-excited states by amines will be given by

$$\Phi_{\text{rad}} = \Phi^* f \beta \tag{10}$$

Where Φ^* stands for the quantum yield of excited state formation in the presence of amine, f is the fraction of excited states quenched by the amine, and β is the fraction of this event that leads to active radicals. Since the radicals can be produced from the interaction of the singlet- and triplet-excited states of the flavin with the amine, the total radical quantum yield is given by

$$\Phi_{\text{rad}} = (\Phi_{\text{rad}})_s + (\Phi_{\text{rad}})_T \tag{11}$$

Thus, an apparent value of radical quantum yield can be calculated from the measured values of the quenching rate constants, the excited state lifetimes, and the triplet-quantum yield. The fraction of excited states of the flavin that are deactivated by the amine can be obtained from the measured values of quenching rate constants (Equation (12)).

$$f = \frac{k_q[Am]}{k_q[\mathrm{Am}] + \sum k^0} \tag{12}$$

The quantum yield of formed triplet has to consider the singlet quenched by the amine,

$$\Phi_\mathrm{T} = \frac{\Phi_\mathrm{T}^0}{\left(1 + {}^\mathrm{S}k_q \tau_\mathrm{S}^o [Am]\right)} \tag{13}$$

Φ_T^0 stands for the triplet-quantum yield in the absence of amine. Then, the apparent radical quantum yield is given by Equation (14).

$$\Phi_\mathrm{rad} = \frac{\beta_\mathrm{S}{}^\mathrm{S}k_q[Am]}{{}^\mathrm{S}k_q[Am] + \sum {}^\mathrm{S}k^0} + \frac{\beta_\mathrm{T}\Phi_\mathrm{T}{}^\mathrm{T}k_q[Am]}{{}^\mathrm{T}k_q[Am] + \sum {}^\mathrm{T}k^0} \tag{14}$$

With Φ_T as given by Equation (13).

Since, singlet- and triplet-excited states of flavins are quenched by amines through an electron transfer mechanism involving a charge transfer intermediate,[13] it is expected that the rate of the process will be highly dependent on the medium properties. This implies to know-how the photochemical behaviour is modified in the polymerization system. Rate constants for the singlet and triplet quenching of Rf by TEOHA, measured in the presence of the monomer, were $3 \times 10^9\ \mathrm{M}^{-1}\ \mathrm{s}^{-1}$ and $8 \times 10^8\ \mathrm{M}^{-1}\ \mathrm{s}^{-1}$, respectively. These values are slightly lower than that obtained in pure solvent. This is consequence of the higher viscosity and the lower polarity of the monomer/methanol mixture. On the other hand, the fluorescence decay does not change in the presence of HEMA, indicating that there is not interaction between the singlet-excited state of Rf and the monomer. The triplet lifetime is shortened by the monomer. At the concentration of HEMA employed in the polymerization mixture (3.8 M), in the absence of amine, 90% of Rf triplet states are quenched by the monomer; however, this process does not lead to polymerization. This implies that the interaction of the triplet with the HEMA does not lead to active radicals. The Rf neutral radicals also were observed in the triplet quenching by the amine, Figure 4, but the yield increases considerably in the presence of monomer. Figure 6 shows that the semireduced Rf-quantum yield at 6 mM TEOHA, where a minimal fraction of excited singlets are intercepted, increases approximately three times in the presence of 1.8 M HEMA. This increase can be related to the addition of amine radicals to the monomer before the radicals escape from the cage and recombine with the flavin radicals.

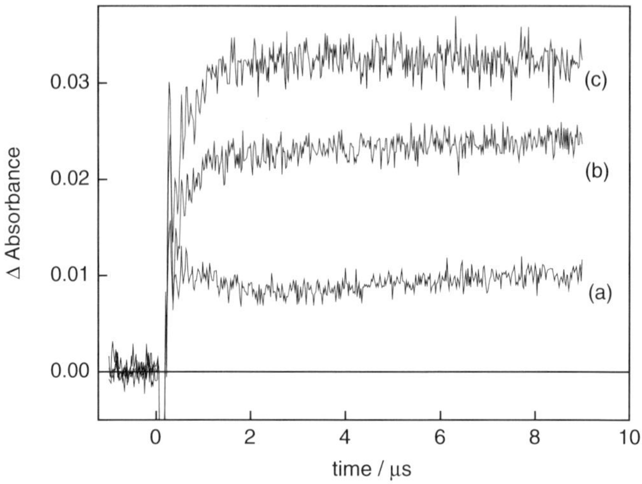

Figure 6. Transient absorption at 570 nm. Riboflavin (1×10^{-5} M) in the presence of TEOHA 6 mM in methanol: (a) in the absence of monomer; (b) in the presence of HEMA 1.14 M; (c) HEMA 1.8 M

When the experimental values of the square of the polymerization rate at different amine concentrations are fitted to the radical yield predicted by Equations (9) and (14), the best fitting for Rf is obtained with $\beta_S = 0.4$ and $\beta_T = 0.75$ (Figure 7). This indicates that the interaction of both, the singlet and the triplet states with amines, leads to neutral radicals. This is surprising since back electron transfer is a very fast process from the singlet radical ion pair. Frequently, it is found with many dyes that the active radicals are only those providing from the triplet interaction.[30,31] With respect to Rf the high photoinitiation quantum yield is in agreement with the polymerization initiated by the radicals formed in the quenching of the singlet and triplet states by the amine. High free radical yields have been also observed in the interaction of flavin singlet state with substrates of biological interest such as ascorbic acid.[32,33]

A mechanism compatible with these results is shown in Scheme 2, where parentheses indicate reactive partners in the solvent cage. The electron transfer from the amine to the Rf-excited states in methanolic solution leads to the Rf anion and amine cation radical pair. Rapid proton transfer, inside the solvent cage, gives the flavin and amine neutral radicals. The later radicals in solvent cage or free are added to the monomer double bond leading to the polymer formation. The addition of the Rf(H$^\cdot$) radicals to the monomer can be disregarded, since spectroscopic measurements of the formed polymer did not reveal incorporation of the flavin moiety.

The same behaviour was hold when irradiation source of 366 nm or 450 nm was used (Figure 5). This shows the applicability of flavins as photoinitiators with near-UV or visible light. When the polymerization was carried out using LC it was obtained a similar, even slightly higher polymerization rate, Figure 5. The fitting of the experimental data of R_p^2 to Equations (9) and (14)

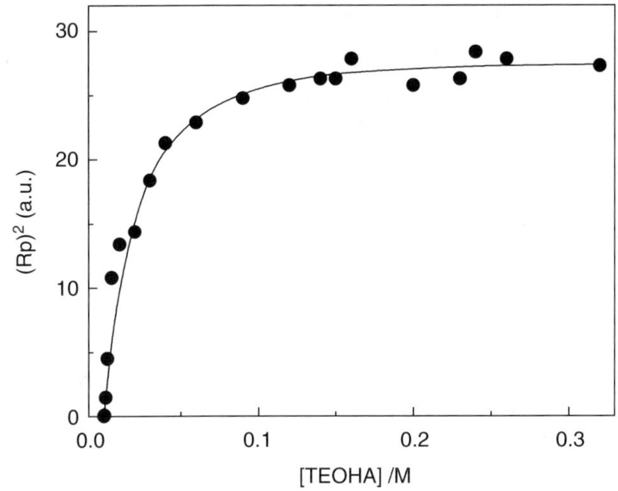

Figure 7. Plot of the square of the polymerization rate in methanol as function of TEOHA concentration. Riboflavin excited at 366 nm, HEMA 4.1 M. The solid line corresponds to the free radical quantum yield calculated with Equations (9) and (14) with $\beta_T = 0.75$ and $\beta_S = 0.4$

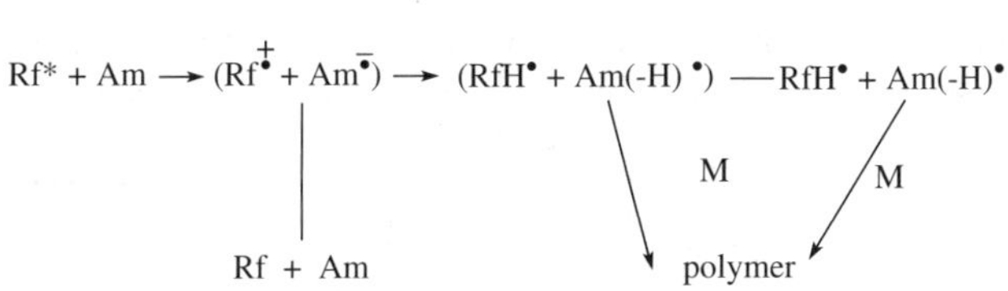

Scheme 2.

also gave an important contribution of active radicals formed in the singlet quenching by the amine ($\beta_S = 0.7$ and $\beta_T = 0.8$). Lumichrome is one of the main product obtained in the anaerobic photodecomposition of Rf. This is interesting because at large irradiation times, where the decomposition of the flavin becomes important, the polymerization could continue by the product.

As expected from the presence of the charge transfer intermediate, the photoinitiation efficiency of these systems is also highly dependent on the amine structure. Figure 8 shows the dependence of R_p with the concentration for different amines. As previously described for TEOHA[25] the polymerization rate increases with the amine concentration reaching a maximum value,[27] but this value is strongly dependent on the amine structure. Only aliphatic tertiary amines are efficient co-initiators, and the higher efficiency was obtained with TEOHA. Secondary and aromatic amines are poor co-initiators. The fractions of the singlet and triplet of Rf quenched by the different amines, calculated from Equation (12) at 0.2 mM amine, are summarized in Table 4. These data show that the quenching efficiencies for both excited states are in the order

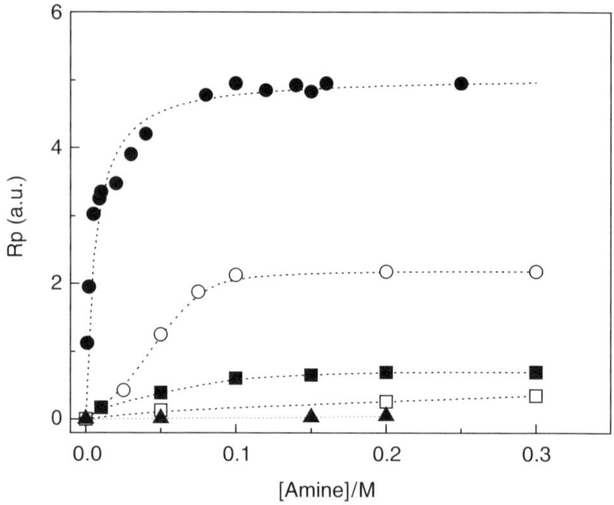

Figure 8. Polymerization rates of HEMA photoinitiated by Rf as a function of amine concentration: (●) TEOHA; (○) triethylamine; (■) N,N′-dimethylaniline; (□) dibutylamine; (▲) aniline. Monomer/methanol: 1/1 vol/vol

Table 4. Square of the polymerization rate, efficiencies of Rf-excited states quenched by the amine and experimental Rf-neutral radical quantum yields at 0.2 M amine concentration

Amine	R_p^2(a.u.)	f_T	f_S	$(\Phi_{rad})_{exp}$
N,*N*′-dimethylaniline	0.036	1	0.92	0.16
Triethanolamine	1.5	0.9	0.68	0.25
Triethylamine	0.32	0.88	0.67	0.14
Dibutylamine	0.004	0.57	0.30	0.05

aromatic amine > tertiary amines > secondary amine, and they are not correlated with the photoinitiation efficiency.

The Rf neutral radical quantum yields at 0.2 M amine, $(\Phi_{rad})_{exp}$, estimated from the absorption at 570 nm (Figure 9) relative to the triplet absorption, are included in Table 4.

The radical yields do not correlate with the amount of excited states quenched by the amine. This suggests that the radical yield is determined mainly by the rate of the proton-transfer process. On the other hand, data of Table 4 show that the radical yield is higher for the TEOHA, this is the amine that behaves as better photoinitiator. The lower value corresponds to the dibutylamine, which gives very low polymerization rate. Even when the initiation quantum yield increases with the radical yield, the correlation as predicted by the polymerization rate law is very poor. This indicates that the reactivity of the radicals towards the monomer double bond, given by β, is an important parameter to consider.

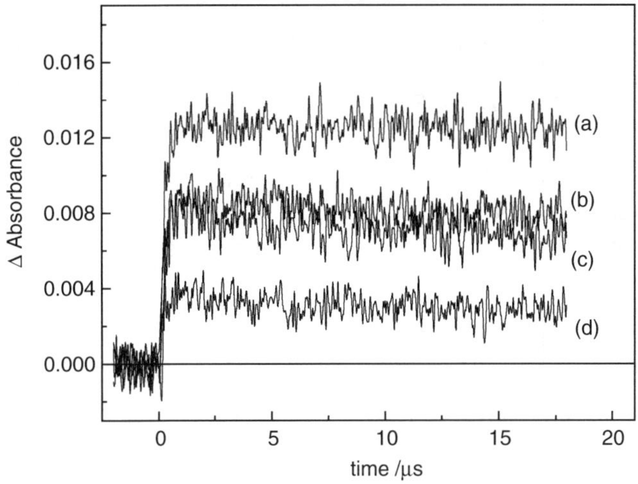

Figure 9. Transient absorption at 570 nm. Riboflavin (5×10^{-5} M in the presence of 0.2 M amine: (a) TEOHA; (b) N,N′-dimethylaniline; (c) triethylamine; (d) dibutylamine

Polymerization of HEMA also has been studied in aqueous solutions.[28] In this case the experiments were carried out at pH 9, where the dominant form of the ground and excited states of Rf is the neutral form.[34,35] Also, at this pH the protonation of the TEOHA is avoided. In this case the transient spectrum measured at total triplet quenching by the amine corresponds to the Rf radical anion, which is consistent with the pK 8.3 reported for the equilibrium between uncharged and monoanionic species of the radical derived from Rf.[34,36] Values of 2.7×10^9 M^{-1} s^{-1} and 1.8×10^8 M^{-1} s^{-1} were determined for the singlet and triplet quenching rate constants in water at pH 9. These values are lower than those measured in methanol. This can be explained in terms of solvent hydrogen bonding with the amine, that protects the amine electron lone pair, decreasing the electron transfer to the flavin.

The polymerization rate increases with the amine concentration, reaching a maximum at 0.01 M TEOHA, further amine addition decreases the rate (Figure 10). This behaviour is different to that before described for the polymerization of HEMA in methanolic solution, where the polymerization rate increases but reaches a constant value. In both media, the 45% of the Rf singlets are quenched by 0.1 M TEOHA. Furthermore, the amine concentration required to reach the maximum photoinitiation efficiency is approximately 8 times lower in the aqueous polymerization than in methanolic solution. These differences indicate that in the polymerization in aqueous medium, the singlet quenching does not lead to active radicals.

Figure 10 shows the fitting of the experimental values of R_p^2 to the radical quantum yield predicted from the quenching rate constants, lifetimes of the excited state, and triplet-quantum yields measured in the monomer/water mixture, according to the Equations (9) and (14). The best fitting was obtained with $\beta_T = 0.4$ and $\beta_S = 0$. This implies that the only active radicals are those

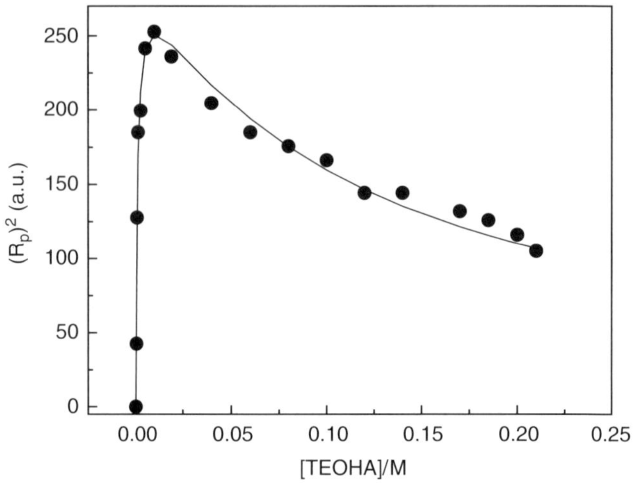

Figure 10. Plot of the square of the polymerization rate in HEMA/water (1/2) pH 9, as a function of TEOHA concentration. The solid line corresponds to the radical quantum yield calculated by Equations (9) and (14) with $\beta_T = 0.4$ and $\beta_s = 0$

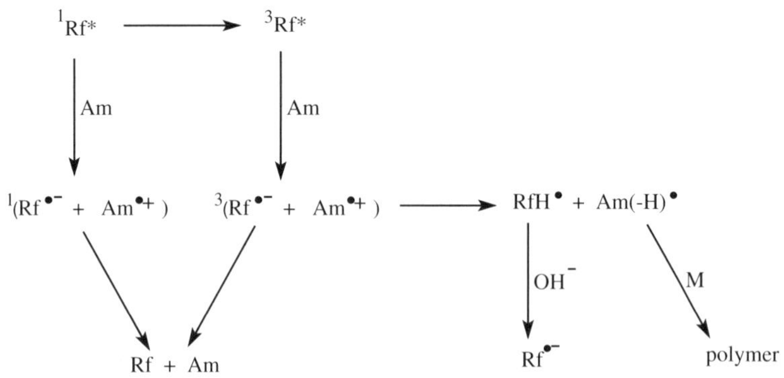

Scheme 3.

originated in the triplet-quenching process. Scheme 3 shows a mechanism consistent with the photochemical behaviour of flavins in the presence of amines and HEMA in aqueous solution at pH 9. The lack of photoinitiation through the singlet-quenching process indicates a fast back electron transfer for the singlet radical ion pair in the aqueous medium.

In conclusion, the studies here presented show that flavins in the presence of aliphatic amines are good initiating systems for the photopolymerization of HEMA. The results show that the nature of the solvent plays an important role on the initiation mechanism of the polymerization of HEMA photoinitiated by Rf, being markedly different when the solvent is changed from water to methanol. Similar results were found with Lf as photoinitiator. In methanol, can be deduced the participation of the radicals arising of both the singlet and triplet states. Meanwhile in water, the singlet quenching leads to the inhibition of the polymerization. The photoinitiation efficiency also is strongly dependent on the structural features of the coinitiator, the amines. This is consequence of differences in the active radical yield generation and in the reactivity of these species towards to the monomer double bond.

References

1. G. Oster and N.L. Yang, Photopolymerization of vinyl monomers, *Chem. Rev.*, 1968, **68**, 125–151.
2. D.F. Eaton, *Dye sensitized photopolymerization*, in *Advances in Photochemistry*, D. Volman, K. Gollnick and G.S. Hammond (eds), **vol 13**, Wiley, New York, 1986, Chapter 4.
3. J. Paczkowski, Z. Kucybala, F. Scigalski and A. Wrzyszczynski, Dyeing photoinitiators. Electron transfer processes in photoinitiating systems, *Trends Photochem. Photobiol.*, 1999, **5**, 79–91.
4. M.A. Slifkin, *Charge Transfer Interactions of Biomolecules*, Academic Press, London, 1971, 132–172.
5. G.R. Penzer and G.K. Radda, The chemistry and biological function of isoalloxazines, *Q.Rev.*, 1967, **21**, 43.
6. P.F. Heelis, The photophysical and photochemical properties of flavins (isoalloxazines), *Chem. Soc. Rev.*, 1982, **11**, 15–39.
7. P.F. Heelis, *The photochemistry of flavins*, in *Chemistry and Biochemistry of Flavins*, F. Muller (ed), **vol 1**, CRC, Boca Raton, 1991.
8. S.A. Naman and L. Tegner, Decay kinetics of the triplet excited state of lumiflavin, *Photochem. Photobiol.*, 1986, **43**, 331–333.
9. P.F. Heelis, B.J. Parsons, G.O. Phillips and J.F. McKellar, The photoreduction of flavins by amino acids and EDTA. A continuous and flash photolysis study, *Photochem. Photobiol.*, 1979, **30**, 343–347.
10. R. Traber, E. Vogelman, S. Schreiner, P. Werner and H.E.A. Kramer, Reactivity of excited states of flavin and 5-deazaflavin in electron transfer reactions, *Photochem. Photobiol.*, 1981, **33**, 41–48.
11. R. Traber, H.E.A. Kramer and P. Hammerich, One and two electron transfer in the photoreduction of flavin, *Pure Appl. Chem.*, 1982, **54**, 1651–1665.
12. A. Karen, N. Ikeda, N. Mataga and F. Tanaka, Picosecond laser photolysis studies of fluorescence quenching mechanisms of flavins: a direct observation of indole-flavin singlet charge transfer state formation in solutions and flavoenzymes, *Photochem. Photobiol.*, 1983, **37**, 495–302.
13. P.F. Heelis, M.A. De la Rosa and G.O. Phillips, A laser flash photolysis study of the photoreduction of the lumiflavin triplet state, *Photobiochem. Photobiophys.*, 1985, **9**, 57–63.
14. C. Lu, W. Lin, W. Wang, Z. Han, S. Yao and N. Lin, Riboflavin (VB_2) photosensitized oxidation of 2′-deoxyguanosine-5′-monophosphate (dGMP) in aqueous solution: a transient intermediates study, *Phys. Chem. Chem. Phys.*, 2000, **2**, 329–334.
15. M.V. Encinas, S.G. Bertolotti and C.M. Previtali, The interaction of ground and excited states of lumichrome with aliphatic and aromatic amines in methanol, *Helv. Chim. Acta*, 2002, **85**, 1427–1438.

16. G. Porcal, S.G. Bertolotti, C.M. Previtali and M.V. Encinas, Electron transfer quenching of singlet and triplet excited states of flavins and lumichrome by aromatic and aliphatic amines, *Phys. Chem. Chem. Phys.*, 2003, **5**, 4123–4128.
17. D. Rehm and A. Weller, Kinetics of fluorescence quenching by electron and H-atom transfer, *Isr. J. Chem.*, 1970, **8**, 259–271.
18. S.L. Murov, I. Carmichael and G.L. Hug, *Handbook of Photochemistry*, Marcel Decker, New York, 1993.
19. R.H. Kayser and R.H. Young, The photoreduction of methylene blue by amines – I. A flash photolysis study of the reaction between triplet methylene blue and amines, *Photochem. Photobiol.*, 1976, **24**, 395–401.
20. R. Ballardini, G. Varani, M.T. Indelli, F. Scandola and V. Balzani, Free energy correlation of rate constants for electron transfer quenching of excited transition metal complexes, *J. Am. Chem. Soc.*, 1978, **100**, 7219–7223.
21. T.N. Inada, K. Kikuchi, Y. Takahashi, H. Ikeda and T. Miyashi, A comparative study on electron-transfer fluorescence quenching by aliphatic and aromatic amines, *J. Photochem. Photobiol. A: Chem.*, 2000, **137**, 93–97.
22. P. Jacques, X. Allonas, D. Burget, E. Haselbach, P.A. Muller, A.C. Sergenton and H. Galliker, Intramolecular multiple Rehm–Weller plots in photoinduced electron transfer: competition between *p*- and *n*-type donor sites in benzylamines, *Phys. Chem. Chem. Phys.*, 1999, **1**, 1867–1871.
23. R.A. Marcus and N. Sutin, Electron transfer in chemistry and biology, *Biochem. Biophys. Acta*, 1985, **811**, 265–322.
24. G.k. Oster, G. Ester and G. Prati, Dye-sensitized photopolymerization of acrylamide, *J. Am. Chem. Soc.*, 1957, **79**, 595–598.
25. S. Caglio and P.G. Righetti, On the pH dependence of polymerization efficiency, as investigated by capillary zone electrophoresis, *Electrophoresis*, 1993, **14**, 554–558.
26. T. Rabilloud, M. Vincon and J. Garin, Micropreparative one- and two-dimensional electropohoresis: improvement with new photopolymerization systems, *Electrophoresis*, 1995, **16**, 1414–1422.
27. S.G. Bertolotti, C.M. Previtali, A.M. Rufs and M.V. Encinas, Riboflavin/triethanolamine as photoinitiator system of vinyl polymerization. A mechanistic study by laser flash photolysis, *Macromolecules*, 1999, **32**, 2920–2924.
28. B. Orellana, A.M. Rufs, M.V. Encinas, C.M. Previtali and S.G. Bertolotti, The photoinitiation mechanism of vinyl polymerization by riboflavin/triethanolamine in aqueous medium, *Macromolecules*, 1999, **32**, 6570–6573.
29. M.V. Encinas, A.M. Rufs, S. Bertolotti and C.M. Previtali, Free radical polymerization photoinitiated by riboflavin/amines. Effect of the amine structure, *Macromolecules*, 2001, **34**, 2845–2847.
30. C.M. Previtali, S.G. Bertolotti, M.G. Neumann, I.A. Pastre, A.M. Rufs and M.V. Encinas, Laser flash photolysis study of the photoinitiator system safranine T-aliphatic amines for vinyl polymerization, *Macromolecules*, 1994, **27**, 7454–7458.
31. J. Jakubiak, X. Allonas, J.P. Fouassier, A. Sionkowska, E. Andrzejewska, L.A. Linden and J.F. Rabek, Camphorquinone-amines photoinitiating systems for the initiation of free radical polymerization, *Polymer*, 2003, **44**, 5219–5226.
32. P.F. Heelis, B.J. Parsons, G.O. Phillips and J.F. McKellar, The flavin sensitized photooxidation of ascorbic acid a continuous and flash photolysis study, *Photochem. Photobiol.*, 1981, **33**, 7–13.
33. H. Kim, L.J. Kirshenbaum, I. Rosenthal and P. Riesz, Photosensitized formation of ascorbate radicals by riboflavin: an ESR study, *Photochem. Photobiol.*, 1993, **57**, 777–784.

34. E.J. Land and A.J. Swallow, One electron reactions in biochemical systems as study by pulse radiolysis. Riboflavin, *Biochemistry*, 1969, **8**, 2117–2125.
35. S. Schreider, U. Steiner and H.E.A. Kramer, Determination of the pK values of the lumiflavin triplet state by flash photolysis, *Photochem. Photobiol.*, 1975, **21**, 81–84.
36. C.E.H. Lmoumene and L. Lindqvist, Stepwise two photon excitation of 1,5-dihydroflavin monucleotide: study of flavosemiquinone properties, *Photochem. Photobiol.*, 1997, **66**, 591–595.

Chapter 4

Riboflavin as a Visible-Light-Sensitiser in the Aerobic Photodegradation of Ophthalmic and Sympathomimetic Drugs

NORMAN A. GARCÍA, SUSANA N. CRIADO AND WALTER A. MASSAD

Departamento de Química, Universidad Nacional de Río Cuarto, Campus Universitario, 5800 Río Cuarto, Argentina

Abstract

This chapter compiles and discusses results published by our group related to the study of the kinetic and mechanistic behavior and, in some cases, the distribution of photoproducts in the Riboflavin (Rf)-promoted photooxygenation of two families of drugs of pharmaceutical relevance, namely sympathomimetic and ophthalmic drugs. The drugs only absorb ultraviolet light, nevertheless, in the presence of Rf, visible-light irradiation under aerobic conditions produce a series of kinetically competitive processes which mainly depend on the relative concentrations of Rf and the drugs, and on the redox properties of the drugs. The overall picture comprises photochemical mechanisms that initiate with the quenching of Rf singlet and triplet excited states, and include the participation of the oxidative species singlet molecular oxygen and superoxide radical anion, which produce the photodegradation of both Rf and the drugs in different extents depending on the characteristics of the drug involved. On the contrast, in particular cases the drugs behave as excellent non-sacrificial (physical) deactivatiors of the oxidative species protecting Rf toward photodestruction.

4.1. Introduction

4.1.1. The Riboflavin-Photosensitising Process

When a substrate is transparent to daylight, but another compound present in the medium absorbs the incident light, sensitised photoreactions can take place. Under aerobic conditions, sensitised photooxidation of the substrate can occur mainly by reaction with photochemically generated reactive oxygenated species (ROS). The sensitised photoprocess of pharmaceutical and bioactive drugs (D), especially under daylight irradiation, has received particular attention in the last decades mainly because the photoreaction can give rise to products with different, null or even undesirable activity, such as drug photosensitivity.[1–3] The mechanism of all these photoprocesses is strongly dependent on the type of sensitiser. This chapter will focus on the photosensitising effects of the natural daylight-absorbing pigment Riboflavin (Rf, vitamin B2) in relation to two important families of bioactive compounds transparent to daylight, namely ophthalmic drugs (OD) and sympathomimetic drugs (SD). The vitamin, which is endogenously present in living organisms[4] and practically in all types of surface waters,[5] is a well-known sensitiser for the visible-light-promoted oxidation of different substrates of biological, medical and environmental interest.[6–8]

The usual mechanism of action of this sensitiser is rather complex, in many cases with the concurrent involvement of the ROS singlet molecular oxygen ($O_2(^1\Delta_g)$) and superoxide radical anion ($O_2^{\bullet-}$). This mechanism can be summarised with a general scheme (Scheme 1), which includes most of the known photochemical processes taking place when a solution containing Rf and a bioactive drug D, is irradiated with visible light in the presence of dissolved oxygen.[9–14]

The initially generated excited singlet state of the sensitiser ($^1Rf^*$, process (1)) can decay to ground state Rf (2), can be quenched by D (3), or can produce

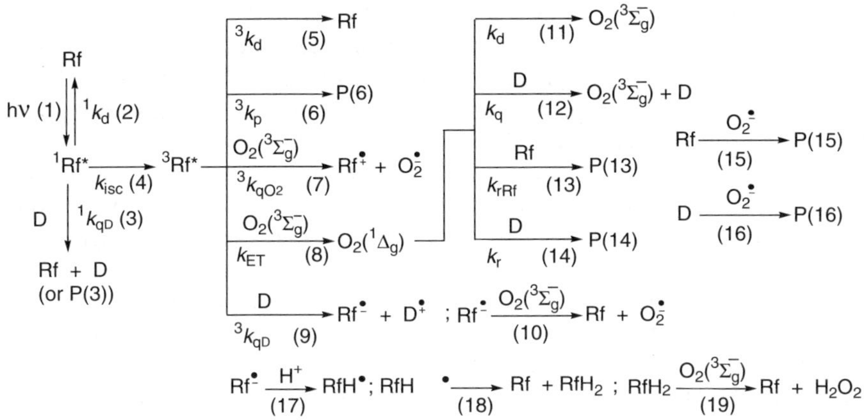

Scheme 1. Main possible interactions in the visible-light aerobic irradiation of bioactive drugs (D) and Riboflavin (Rf). P denotes a reaction product

excited triplet Rf ($^3Rf^*$) (4). $^3Rf^*$ can decay to ground state Rf (5), it can yield products (6), it can be quenched by ground state oxygen, $O_2(^3\Sigma_g^-)$, generating $O_2^{\cdot -}$ by electron transfer (7) and $O_2(^1\Delta_g)$ by energy transfer (8), or it can accept an electron from D, yielding semireduced Rf ($Rf^{\cdot -}$) (9). $Rf^{\cdot -}$ can also generate $O_2^{\cdot -}$ by electron transfer to oxygen (10). On the other hand, $O_2(^1\Delta_g)$ can be physically quenched either by the medium (11) or by D (12), or can react either with Rf (13) or with D (14). Reaction (14), with reactive rate constant k_r, is the main pathway of disappearance of D in $O_2(^1\Delta_g)$-mediated processes.

In parallel, the generated $O_2^{\cdot -}$ can chemically react with Rf (15) or with D (16). $O_2(^1\Delta_g)$ (process (8)) and $O_2^{\cdot -}$ (process (7)) are photogenerated, with reported quantum yields in water of 0.49 [15] and 0.009,[16] respectively, although the generation of $O_2^{\cdot -}$ via process (7) is practically negligible in kinetic terms. Rf is also a moderate quencher of $O_2(^1\Delta_g)$, with an overall rate constant, $k_{tRf} = k_{qRf} + k_{rRf}$, for the physical and chemical quenching (process (12), with Rf instead of D) of $6 \times 10^7\ M^{-1}\ s^{-1}$.[17] Nevertheless, in the presence of an adequate substrate D, $^3Rf^*$ can give rise to $O_2^{\cdot -}$ (process (10)) via $Rf^{\cdot -}$ (process (9)). Semireduction of $^3Rf^*$ to $Rf^{\cdot -}$ has been proposed in order to explain the photoinitiation of vinyl polymerisations with Rf,[18] as well as the interaction of ascorbic acid with the Rf-chromophore radical from flavin-mononucleotide.[19] In the presence of proton donating species, the generation of Rf neutral radical ($RfH^{\cdot}$, process (17)) should occur, with a pK value of 8.3.[13,14] Nevertheless, although it is known that the reaction of ground state oxygen with $RfH^{\cdot}$ is much slower than with $Rf^{\cdot -}$ (process (10)), the bimolecular decay of $RfH^{\cdot}$ through a disproportionation reaction can yield Rf and fully reduced Rf (RfH_2) (process (18)), which in the presence of $O_2(^3\Sigma^-_g)$ can be reoxidised, giving Rf and H_2O_2 (process (19)).[13]

Regarding $O_2(^1\Delta_g)$-mediated photooxidations (process (14)), their efficiency is correctly evaluated through the determination of the photooxidation quantum efficiency, $\varphi_r = k_r\,[D]/(k_q + k_t\,[D])$.[20,21] The φ_r value so determined takes into account the simultaneous effect of the physical and chemical interactions, being the k_q contribution usually interpreted in practical terms as a form of self-protection against $O_2(^1\Delta_g)$-mediated photooxidations. Nevertheless, the φ_r value depends on the concentration of the photooxidizable substrate ([D] in the present case), which is particularly difficult to estimate in complex systems such as biological environments. On the other hand, no relevant information about these photoreactions can be obtained from the straightforward analysis of isolated k_t and k_r values.[21] A simpler and more useful approach is the evaluation of the k_r/k_t ratio, which can be envisaged as the fraction of the overall interaction $O_2(^1\Delta_g)$-substrate that leads to effective chemical transformation. In this chapter the k_r/k_t ratio will be used for the discussion of results.

A positive effect that can arise from the interaction of D with some photogenerated oxidative species is the eventual scavenging of ROS (reactions (10–12)). These species are known to produce pernicious effects in living organisms.[2,22] In this context, the prevalence of the physical process (11) is a desirable possibility since the final result is the elimination of the oxidative species without considerable loss of the scavenger D. These actions (processes (10–12)) represent a protection for proteins, DNA and other cell matrix components, which, due to their known reactivity and high local concentration, constitute the primary targets for the attack of ROS or reactive species generated upon Rf-photosensitisation in biological environments.[12,23]

4.1.2. Sympathomimetic and Ophthalmic Drugs

For this chapter we selected some members of the biologically and pharmaceutically relevant families of SD and OD, to describe their interaction with ROS generated by visible-light-promoted Rf-photosensitisation. The knowledge of the kinetics and mechanism of these photoprocesses can be of interest for the understanding, evaluation of its feasibility and prevention of photosensitivity and unexpected therapeutic behaviors of potentially photodegradable drugs and medicaments.

The SD studied, here represented by Phen and Iso, consist of a series of compounds, with phenolamine or catecholamine structures and properties resembling the neurotransmitters epinephrine, norepinephrine and dopamine[24] (Scheme 2). Phen is a phenolamine, and as it is well known,[21] several compounds with phenol-like structures have been reported to possess remarkable antioxidative properties, among which the naturally-occurring α-tocopherol (vitamin E) is a paradigmatic example.[25]

Iso is a synthetic catecholamine (Scheme 2), whose photooxidation upon absorption of UV irradiation has been extensively described,[26,27] whereas only very recently some results have been published about its dye-sensitised phototransformations.[11]

The OD, and very especially those of topical administration, obviously constitute a family of particular photochemical risk and deserve attention. The diversity of structures in potentially-photosensitive OD (Scheme 2)

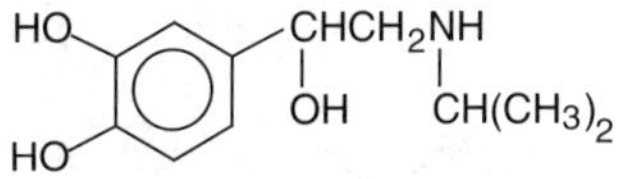

Isoproterenol (Iso)

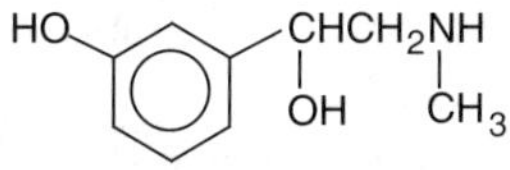

Phenylephrine (Phen)

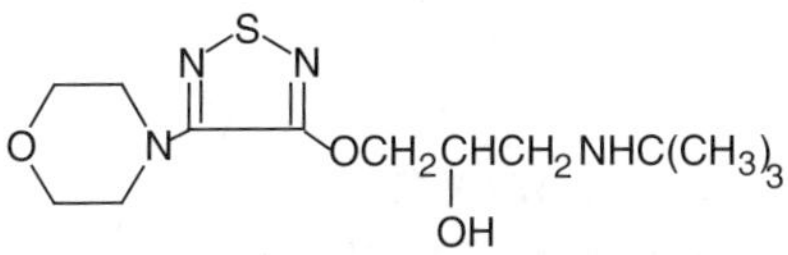

Timolol (Tim)

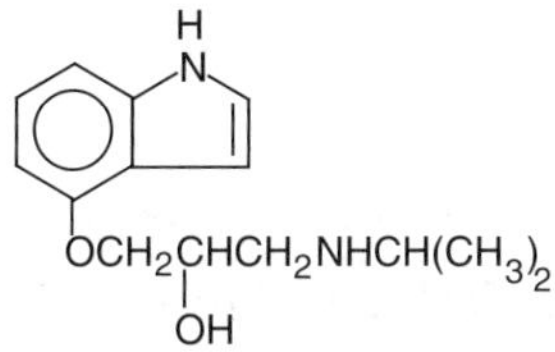

Pindolol (Pin)

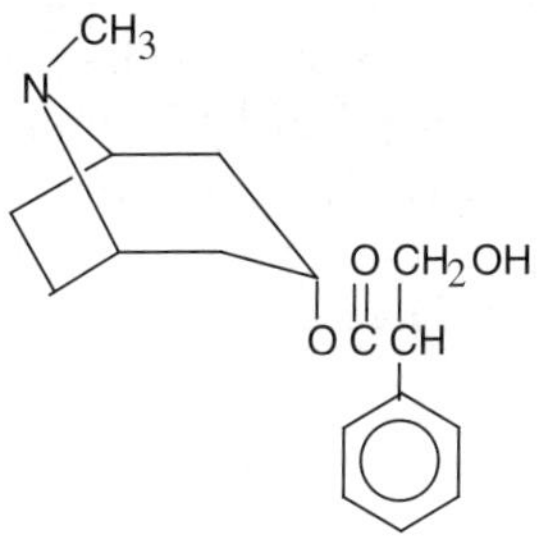

Atropine (Atr)

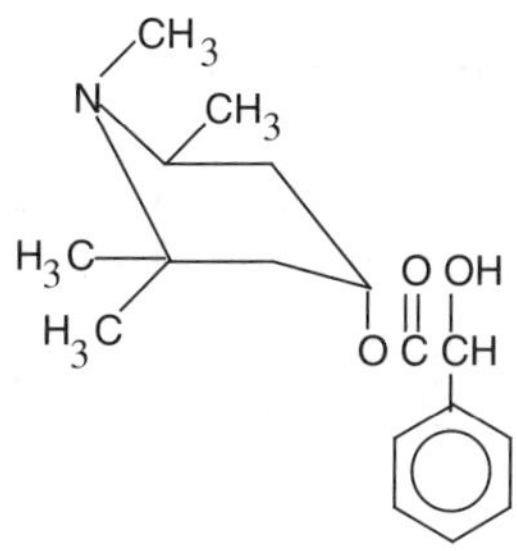

Eucatropine (Euc)

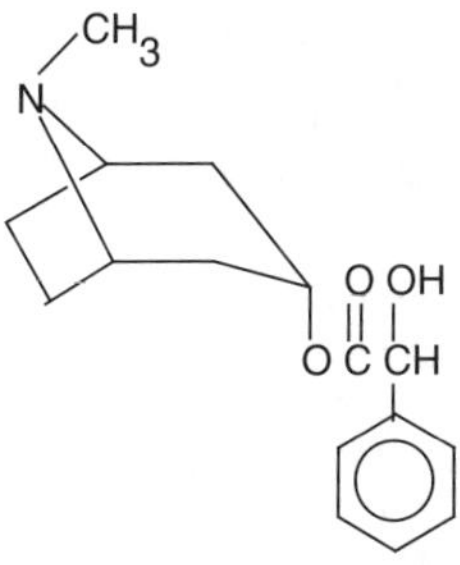

Homatropine (Hom)

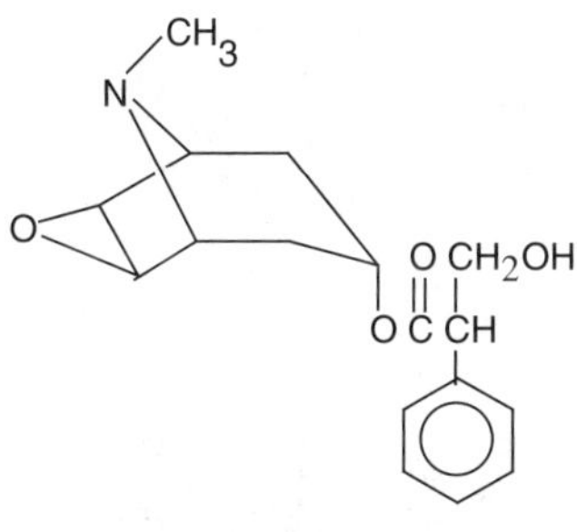

Scopolamine (Sco)

Scheme 2. Chemical structures of Isoproterenol, Phenylephrine, Timolol, Pindolol, Atropine, Eucatropine, Homatropine and Scopolamine

indicates that different mechanisms are likely to be responsible for initiating photoreactions in the eye.

We have recently studied[9] the kinetics and mechanism of Rf-sensitised photooxidation of two pharmaceutical OD families: the antiglaucoma drugs timolol (Tim) and pindolol (Pin) and the family of atropinic compounds (AT), represented by atropine (Atr), eucatropine (Euc), homatropine (Hom) and scopolamine (Sco).

The first family, constituted by β-adrenergic receptor antagonists, is widely used for the treatment of open-angle glaucoma and ocular hypertension.[28,29] The atropine family, employed also through topical applications, is a class of compounds with antimuscarinic action, utilised in the therapy of some kinds of myopia.[30]

4.2. Riboflavin-Sensitised Photodegradation of the Sympathomimetic Drugs Isoproterenol and Phenylphrine

4.2.1. Dark Complexation and Rf-Photosensitised Processes

In aqueous solutions, Rf forms dark complexes with Iso and Phen, which are only detectable at relatively high concentrations of the SD, typically higher than 15 mM. The behavior of Iso and Phen in the presence of Rf is similar to that found for phenols, catechols and indoles,[31] the interaction being attributed to a weak charge-transfer process mainly driven by hydrophobic forces. The set of difference spectra, shown for the case of Iso (Figure 1), reveals the formation of the dark complex (process (20)).

$$\mathrm{Rf} + \mathrm{SD} \underset{}{\overset{Kass}{\rightleftarrows}} [\mathrm{Rf} - \mathrm{SD}] \tag{20}$$

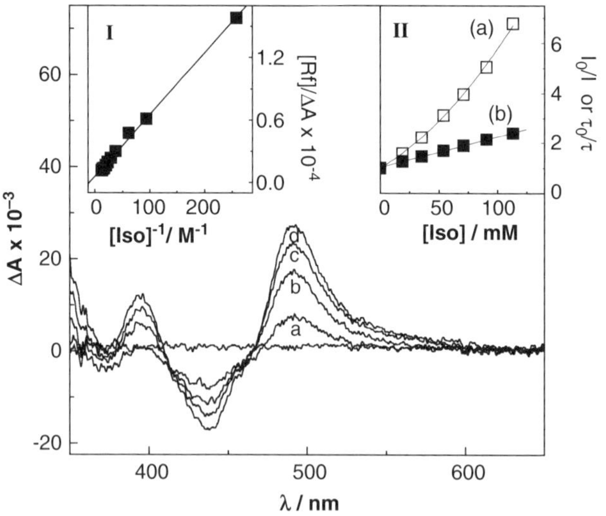

Figure 1. Difference spectra of Riboflavin+Isoproterenol vs. Isoproterenol, in water. [Dye] 0.03 mM. [Isoproterenol]: (a) 16.24 mM; (b) 43.35 mM; (c) 61.35 mM; (d) 81.18 mM. Inset I: Benesi-Hildebrand plot for the association Riboflavin- Isoproterenol, in water. Inset II: Stern-Volmer plots for the steady-state (□) and time-resolved (■) fluorescence quenching of the triplet excited Riboflavin by Isoproterenol

The values for the apparent association constants (K_{ass}), determined by the Benesi-Hildebrand method (Figure 1, inset I), are 10 and 5.5 M^{-1} (Unpublished results by Massad and García) for Iso and Phen, respectively. These low values denote a weak interaction and indicate that the fractions of complexed pigment and complexed SD can be ignored under sensitizing conditions, *i.e.* Rf *ca.* 0.02 mM and SD in the sub-mM concentration range.

The photoirradiation of aqueous mixtures of Rf (*ca.* 0.02 m mM plus Iso or Phen 0.1–0.4 mM) with visible light produces oxygen consumption and spectral changes in the absorption spectra of the mixtures. The evolution of the absorption spectra of Iso and Phen reflect different behaviors with respect to the photodecomposition of the individual SD. Whereas the changes in UV–Vis spectra of irradiated mixtures of Rf-Iso can be attributed to transformations on both components of the mixture (see Figure 2, main) the spectral changes in the Rf–Phen mixture correspond to transformations in Rf and, to a lesser extent, in Phen, as shown in the difference spectra Rf+Phen *vs.* Rf before and after photoirradiation (Figure 2, inset).

The anaerobic and aerobic photodegradation rates of Rf, processes that are well known to occur from $^3Rf^*$ (anaerobically)[32] and/or from $^3Rf^*$ plus autosensitisation via $O_2(^1\Delta_g)$ (aerobically),[8] suffer a decrease in the presence of either of the SD in concentrations in the range of 0.1 mM. All these pieces of experimental evidence clearly indicate an interaction between SD and

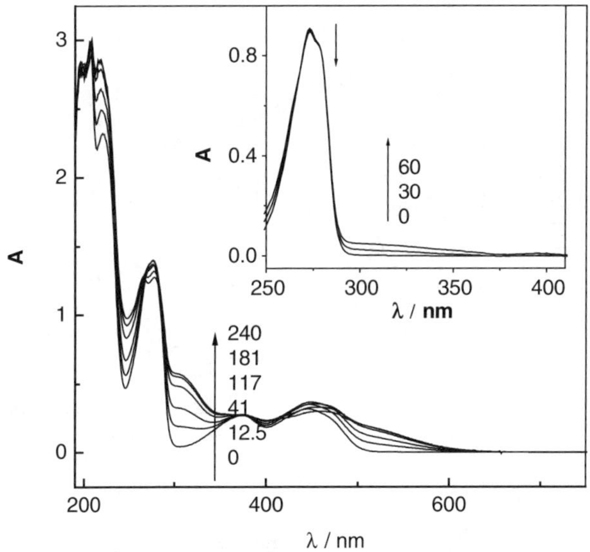

Figure 2. Changes in the UV–Vis absorption spectra of Riboflavin (0.027 mM)+Isoproterenol (0.33 mM). Inset: changes in the UV–Vis absorption spectra of Riboflavin (0.024 mM)+Phenylephrine (0.51 mM) vs. Riboflavin (0.024 mM). In both cases irradiation at 445 nm under air-saturated conditions was employed. Numbers in the spectra represent the irradiation time in minutes

electronically excited singlet and/or triplet states of Rf with the concomitant participation of ROS.

4.2.2. *Quenching of $O_2(^1\Delta_g)$ and $^1Rf^*$ by isoproterenol and Phenyephrine*

The values for overall quenching constants k_t (Scheme 1) evaluated through a Stern-Volmer treatment (see Figure 3, inset for a typical case) by time-resolved phosphorescence detection of $O_2(^1\Delta_g)$ (TRPD) for each SD in aqueous solution are shown in Table 1. This experiment unambiguously demonstrates an interaction between $O_2(^1\Delta_g)$ and D, which may be merely physical in nature (process (12)), purely reactive (process (13)) or a composition of both mechanisms simultaneously. Table 1 also collects the k_r values, independently determined by means of oxygen uptake employing Rose Bengal, RB, as a dye-sensitiser (Figure 3). This dye, possibly the most widely employed in $O_2(^1\Delta_g)$ research (15), generates $O_2(^1\Delta_g)$ with a quantum yield of 0.75 (25).

In the case of Iso, the first order representation (Figure 3) clearly shows a curvature when Rf was replaced by RB in the oxygen uptake experiments. This fact indicates that some other/others mechanisms, besides the $O_2(^1\Delta_g)$-mediated one, are operating. The initial rate of oxygen consumption upon Rf-photosensitisation was considerably diminished by the presence of 1 μg ml^{-1} SOD. This enzyme is a specific quencher for the species $O_2^{\bullet -}$, and is frequently employed

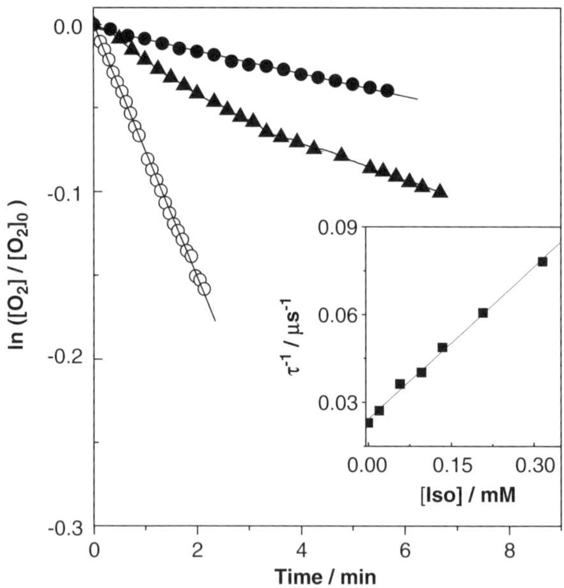

Figure 3. First-order plots for oxygen uptake upon visible-light irradiation in aqueous solutions containing: (○) Rose Bengal (Absorbance at 549 nm = 0.552)+Furfuryl Alcohol (0.69 mM); (○) Rose Bengal (Absorbance at 549 nm = 0.552)+Isoproterenol (0.78 mM); (▲) Riboflavin (0.045 mM)+Isoproterenol (15 mM). Inset: Stern-Volmer plot for the $O_2(^1\Delta_g)$ quenching by Isoproterenol

Table 1. Rate constants for the fluorescence quenching of Riboflavin, determined by dynamic methods (^1k_q) and by static methods (apparent $^1k_{q\ App}$), for the quenching of riboflavin excited triplet state (^3k_q) and for the overall (k_t) and reactive (k_r) quenching of singlet molecular oxygen by phenylephrine, isoproterenol, timolol, pindolol and atropines

Compound	$^1k_q \times 10^{-9}$ ($M^{-1}\,s^{-1}$)	$^1k_{q\ App} \times 10^{-9}$ ($M^{-1}\,s^{-1}$)	$^3k_q \times 10^{-9}$ ($M^{-1}\,s^{-1}$)	$k_t \times 10^{-8}$ ($M^{-1}\,s^{-1}$)^b	$k_r \times 10^{-8}$ ($M^{-1}\,s^{-1}$)
Isoproterenol[11]	2.5	nd	1.5	1.7	0.08
Phenylephrine	3.33	nd	1.6	2.2	0.003
Timolol[9,53]	4.6^a	nd	2.0^a	1.5	0.23
Pindolol[9,53]	7.2^a	nd	1.1^a	2.1	1.1
Atropine[54]	0.97	8.7	nq	0.09	0.01
Eucatropine[54]	0.20	0.30	nq	0.01	0.008
Homatropine[54]	2.5	2.5	0.07	0.005	$<$ 0.001
Scopolamine[54]	2.6	2.7	0.07	0.02	0.02

The measurements were done in water solutions, unless other solvent specified. ^{a}MeOH as a solvent. ^{b}D_2O as a solvent; nd: not determined; nq: no quenching of $^3Rf^*$ was observed.

to confirm/discard the participation of the oxygenated species in a given event.[14,21,33,34]

Experiments of oxygen uptake upon Rf-photosensitised irradiation of Phen, trolox and furfuryl alcohol (FFA), are shown in Figure 4. Oxygen consumption was scarcely detected for the system containing the sensitiser plus Phen 0.2 mM, whereas substantially higher rates of oxygen uptake could be observed when Phen was replaced by trolox or FFA in said concentration. Trolox is a known phenolic antioxidant considered the water-soluble analogue of α-tocopherol with regard to its properties as ROS scavenger.[35] FFA reacts with $O_2(^1\Delta_g)$ and is currently used as a reference compound for kinetic determinations in photodynamic action.[36,37]

Regarding the Rf fluorescence quenching by Iso and Phen, the presence of SD produces a decrease in the intensity of the steady-sate emission of Rf without any change in the shape of the emission spectrum. In parallel, the time decay of $^1Rf^*$ was evaluated in the presence and in the absence of SD by means of the SPC technique. In both cases, $K_{SV} > {}^1k_q{}^1\tau_0$, being K_{SV} the Stern-Volmer constant and $^1\tau_0$ the experimentally determined lifetime of $^1Rf^*$ (4.8 ns, in excellent agreement with data in the literature,[32] (Table 1). This behavior, shown in the Stern-Volmer plots of Figure 1, inset II, observed for Iso and Phen, corresponds to the typical case in which a fluorescent probe is simultaneously quenched by dark association with its ground state and by collisional interaction with its excited state.

It can be observed that although the quenching of $^1Rf^*$ by SD (process (3)) occurs with a rate constant close to the diffusion limit, only ~0.5% of the Rf excited singlet states are quenched in the concentration range of SD ~0.4 mM, employed as an upper limit in the photosensitised experiments. This suggest that the kinetic analysis should be focalised on those processes that involve $^3Rf^*$ and the ROS produced from this state.

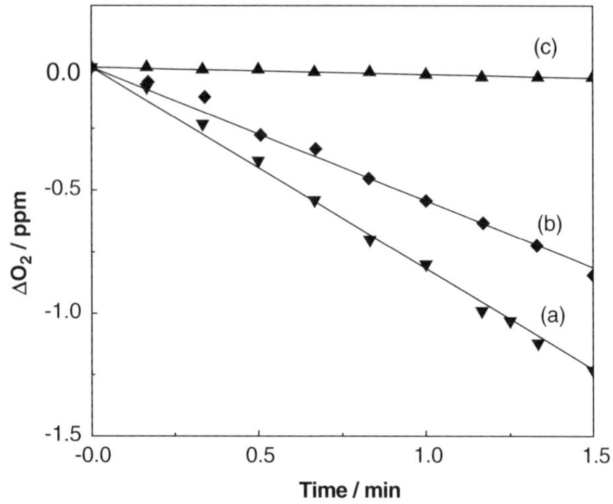

Figure 4. Rates of oxygen uptake upon visible-light irradiation (cut off 400 nm) of aqueous solutions containing: (a) Riboflavin (0.04 mM)+Phenylephrine (0.2 mM); (b) Riboflavin (0.04 mM)+Furfuryl Alcohol (0.2 mM) (c) Riboflavin (0.04 mM))+Trolox (0.2 mM)

4.2.3. The Interaction of Phenylephrine and Isoproterenol with $^3Rf^*$ and the Generation of $O_2^{\bullet-}$

The reduction of ^{3}Rf* lifetime in the presence of SD, demonstrates the occurrence of an interaction between the drugs and ^{3}Rf*. The bimolecular rate constants for this process (process (7)) are shown in Table 1. Plot II in Figure 8 shows the known transient absorption spectra of ^{3}Rf* obtained after the laser pulse[38–40] (trace (a)). The shape of the long-lived absorption, obtained in the presence D (*ca.* 95% ^{3}Rf* quenched by SD), is in good agreement with that reported for the semiquinone radical, $RfH^{\bullet}$.[18,41]

Quantum yields of 0.27 ± 0.03 (Phen) and 0.40 ± 0.04 (Iso) were obtained for $RfH^{\bullet}$ generation ($\Phi_{RfH^{\bullet}}$). As shown in Scheme 1, reactions (10) and (19) constitute pathways for $O_2^{\bullet-}$ and H_2O_2 production, being both oxygenated species good candidates to react with SD.

4.2.4. Photooxidation Products of Iso

As it is well known, catecholamines in general and Iso in particular, are photochemically unstable, upon direct UV irradiation, being aminochromes the main photoproducts.[26,42–44] Unpublished results by Massad and García, employing Rf as dye-sensitiser and through HPLC analysis, indicate that N-isopropylnoradrenochrome is the main product of Iso photooxidation.

4.2.5. Discussion

Although an upper limit of 0.4 and 0.28 was calculated for the quantum yield of $O_2^{\cdot-}$ production by Iso and Phen, respectively, (process (9) followed by (10)), the yield of such a mechanism could be appreciably diminished under air-saturated conditions, because of the competition between Iso (process (9)) and $O_2(^3\Sigma_g^-)$ (process (8)) toward $^3Rf^*$. Besides, in that case, although the species $O_2^{\cdot-}$ would not be formed through process (10), (reported $k_{10} = 1.4 \times 10^8$ M^{-1} s^{-1})[12,14] because of the protonation step (17), reactions (10) and (19) constitute pathways for $O_2^{\cdot-}$ and H_2O_2 production and Rf regeneration. The recovery of the pigment -through processes (18) and/or (19)- represents a crucial step in living organisms, in which it is well known that $O_2^{\cdot-}$ is a key intermediate in the oxygen redox chemistry.[33]

Within the typical sensitising concentrations employed, *ca.* 0.4 mM for Iso and Phen and approximately the same value for dissolved oxygen in water,[45] employing the determined values for ^3k_q and assuming a value $k_{ET} = 0.7\ 10^9$ M^{-1} s^{-1}(equivalent to 1/9[46] of the diffusion-controlled value in water,[47] results in ^3k_q [SD] $\cong 3\ k_{ET}\ [O_2(^3\Sigma_g^-)]$. This kinetic balance indicates that the generation of $Rf^{\cdot-}$, the $O_2^{\cdot-}$- precursor species, is, in the presence of either SD, *ca.* 3 times faster than the generation of $O_2(^1\Delta_g)$ and suggests that the $O_2^{\cdot-}$-mediated mechanism is the predominant pathway in the Rf-sensitised photooxidation of SD. The inhibition of oxygen uptake in the Rf-sensitised irradiation of Iso in the presence of SOD confirms the involvement of $O_2^{\cdot-}$ in the photoprocess. Degradation of Iso via the $Iso^{\cdot+}$ production (process (9)), magnified in anaerobic conditions, could also be operative in the presence of air, whereas $Rf^{\cdot-}$ is transformed through process (10) to ground state Rf. A rate constant value of 1.4×10^8 M^{-1}s^{-1} has been reported for such a process.[16] This is an important pathway in living organisms, since it constitutes a source of Rf recovery from the semireduced species.[33] Besides, results on Iso photoproducts upon Rf sensitisation (Section 4.2.4), confirm that $O_2^{\cdot-}$-mediated oxidation is in the dominant degradation pathway.

Regarding the interaction with $O_2(^1\Delta_g)$, the Iso molecule presents two centers (Scheme 1) susceptible to attack by the oxidative species: the catechol moiety and the isopropylamine group. A k_t value of 5.4×10^7 M^{-1} s^{-1} has been reported for catechol in D_2O,[48] not so far from the k_t value of 1.7×10^8 M^{-1} s^{-1} determined for Iso in the same solvent. Besides, a k_t value of 1.8×10^6 M^{-1} s^{-1} was reported for the quenching of $O_2(^1\Delta_g)$ by isopropylamine in $CHCl_3$[49] and it is known that the quenching of $O_2(^1\Delta_g)$ by aliphatic amines also proceeds by a dominant physical mechanism.[50] Although Iso also quenches $O_2(^1\Delta_g)$ with a reactive component, the ratio $k_r/k_t \sim 0.05$ is relatively low. This indicates that most of the deactivation of $O_2(^1\Delta_g)$ proceeds by a physical pathway, and that even under competitive conditions favorable to process (8), the $O_2(^1\Delta_g)$ mechanism can constitute only a minor contribution within the whole process of Rf-sensitised photooxidation of Iso because of the kinetic reasons already discussed.

As estimated by oxygen uptake and UV absorption experiments (Figure 2 inset and Figure 4), the photoreaction of Phen employing Rf as a photosensitiser, is extremely low and, according to experimental evidence, the mechanism should be due to one or more of the following processes: oxidation because of electron transfer mechanism (9), or oxygenation through reactions (14) and/or (16). The fate of $Rf^{\cdot -}$ and $Phen^{\cdot +}$ in the absence of $O_2(^3\Sigma^-{}_g)$ is possibly back electron transfer, as suggested by the results of Rf photoprotection exerted by Phen upon photoirradiation of the pigment.

Data shown in Figure 4 agree with that of absorption spectroscopy, in the sense that limited oxygen is consumed by Phen upon Rf-sensitisation. This scarce Phen degradation, in fact, not only confirms its poor proclivity toward a $O_2(^1\Delta_g)$-mediated oxidation, but also discards, in practice, a substantial chemical reactivity toward $O_2^{\cdot -}$. In other words, Phen complies with the requirements for a photoprotective agent provided that the Phen deactivates ROS without significant degradation.

Phen molecule presents two centers (Scheme 1) susceptible to attack by the oxidative species $O_2(^1\Delta_g)$: the phenolic moiety and the isopropylamine group. Literature reports indicate k_t values in the order of 10^7 M^{-1} s^{-1} for substituted phenols in water[21] and k_t values up to the order of 10^8 M^{-1} s^{-1} for different aliphatic amines in several solvents.[50] For both families of compounds it is currently accepted that bimolecular reactions with $O_2(^1\Delta_g)$[21] can take place through a mechanism that involves an polar encounter complex, as illustrated in Scheme 3. Physical quenching results from the intersystem crossing (ISC) within the encounter complex. In the case of Phen, the balance between physical quenching and chemical reaction, an event sensitive to spin-orbit coupling and entropy factors,[34] seems to be dominated by the ISC pathway, resulting in a practically exclusive physical deactivation of $O_2(^1\Delta_g)$.

The antioxidant properties of Phen may be estimated by comparison to trolox, a synthetic compound extensively recognised as an antioxidant in processes mediated by $O_2(^1\Delta_g)$ and radical species.[35] Although a k_t value of 3.5×10^8 M^{-1} s^{-1} has been determined in our laboratory for trolox in water,[35] very similar to the corresponding one for Phen, the k_r/k_t quotient, a measure of the degradability of a given substrate due to photooxidation by the species $O_2(^1\Delta_g)$, reaches a value of 0.63 for trolox,[35] as compared to the value $k_r/k_t = 0.001$ obtained for Phen. This photoprotective effect of Phen has been tested toward a vitally important photooxidisable biological target such as Trp. A concentration of 0.1 mM Phen reduces *ca.* 80% the rate of the RB-sensitised

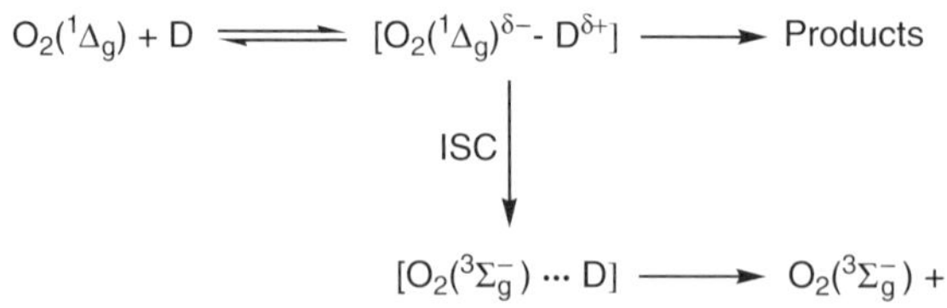

Scheme 3. Possible reaction pathways in the interaction of $O_2(^1\Delta_g)$ with bioactive drugs (D)

photooxidation of Trp, an amino acid frequently employed to monitor the extent of photooxidation of Trp-containing proteins.[51,52]

4.3. Riboflavin-Sensitised Photodegradation of the Ophthalmic Drugs Timolol, Pindolol and Atropines

4.3.1. Dark Complexation with Riboflavin and Riboflavin-Sensitised Photoconsumption of the Ophthalmic Drugs

Flavins in general, and very especially Rf, are probably the most extensively studied biomolecules with respect to their complexation ability with other molecules of biological and environmental relevance.[31] Nevertheless, no dark association complexes of Rf with Tim, Pin,[53] and the atropines Hom and Sco could be spectrophotometrically detected,[54] up to OD concentrations *ca.* 1 mM. The compounds Atr and Euc complex Rf ground state. The value for the apparent association constant Rf-Atr, reported by Criado *et. al.*,[54] was 3200 $M^{-1} \pm 5\%$ in water and $100 \pm 10\%$ M^{-1} in water-MeOH 1:1 (v/v). In the case of Euc the spectral perturbations were hardly detected and the respective values for ground state complexation with Rf were not evaluated.[54]

The solvent effect on the value of the association constant for the dark-interaction between Rf and Atr, suggests the presence of a charge-transfer-mediated complex. This behavior has already been observed for dark associations of Rf with aromatic compounds[31] and amino acids,[55] being the interaction attributed to a combination of a weak charger-transfer process and hydrophobic forces.

The fact that structurally similar AT show very different tendency to form dark complexes with Rf is, in principle, an unexpected result. Nevertheless, as emphasised by Slifkin,[31] the characteristic of the charge-transfer interactions are highly dependent on the molecular orientation of the donor and acceptor, and on the contribution of other weak binding forces such as hydrophobic forces. Hence, the characteristics of the charge-transfer interaction of both the donor and the flavin may be different for apparently similar compounds with minor structural differences, as the highly methylated (hydrophobically enhanced) compound Hom or the presence of the ethylene oxide group in compound Sco. It is remarkable that both AT forming dark-complexes with Rf have practically the same half-wave oxidation potential.[54]

The visible/UV absorption spectrum of the systems Tim or Pin /Rf/oxygen (air), in water-MeOH, (9:1 v/v) and AT/Rf/oxygen (air), suffer modifications when the mixtures are irradiated with light of wavelength higher than 400 nm, indicating chemical changes in both the OD and the dye (Figure 5).

Additionally, oxygen uptake experiments indicate the participation of reactive oxygen species in the photopromoted process. In all cases the rate of oxygen consumption was reduced in the presence of sodium azide (1 mM). Only in the cases of Tim and Pin did the presence of SOD, (1 mg per 100 ml) produce

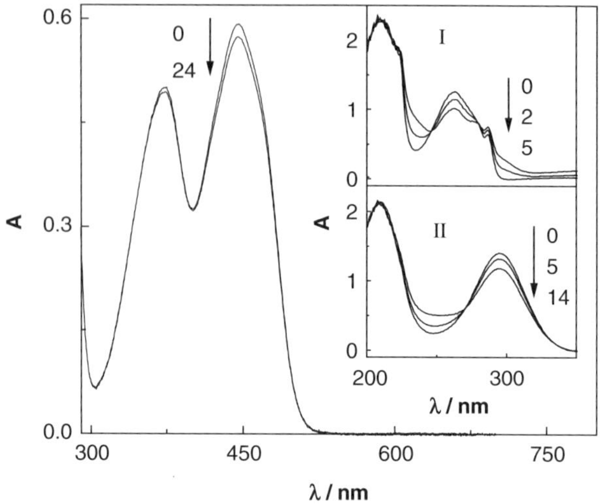

Figure 5. Spectral changes in the visible spectral range of Rf (0.06 mM) upon photoirradiation (cut off 400 nm). Inset I: Pindolol (0.2 mM)+Riboflavin (0.06 mM) vs. Riboflavin (0.06 mM). Inset II: Spectral evolution of Timolol (0.44 mM)+Riboflavin (0.06 mM) vs. Riboflavin (0.06 mM). Solvent water-MeOH (9:1, v/v). Numbers in the spectra represent the irradiation time in minutes

a delay in the rate of oxygen uptake (Figure 6). Sodium azide is a well known[50] selective physical quencher of O_2 ($^1\Delta_g$), although the salt also efficiently reacts with excited triplet Rf by an electron-transfer mechanism.[13]

The rate of the photoxidation of Rf (0.06 M), in N_2-saturated aqueous solutions, decrease in the presence of Tim and Pin in the sub-mM concentration range. This fact suggests that specific interactions of OD and excited triplet Rf can occur (process (9)). The effect is much less intense for Hom and Sco and imperceptible for the cases of Atr and Euc.

4.3.2. The Interaction of Ophthalmic Drugs with $^1Rf^*$, $^3Rf^*$ and the Generation and Quenching of $O_2^{\bullet-}$ and $O_2(^1\Delta_g)$

In air-equilibrated methanolic solution, Rf exhibits an intense fluorescence emission centered at 525 nm. In the presence of $\geq$ 10 mM OD, the fluorescence lifetime of ^{1}Rf* decrease. Figure 7 shows the fluorescence decay of Rf (0.03 mM) in MeOH in the absence and in the presence of Pin 10 mM. Table 1, includes the rate constants for the interaction ^{1}Rf*-OD (process (3)), determined by Criado *et al.*[53] Again as in the case of SD, although the interaction ^{1}Rf*-OD occurs with a rate constant close to the diffusion limit, relatively high concentrations of OD are necessary to appreciably hinder Rf triplet state population (process (4)).

The lifetime of ^{3}Rf* experiments a considerable decrease in the presence of Tim and Pin in the mM range of concentrations, as shown, in Figure 8, plot I,

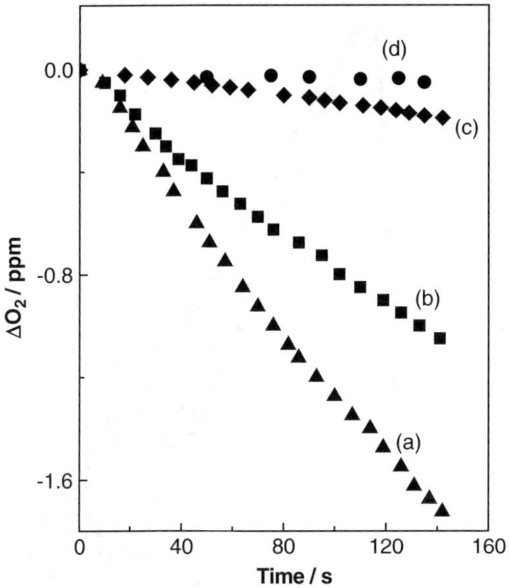

Figure 6. Oxygen consumption upon Riboflavin (0.06 mM)-sensitised photoirradiation of: (a) Pindolol 0.5 mM; (b) Pindolol (0.5 mM)+Superoxide Dismutase (1 mg/100 ml); (c) Timolol (0.5 mM); (d) Pindolol (0.5 mM)+Sodium Azide (1 mM). Solvent water-MeOH (9:1, v/v)

for a typical case. The bimolecular rate constants for the deactivation of $^3Rf^*$ (process (9)) are shown in Table 1. Sco and Hom also quench $^3Rf^*$, with much lower rate constants, whereas the presence of Atr or Euc, even at relatively high concentration, do not modify at all triplet Rf lifetime (Table 1).

The transient absorption spectrum of the Rf solution 0.1 μs after the laser pulse (Figure 8, plot II, trace (a)) is similar to that reported for the Rf neutral triplet state in water or MeOH.[18] In the presence of Tim, Pin, Sco and Hom 40 μs after the laser pulse, a different spectrum was observed[53,54] (Figure 8, plot II, trace (b)) with a shape similar to that habitually identified with the well known species $RfH^{\cdot}$(see Section 4.2.3).

The $^3Rf^*$ quenching by OD is due to an electron transfer process from OD to the pigment, with the concomitant production of the semireduced pigment (reaction (9), Scheme 1). This reaction has already been described for Rf-photosensitised processes toward different substrates.[18,56–58] The potential contribution of $O_2(^1\Delta_g)$ in the Rf-photosensitised degradation of OD, was evaluated, employing the sensitisers RB for the group of the AT [9] and perinaphthenone-2-sulfonic acid (PNS), an exclusive $O_2(^1\Delta_g)$-generator, for Tim and Pin.[9] The rate constants k_t and k_r accounting for the respective processes (12)+(14) and (14) are shown in Table 1. A pronounced increase of k_t going from MeOD to D_2O:MeOD (9:1 v/v) was observed for Tim and Pin.[9]

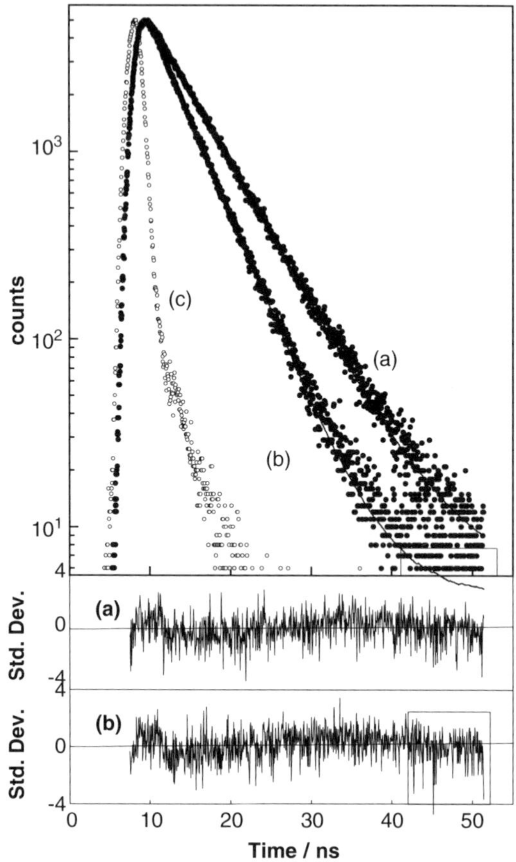

Figure 7. Fluorescence decay signals (SPC). Signal (a): Riboflavin 0.03 mM; signal (b) Riboflavin 0.03 mM+Pindolol 10 mM; signal (c) the lamp. Excitation and emission wavelengths of 445 and 515 nm, respectively. Solvent: MeOH

4.3.3. Photoproducts in the O_2 ($^1\Delta_g$)-Mediated Oxidation of Pindolol and Timolol

A mass spectrometry study of the photooxidation products of Pin and Tim, indicate that the mechanistic pathways involve $O_2(^1\Delta_g)$ attack to the Pin indole ring and oxidation of the pindolol isopropyl or timolol terbutyl methyl groups.[10]

The final photooxidation products for Pin are salicylaldehyde, aminosalycylaldehyde and o-substituted derivatives. A similar mechanism, namely the breakage of the indolic group of tryptophan through reaction leading to the formation of N-formyl kynurenine and L-kynurenine was also proposed by Langlois *et al.*[59]

In the case of Tim only one photoproduct retains the morpholine-thiadiazol structure and the other two products are aliphatic amino acids. They are the

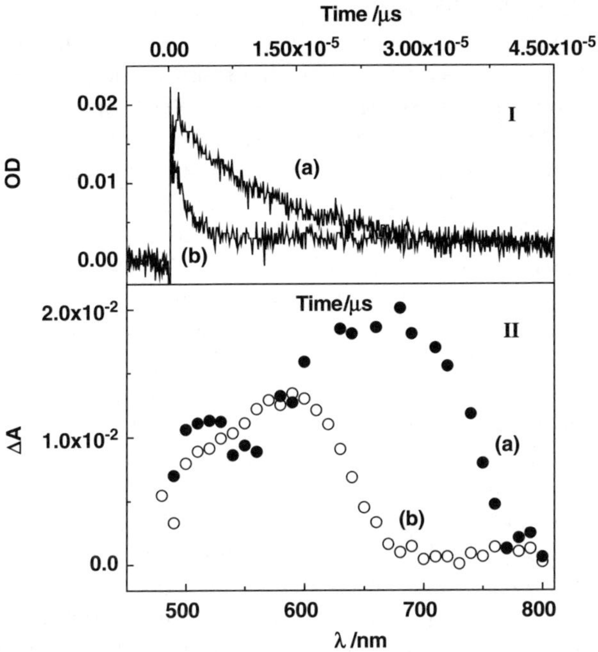

Figure 8. Plot I: Absorption-time profiles of triplet excited Riboflavin in the absence (a) and in the presence (b) of Timolol 0.08 mM. Plot II: Transient absorption spectra of triplet excited Riboflavin 0.1 μs after the laser pulse (trace (a)) and 40 μs after the laser pulse, in the presence of Timolol 0.08 mM (trace (b)). Solvent MeOH

result of the oxidation of the terbutyl moiety. This step would render only one carboxyl group which would inhibit further oxidation of the aldehyde groups.[10]

4.3.4. Discussion

From ^{3}Rf* four main competitive pathways can occur[56,60] (Scheme 1). Two of them (reactions (6 and 9)) implying Rf and/or OD consumption, while the other two steps (5 and 8) are merely physical processes. The dominant mechanism depends on the competition between $O_2(^3\Sigma_g{}^-)$ and OD for the quenching of ^{3}Rf*.

Employing a similar kinetic analysis as for SD, it can be demonstrated that in the cases of Tim and Pin the quenching of ^{3}Rf* by OD prevails against the quenching by O_2 $(^3\Sigma_g{}^-)$, prevailing also the generation of Rf $^{\cdot-}$ through processes (9) against the generation of $O_2(^1\Delta_g)$.

On the contrary, for the AT that effectively quench ^{3}Rf* state, the generation of another reactive oxygen species, such as O_2 $({}_1\Delta_g)$ (process (8)), favorably competes with reaction (9). For identical concentrations of AT and dissolved molecular oxygen, the rate of ^{3}Rf* quenching by AT will be two orders of magnitude smaller than the rate of $O_2({}_1\Delta_g)$ production. In other words: oxygen consumption upon Rf photosensitised irradiation of AT is mainly because of an $O_2(_1\Delta_g)$-mediated process, and practically no contribution of $O_2^{\cdot -}$ mediated reactions operates. This is a very interesting result in the context of photopromoted interactions in living systems for which the $O_2^{\cdot -}$-mediated reactions are believed to be one of the most important sources of environmental photodegradation. The whole process of $O_2^{\cdot -}$ generation is inhibited in the presence of Hom and Sco, whereas Atr and Euc do not quench at all $^3Rf^*$, the electronic state precursor of $O_2^{\cdot -}$. In this case the degradation of the Rf can occur through reaction (13) for which a value of $6 \times 10^7\ M^{-1}\ s^{-1}$ has been reported.[17]

It was reported[9]for Tim and Pin that the respective k_r/k_t values do not parallel the values for the relative rates of oxygen uptake in the Rf-sensitised process. This fact indicates that when Rf is the sensitiser a second source of oxygen consumption is operating, besides the $O_2(_1\Delta_g)$-mediated mechanism. This additional source can be a $O_2^{\cdot -}$ as supported by the inhibitory effect in oxygen consumption exerted by SOD and also by the spectral evidence for the generation of the transitory species $Rf^{\cdot -}$. The fate of this species $Rf^{\cdot -}$ in the presence of Tim and Pin but in the absence of O_2 ($^3\Sigma^-{}_g$) is possibly back electron transfer, as suggested by the results of photoprotection exerted by OD upon anaerobic photoirradiation of Rf. The solvent effect on k_t for Pin and Tim[9] confirms the participation of a polar encounter complex O_2 ($^1\Delta_g$)-OD (Scheme 3). This fact has already been reported[61] for a set of substituted indoles for which a series of five solvents of different polarity were tested. The efficiency of the process $O_2(^1\Delta_g)$-mediated is significantly higher for Pin, whereas Tim mainly behaves as a physical O_2 ($^1\Delta_g$) deactivator.

Regarding the O_2 ($^1\Delta_g$)-mediated- photooxidation products of Tim and Pin, the high degree of oxidation observed in the identified products suggests that important thermal reactions following the primary interaction between the ROS and OD take place.[10]

Acknowledgments

Financial support from Consejo Nacional de Investigaciones Científicas y Técnicas (CONICET), Agencia Nacional de Promoción Científica y Tecnológica (ANPCyT) and Secretaría de Ciencia y Técnica de la Universidad Nacional de Río Cuarto (SECyT UNRC), all from Argentina, is gratefully acknowledged.

References

1. H.H. Tonnesen and D.E. Moore, Photochemical stability of biologically active compounds. III. Mefloquine as a photosensitizer, *Int. J. Pharm.*, 1991, **70**, 95–101.
2. R.C. Straight and J.D. Spikes, *Photosensitised oxidation of biomolecules*, In: *Singlet Oxygen*, A.A. Frimer, (ed.), **vol. 4**. CRC Press, Boca Raton, 1985, 91–143.
3. A. Albini and E. Fasani, *Drugs: Photochemistry and Photostability*, The Royal Society of Chemistry, Cambridge, 1998.

4. A.M. Edwards and E. Silva, Effect of visible light on selected enzymes vitamins and amino acids, *J. Photochem. Photobiol. B: Biol.*, 2001, **63**, 126–131.
5. A. Momzikoff, R. Santus and M. Giraud, A study of the photosensitizing properties of seawater, *Mar. Chem.*, 1983, **12**, 1–14.
6. A. Posadaz, E. Sánchez, M.I. Gutiérrez, M. Calderón, S. Bertolotti, M.A. Biasutti and N.A. García, Riboflavin and rose bengal sensitized photooxidation of sulfathiazole and succinylsulfathiazole. Kinetic study and microbiologiacal implications, *Dyes Pigm.*, 2000, **45**, 219–228.
7. A. Pajares, J. Gianotti, G. Stettler, S. Bertolotti, S. Criado, A. Posadaz, F. AmatGuerri and N.A. Garcia, Modelling the natural photodegradation of water contaminants – a kinetic study on the light-induced aerobic interactions between riboflavin and 4-hydroxypyridine, *J. Photochem. Photobiol. A: Chem.*, 2001, **139**, 199–204.
8. J.N. Chacon, J. McLearie and R.S. Sinclair, Singlet oxygen yields and radical contributions in the dye-sensitised photo-oxidation in methanol of esters of polyunsatured fatty acids (oleic, linoleic, linolenic and arachidonic), *Photochem. Photobiol.*, 1988, **47**, 647–656.
9. S. Criado, D.O. Mártire, P.E. Allegretti, J. Furlong, S.G. Bertolotti, E. La Falce and N.A. García, Singlet molecular oxygen generation and quenching by the antiglaucoma ophthalmic drugs, *Timolol and Pindolol, Photochem. Photobiol. Sci.*, 2002, **10**, 788–792.
10. S. Criado, D. Mártire, P. Allegretti, J. Furlong, S. Bertolotti, E. La Falce and N.A. García, Mass spectrometric study of the photooxidation of the ophtalmic drugs timolol and pindolol, *Pharmazie*, 2003, **58**, 551–553.
11. W.A. Massad, S. Bertolotti and N.A. García, Kinetics and Mechanism of the Vitamin B_2-sensitized Photooxidation of Isoproterenol, *Photochem. Photobiol.*, 2004, **79**, 428–433.
12. C.-Y. Lu, W.-F. Wang, W.-Z. Lin, Z.-H. Han, S.-D. Yao and N.-Y. Lin, Generation and photosensitization properties of the oxidized radicals of riboflavin: a laser flash photolysis study, *J. Photochem. Photobiol. B Biol.*, 1999, **52**, 111–116.
13. C.Y. Lu, W.Z. Lin, W.F. Wang, Z.H. Han, S.D. Yao and N.Y. Lin, Riboflavin (VB_2) photosensitized oxidation of 2′-deoxyguanosine-5′-monophosphate (dGMP) in aqueous solution: a transient intermediates study, *Chem. Phys.*, 2000, **2**, 329–334.
14. C. Lu, G. Bucher and W. Sander, Photoinduced interactions between oxidized and reduced lipoic acid and Riboflavin (Vitamin B2), *Phys. Chem.*, 2004, **5**, 47–56.
15. F. Wilkinson, W.P. Helman and A.B. Ross, Quantum yields for the photosensitized formation of the lowest electronically excited singlet state of molecular oxygen in solution, *J. Phys. Chem. Ref. Data*, 1993, **22**, 113–262.
16. C.M. Krishna, S. Uppuluri, P. Riesz, J.S. Zigler and D. Balasubramanian, A study of the photodynamic efficiencies of some eye lens constituents, *Photochem. Photobiol.*, 1991, **54**, 51–58.
17. J.N. Chacon, J. McLearie and R.S. Sinclair, Singlet oxygen yields and radical contributions in the dye-sensitised photo-oxidation in methanol of esters of polyunsatured fatty acids (oleic, linoleic, linolenic and arachidonic), *Photochem. Photobiol.*, 1988, **47**, 647–656.
18. S.G. Bertolotti, C.M. Previtali, A.M. Rufs and M.V. Encinas, Riboflavin/triethanolamine as photoinitiator system of vinyl polymerization. A mechanistic study by laser flash photolysis, *Macromolecules*, 1999, **32**, 2920–2924.
19. P.F. Heelis, B.J. Parsons, G.O. Phillips and J.F. McKellar, The flavin sensitised photooxidation of ascorbic acid – A continuous and flash photolysis study, *Photochem. Photobiol.*, 1981, **33**, 7–13.

20. G. Bocco, M. Luiz, M.I. Gutiérrez and N.A. Garcia, Influence of the nuclear substitution on the Sensitized Photooxidation of model compounds for phenolic-type pesticides, *J. Prakt. Chem.*, 1994, **336**, 243–246.
21. N.A. García, Singlet molecular oxygen-mediated photodegradation of aquatic phenolic pollutants, *J. Photochem. Photobiol. B Biol.*, 1994, **22**, 185–196.
22. I. Fridovich, *Free Radicals in Biology*, **vol 2**, Academic Press, New York, 1972.
23. J.D. Spikes, H.-R. Shen, P. Kopeccková and J. Kopecek, Photodynamic crosslinking of proteins. III. Kinetics of the FMN- and Rose Bengal-sensitized photooxidation and intrermolecular crosslinking of model tyrosine-containing N-(2-hydroxypropyl) methacrylate copolymers, *Photochem. Photobiol.*, 1999, **70**, 130–137.
24. M. Litter, *Farmacología Experimental y Clínica*, El Ateneo, Buenos Aires, 1988.
25. A. Kamal-Eldin and L.-A. Appelqvist, The chemistry and antioxidant properties of tocopherols and tocotrienols, *Lipids*, 1996, **31**, 671–701.
26. I. Kruk, The identification by electron spin resonance spectroscopy of singlet oxygen formed in the photooxidation of catecholamines, *Z. Phys. Chemie, Leipzig*, 1985, **266**, 1239–1242.
27. N.J.D. Mol, G.M.J. Beijersbergen Van Henegouwen and K.W. Gerritsma, Photochemical decomposition of catecholamines-II. The extent of aminochrome formation from adrenaline, isoprenaline and noradrenaline induced by ultraviolet light, *Photochem. Photobiol.*, 1979, **29**, 479–482.
28. P. Watson and J. Stjernnschantz, The latanoprost study group, a six-month, randomized, double-masked study comparing latanoprost with timolol in open-angle glaucoma and ocular hypertension, *Ophthalmol.*, 1996, **103**, 126–137.
29. M. Araie, M. Sekine, Y. Suzuki and N. Koseki, Factors contributing to the progression of visual field damage in eyes with normal-tension glaucoma, *Ophthalmol.*, 1994, **101**, 1440–1444.
30. A.C. Chou, Y.F. Shih, T.C. Ho and L.L.K. Lin, The effectiveness of 0.5 percent atropine in controlling high myopia in children, *J. Ocular Pharm. Ther.*, 1997, **13**, 61–67.
31. M.A. Slifkin, *Charge Transfer Interactions of Biomolecules*, Academic Press, London, 1971.
32. P.F. Heelis, The photophysical and photochemical properties of flavins (isoalloxazines), *Chem. Soc. Rev.*, 1982, **11**, 15–39.
33. J.R. Kanofsky, Singlet oxygen production from the reactions of superoxide ion in aprotic solvents: Implications for hydrophobic biochemistry, *Free Rad. Res. Comms.*, 1991, **12–13**, 87–92.
34. A.A. Gorman, I.R. Gould, I. Hamblett and M.C. Standen, Reversible exciplex formation between singlet oxygen, $^1\Delta_g$, and vitamin E. Solvent and temperature effects, *J. Am. Chem. Soc.*, 1984, **106**, 6956–6959.
35. S. Nonell, L. Moncayo, F. Trull, F. Amat-Guerri, E.A. Lissi, A.T. Soltermann, S. Criado and N.A. Garcia, Solvent influence on the kinetics of the photodynamic degradation of trolox, a water-soluble model compound for vitamin E, *J.Photochem. Photobiol. B Biol.*, 1995, **29**, 157–162.
36. W.R. Hagg and J. Hoigné, Singlet oxygen in surface waters. 3. Photochemical formation and steady-state concentrations in various types of waters, *Environ. Sci. Technol.*, 1986, **20**, 341–348.
37. G. Jori, Photosensitized reactions of amino acids and proteins, *Photochem. Photobiol.*, 1975, **21**, 463–467.
38. A. Pajares, J. Gianotti, G. Stettler, S. Bertolotti, S. Criado, A. Posadaz, F. AmatGuerri and N.A. Garcia, Modelling the natural photodegradation of water

contaminants – a kinetic study on the light-induced aerobic interactions between riboflavin and 4-hydroxypyridine, *J. Photochem. Photobiol. A: Chem.*, 2001, **139**, 199–204.
39. E. Haggi, S. Bertolotti, S. Miskoski, F. Amat-Guerri and N.A. Garcia, Environmental photodegradation of pyrimidine fungicides. – Kinetics of the visible-light-promoted interactions between riboflavin and 2-amino-4-hydroxy-6-methylpyrimidine, *Can. J. Chem.*, 2002, **80**, 62–67.
40. I. Gutiérrez, S. Criado, S. Bertolotti and N.A. García, Dark and photoinduced interactions between Trolox, a polar-solvent-soluble model for vitamin E, and riboflavin, *J. Photochem. Photobiol. B Biol.*, 2001, **62**, 133–139.
41. M.V. Encinas, A.M. Rufs, S. Bertolotti and C.M. Previtali, Free radical polymerization photoinitiated by riboflavin/amines. Effect of the amine structure, *Macromolecules*, 2001, **34**, 2845–2847.
42. N.J.D. Mol, G.M.J. Beijersbergen Van Henegouwen and K.W. Gerritsma, Photochemical decomposition of catecholamines-II. The extent of aminochrome formation from adrenaline, isoprenaline and noradrenaline induced by ultraviolet light, *Photochem. Photobiol.*, 1979, **29**, 479–482.
43. R.A. Heacock and B.D. Scott, The Chemistry of the "aminochromes". Part IV. Some new aminochromes and their derivatives, *Can. J. Chem.*, 1960, **38**, 516–527.
44. L.S. Jahnke and A.W. Frenkel, Photooxidation of epinephrine sensitized by methylene blue-evidence for the involvement of singlet oxygen and of superoxide, *Photochem. Photobiol.*, 1978, **28**, 517–523.
45. S.L. Murov, I. Carmichael and G.L. Hug, *Handbook of Photochemistry*, Marcel Dekker Inc., New York, 1973.
46. M. Koizumi, S. Kato, N. Mataga, T. Matsuura and I. Isui, *Photosensitized Reactions*, Kagakudogin, Kyoto, 1978.
47. J. Calvert and J. Pitts Jr, *Photochemistry*, Wiley, New York, 1966.
48. D.O. Mártire, S.E. Braslavsky and N.A. García, Sensitised photooxidation of dihydroxybenzenes and chlorinated derivatives. A kinetic study, *J. Photochem. Photobiol. A: Chem.*, 1991, **61**, 113–124.
49. B.M. Moore, Quenching of singlet oxygen by aliphatic amines, *J. Phys. Chem.*, 1991, **81**, 1681–1684.
50. F. Wilkinson, W.P. Helman and A.B. Ross, Rate constants for the decay and reactions of the lowest electronically excited state of molecular oxygen in solution. An extended and revised compilation, *J. Phys. Chem. Ref. Data*, 1995, **24**, 663–1021.
51. A. Biasutti, A.T. Soltermann and N.A. Garcia, Photodynamic Effect in Lysozime. A kinetic study in different micellar media., *J. Pept. Res.*, 2000, **55**, 41–50.
52. M.A. Biasutti, A. Posadaz and N.A. Garcia, A comparative kinetic study on the singlet molecular oxygen-mediated photooxidation of α- and β-chimotrypsines., *J. Pept. Res.*, 2003, **62**, 11–18.
53. S. Criado and N.A. Garcia, Vitamin B2-sensitised photooxidation of the ophthalmic drugs Timolol and Pindolol. Kinetics and mechanism., *Redox Rep.*, 2004, **9**, 291–297.
54. S. Criado, C. Guardianelli, J. Tuninetti, P. Molina and N.A. Garcia, Scavenging of photogenerated oxidative species by antimuscarinic drugs: Atropine and derivatives, *Redox Rep.*, 2002, **7**, 358–394.
55. N.A. Garcia, J. Silber and C. Previtali, Interaction of aliphatic amino acids with riboflavin, *Tetrahedron Lett.*, 1977, **24**, 2073–2076.
56. P.F. Heelis, *The photochemistry of flavins*, in *Chemistry and Biochemistry of Flavoenzymes*, F. Muller, (ed.), **vol 1**, CRC Press, Boca Ratón, 1991, pp. 171–193.

57. S. Miskoski and N.A. Garcia, Effect of cholorphenolic pesticides on the photochemistry of riboflavin, *Toxicol. Environ. Chem.*, 1989, **25**, 33–43.
58. S. Miskoski and N.A. Garcia, Dark and photoinduced interactions between riboflavin and indole auxins, *Collect. Czech. Chem. Commun.*, 1991, **56**, 1838–1849.
59. R. Langlois, H. Ali and J.E. Van Lier, Réaction du L-tryptophane avec le radical peroxyle du trichlorocarbone et l'oxygène singulet: formation de diffèrents paires d'hydroperoxydes isomèriques, *J. Chim. Phys.*, 1993, **90**, 985–999.
60. C.S. Foote, Definition of Type I and Type II Photosensitized oxidation, *Photochem. Photobiol.*, 1991, **54**, 659.
61. M.C. Palumbo, N.A. García and G.A. Argüello, The interaction of singlet molecular oxygen, $O_2(^1\Delta_g)$, with indolic derivaties. Distinction between physical and reactive quenching, *Photochem. Photobiol.*, 1990, **7**, 33–42.

Chapter 5

The Antiviral and Antibacterial Properties of Riboflavin and Light: Applications To Blood Safety and Transfusion Medicine

RAYMOND P. GOODRICH,[a] RICHARD A. EDRICH,[a] LAURA L. GOODRICH,[a] CYNTHIA A. SCOTT,[a] KEITH J. MANICA,[a] DENNIS J. HLAVINKA,[a] NICK A. HOVENGA, ERIC T. HANSEN,[a] DEANNA GAMPP,[a] SHAWN D. KEIL,[a] DENISE I. GILMOUR,[a] JUNZHI LI,[a] CHRISTOPHER B. MARTIN[b] AND MATTHEW S. PLATZ[c]

[a] Navigant Biotechnologies, Inc., Lakewood, CO, USA
[b] Lamar University, Beaumont, TX, USA
[c] The Ohio State University, Columbus, OH, 43210, USA

5.1. Background

In the developed world, blood products are extensively tested for the presence of pathogens prior to administration. Nevertheless, there exists a small, but finite risk of transmission of infectious agents in transfusion medicine.[1] The risks of viral infection are due to the "window period"; the period of time between the infection of a donor and the development of detectable levels of antibodies.[2,3] Nucleic Acid Amplification Testing (NAT) was introduced for Human Immunodeficiency Virus (HIV) and Hepatitis C Virus (HCV) in the United States in 1998 and has shortened the window period and further decreased the incidence of transfused units of blood products containing pathogens.[4,5] Nevertheless, research efforts to develop a methodology for the photochemical pathogen reduction of blood products remains an active field. Psoralens,[6] methylene blue,[7–9] porphyrins,[10,11] and more recently riboflavin[12] are under study, as photosensitizers. The goal of such programs is both to decrease the window period for known and tested diseases even further and to protect against emerging pathogens for which testing is not presently a viable alternative. Examples of concerns in this area include emergence of parasitic disease-causing agents, such as leishmaniasis, chagas, and babesia into the general population.

Riboflavin (RB, vitamin B2) is a vitamin essential to humans. It is present in aerobic organisms and is found in many foodstuffs, such as milk, beer, eggs, yeast, and leafy vegetables. It is also the precursor for flavin mononucleotide (FMN) and flavin adenine dinucleotide (FAD), which are major coenzymes that participate in a number of one-electron processes in the human body.[13] Its ubiquitous nature, constant systemic exposure through foodstuffs, and the extensive information regarding its toxicological profile make it an ideal candidate for application in a blood safety process.

Riboflavin has absorption maxima at 220, 265, 375, and 446 nm in water and is yellow-orange in color. When aqueous solutions containing RB are exposed to sunlight, RB is converted into lumichrome (LC) under neutral conditions, and into lumiflavin (LF) in alkaline solutions.[13,14] LC is also a known metabolic breakdown product of RB in the human body.[15] Flavin systems are known to be photochemically active, and the products of flavin photochemistry are known.[13,15,17]

A great deal of information on the photochemistry of riboflavin has been reported recently. Navigant Biotechnologies™ has described the use of riboflavin-sensitized eradication of pathogens present in blood products.

This chapter attempts to summarize this work. The mechanism of pathogen reduction using riboflavin likely involves three potential pathways: Type I Photochemistry,[18,19] Type II Photochemistry,[20] and the effects of UV light alone.[21] The contribution of each of these three pathways to the MIRASOL PRT System pathogen reduction process is currently being studied by NAVIGANT BIOTECHNOLOGIES™.

The reported mode of action of riboflavin in the reduction of pathogens is postulated to be based on the ability of riboflavin to interact with nucleic acids and to undergo chemistry with those nucleic acids upon exposure to light. This chemistry is believed to involve both oxygen-dependent and oxygen-independent (electron transfer) processes. It has been described thoroughly in the chemical literature over the past several decades.[22–29] The use of UV light with platelets and plasma also affords a third contributor to pathogen kill *via* the direct action of light.

The potential application of riboflavin to pathogen inactivation has several appealing features. Blood products are used extensively in medical practice. On a worldwide basis, over 40 million units of blood are collected and transfused annually. Despite years of research, no suitable substitute for blood has been identified. Its routine use in medicine also gives it the potential to act as a vehicle for the introduction and transmission of pathogens. The example of HIV transmission through blood components, which occurred from the early to late 1980s, posed a significant issue to the healthcare community. In the days prior to identification of the causative agent for AIDS and suitable tests for detection, many cases of transfusion-related disease transmission occurred. The ramifications from this episode still reverberate today throughout the world.

The introduction of methods, such as solvent-detergent treated plasma for fractionated plasma and plasma used for direct transfusion (Fresh Frozen Plasma (FFP)) has effectively eliminated the transmission of HIV, HCV, and Hepatitis B Virus (HBV) *via* these treated blood products. These methods are limited, however, owing to the nature of their mode of action in destroying viral envelopes. Such chemistry is not possible with cellular blood components, such as platelets and red cells because these cells would also be destroyed in the treatment process. A photochemical-based method that can inactivate pathogens while preserving cell and protein functionality would be highly desirable and could afford the same levels of protection against disease transmission that methods for plasma and plasma-derived products enjoy today.

5.2. The Action Spectrum of Riboflavin Photosensitized Inactivation of λ Phage

The action spectra (AS) of the inactivation of λ phage in PBS with 0, 50, 100, 150, and 200 μM riboflavin was determined between 266–575 nm to gain a better appreciation of the molecular mechanism of action of the sensitizer. Each irradiated sample contained 2.5 mL of solution in a 3.0 mL quartz cuvette.

Each data point was performed in triplicate and the reported value reflects the average of these three values.

AS: 266–304 nm. It is well known that UV radiation with wavelengths of less than or equal to 304 nm is lethal to cells.[30,31] Thus, the AS for the inactivation of λ phage in PBS from 266–304 nm required only a small amount of incident energy, 0.1 J mL^{-1}, to achieve a maximum phage reduction of 2.5 logs in this region. As the AS approaches 304 nm, the reduction of phage drops off sharply as the absorption curve of DNA approaches zero (Figure 1).

Phage inactivation is dominated by direct light absorption by the phage from 266–304 nm. Not only does the presence of RB fail to increase the extent of phage inactivation, but also the presence of RB reduces the amount of phage inactivation by acting as a screening agent. This occurs because of the decrease in the amount of light that is absorbed by the phage because riboflavin is absorbing this incident light. This screening effect is proportional to the concentration of RB in solution. Because the competitive absorption of RB dominates any sensitization of the phage, it is concluded that direct absorption of light by the phage is much more efficient at phage inactivation compared to sensitized inactivation in this region.

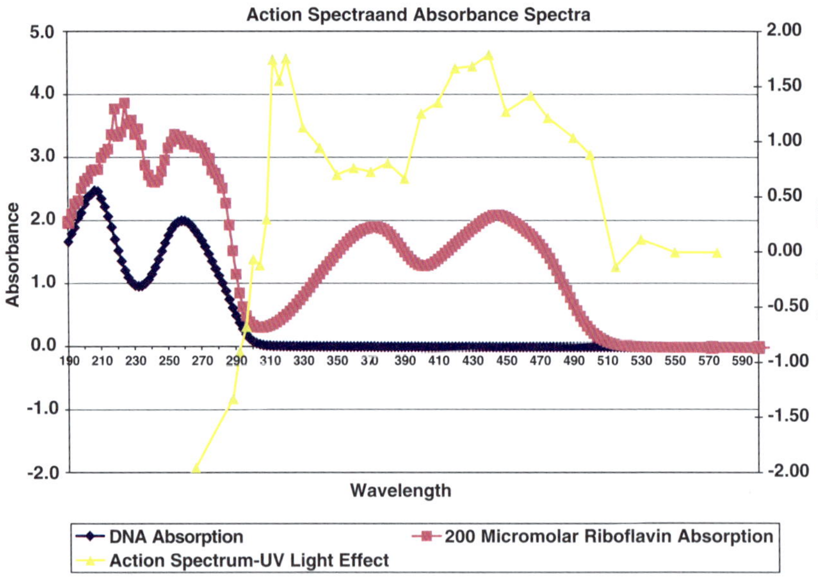

Figure 1. Action spectrum of riboflavin and lambda phage minus light alone (yellow) over-laid with the absorbance profile of riboflavin in PBS (pink) and absorbance of DNA in PBS (blue). At wavelengths lower than 300 nm, riboflavin acts to shield the effects on DNA due to the direct action of UV light. Greater levels of inactivation in the presence of riboflavin occur at wavelengths between 300 and 350 nm compared to the prediction due to the absorbance profile of free riboflavin in solution. This is also observed for wavelengths higher than 500 nm

AS: 308–575 nm. To achieve the same magnitude of log reduction that was observed between 266 and 304 nm, the energy required for the experiments between 308 and 575 nm was increased 50-fold from 0.1 to 5.0 J mL^{-1}. There is no inactivation of λ with light of wavelength ≥ 330 nm in the absence of riboflavin (Figure 1).

The AS do not correlate perfectly with the absorption curve of either riboflavin or LC in PBS over the entire wavelength regime. There appears to be essentially an identical amount of viral inactivation at 355 and 500 nm, although the optical densities (of solutions containing the same concentration of RB) differ by a factor of five at these two wavelengths. The phage reduction obtained at 320 nm and at 500 nm is greater than that expected based on the absorption spectrum of RB in PBS at these wavelengths. The effect is clearly seen in Figure 1, which plots λ reduction achieved in the presence of RB at various concentrations minus that realized in its absence.

There is another deviation of λ inactivation from the absorption spectrum of RB in PBS. The inactivation obtained at 375 nm, where RB has an absorption maximum, is lower than expected. We have hypothesized that RB is largely free in PBS solution and only a small amount of flavin (not detectable by absorption spectroscopy) is bound to phage. We have speculated that the absorption spectrum of bound RB will differ from that of free RB. We hypothesized that photolysis of the RB-phage complex is orders of magnitude more efficient at inactivating λ phage than photolysis of free RB. This can be responsible for differences between the action spectrum and the absorption spectrum of free RB as regions of enhanced absorption of bound flavin that will appear as regions of enhanced pathogen reduction.

The aptamer sequence (GGC GUG UAG GAU AUG CUU CGG CAG AAG GAC ACG CC) is known[32] to have a high affinity for FMN. We used this aptamer to observe potential changes in the absorbance profile of riboflavin following association with a nucleic acid. Differences in spectra were obtained with 200 μM, each of RNA aptamer and FMN (Figure 2). A shift in the absorbance of riboflavin in the presence of aptamer was observed. This shift resulted in increased absorbance of riboflavin above 500 nm compared to that normally observed for riboflavin in Phosphate Buffered Saline (PBS). The shift in absorbance corresponded with increased inactivation of phage above 500 nm in the presence of riboflavin (Figure 1).

The negative values in the difference spectrum (*ca.* 370 and 445 nm) are likely due to the lower molar absorbtivity of the bound flavin compared to the free flavin. It is obvious that FMN bound to the RNA aptamer has increased absorbance at 500 nm. This is consistent with a possible increased inactivation of λ by a small quantity of RB bound to phage nucleic acid. The difference spectrum observed in the presence of aptamer does not explain the increased efficiency of inactivation of λ phage between 320 and 340 nm and the lower than expected amount of inactivation observed between 350 and 400 nm. However, the effect of solvent polarity on the absorption spectrum of RB mirrors results in the deviation of the Action Spectrum of RB in PBS. In 50% dioxane-water, there is enhanced absorption of RB between 320 and 350 nm

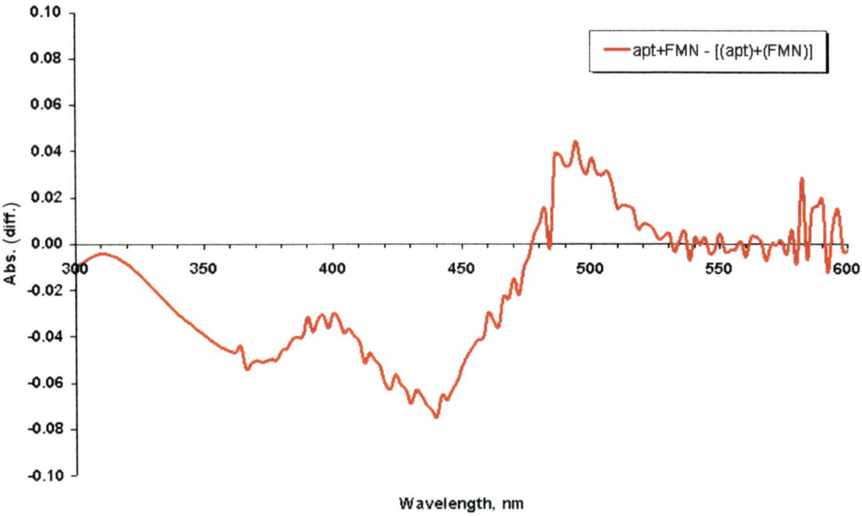

Figure 2. The difference spectrum of absorbance of riboflavin in the presence and absence of flavin-binding aptamer. The positive values observed at 500 nm indicate higher absorbance when riboflavin is mixed with aptamer than when riboflavin is measured free in solution in the absence of aptamer. This increased absorption corresponds to a region where higher levels of phage inactivation occur in the action spectrum compared to predictions based on the absorbance of free riboflavin alone

and decreased absorption between 360 and 410 nm, relative to water (Figure 3).[33] Thus, it seems likely that the enhanced pathogen reduction observed at 320 nm in the AS is because of the binding of RB to regions of the phage resembling 50% dioxane-water. This is consistent with, but not limited to, the polarity observed for the major groove region of DNA.

An examination of the increase in the log reduction of λ beyond the predicted inactivation based upon the excitation scan (and photobleaching) of RB reveals two interesting features. There is *significantly* more phage inactivation at 350 nm and shorter wavelengths than would be predicted by examining the absorption spectrum of RB. This additional amount of inactivation is attributed to direct absorption by the phage for irradiation wavelengths of less than 320 nm. When the irradiation wavelength is between 320 and 350 nm, the inactivation is clearly because of the addition of RB because there is no inactivation in the absence of the sensitizer. Because the inactivation in this range is also beyond that predicted for the bulk photochemistry of RB, it is reasonable to conclude that the inactivation originates from the direct absorption of a phage-RB complex in an environment resembling 50% dioxane-water.

As mentioned previously, there is less inactivation obtained between 350 and 400 nm then from 400 to 500 nm as expected based on the absorption spectrum of RB in PBS. The formation of LC (that absorbs in this region) is probably responsible for the difference between these two wavelength regimes. As RB is

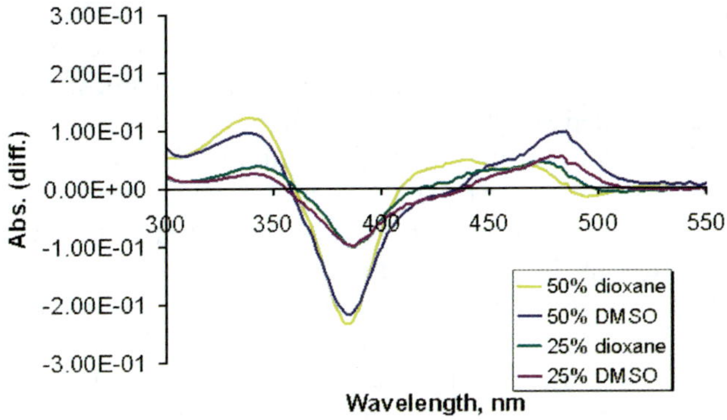

Figure 3. Difference spectra of riboflavin in the presence of various concentrations of organic solvents. Increased levels of dioxane and Dimethyl Sulfoxide (DMS) in water mimic the polarity of the DNA major groove. The difference spectra indicates higher absorbance in the region between 300–350 nm and at 480–500 nm that are higher in the presence of solvents than when riboflavin is free in aqueous solution

photo-bleached, the optical density decreases in the region beyond where the photoproduct, LC, absorbs (wavelength >400 nm). As the absorbance decreases, the inner filter effect will also decrease and photochemistry can proceed more deeply in the cuvette. Conversely, the photobleaching of RB does not lead to a reduction in the optical density of the solution from 350 and 400 nm because the conversion of RB leads to the formation of LC which absorbs in this region. This maintains the optical density in this region resulting in decreased photochemistry due to the inner filter effect. This behavior provides a possible explanation for the shape broadening of the action spectrum curve at the edges and the deviations from expected values of pathogen kill in these regions based on riboflavin absorbance alone.

It is also important to note that the dioxane-water spectrum for riboflavin in the region of 350–400 nm goes negative (Figure 3). This suggests that when riboflavin enters a less hydrophilic environment, the change in spectrum that is induced causes a decrease in absorbance of the molecule in this wavelength region. Hence, movement of riboflavin into a less polar media leads to increase in absorbance at 490–510 nm and 310–350 nm with a decrease in the absorbance between 350 and 400 nm. These observations are consistent with the behavior of the action spectrum of riboflavin where light wavelength effects are plotted against observed virus inactivation of λ phage. We believe that these observations are consistent with a hypothesis of riboflavin association with virus, most likely through nucleic acid binding.

5.3. Riboflavin Sensitized Modification of Nucleic Acids

Navigant Biotechnologies has conducted several studies (Kumar *et al.*)[34] to examine the ability of riboflavin sensitization to modify the nucleic acids of microorganisms. These studies have evaluated the effects of riboflavin and light exposure on DNA in white cells, bacteria, and viruses. In several studies, these effects have been contrasted to those obtained under identical conditions, but in the absence of added riboflavin (*i.e.*, UV light alone).

The DNA fragmentation studies in white cells and bacteria utilized chemical agents that bind to portions of the DNA strand, which have been severed or broken because of chemical modification (Figure 4). The fragments that are produced leave regions that can be chemically tagged with a fluorescence probe and subsequently measured to provide an estimate of the extent of fragmentation that has occurred. Single-strand breaks throughout the nucleic acid sequence can be identified in this way. More complete breaks leading to denaturation of the nucleic acid can also be monitored by gel electrophoresis. In the latter case, the complete denaturation of the nucleic acid can be followed by examining migration patterns on polyacrylamide gels (Figure 5).[34,35] This assay looks for much more severe and complete nucleic acid degradation than single-strand breaks.

In one set of studies, the level of DNA fragmentation occurring in white cell DNA was determined using a flow cytometric assay (Trevigen Apoptotic Cell System (TACS) assay; Figure 4). The level of DNA fragmentation obtained was significantly increased in the presence of riboflavin.[34] Similar observations were made for samples of plasmid DNA and for DNA isolated from *Escherichia coli* following treatment in the presence and absence of riboflavin (Figure 5).[34] These combined studies demonstrate a sensitizing effect, with respect to nucleic acid damage, which riboflavin imparts to samples treated with UV light. These observations are consistent with literature reports for riboflavin.

Cadet and co-workers have evaluated the chemistry involved in the formation of specific lesions induced in nucleic acids by riboflavin and light. These lesions differ from those induced by exposure to light alone in that chemically distinct oxidized species of guanine where residues are formed. This chemistry

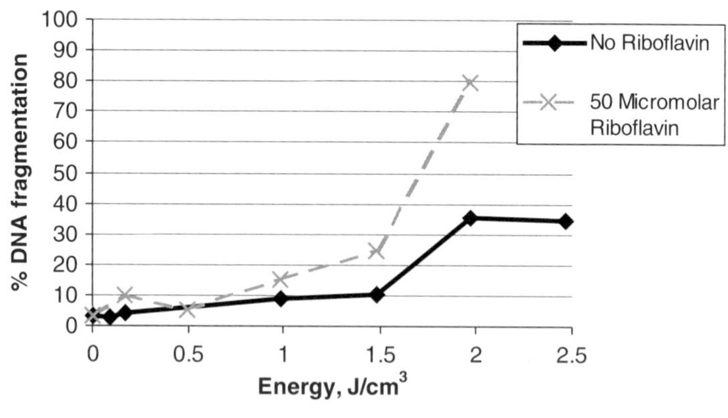

Figure 4. DNA fragmentation in isolated white cell DNA following exposure to light in the presence and absence of riboflavin

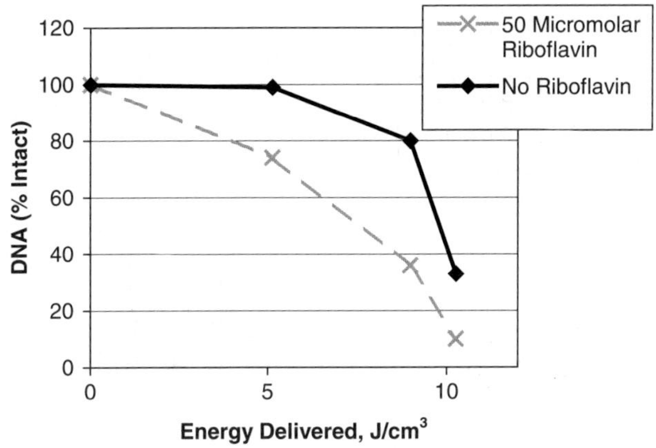

Figure 5. Degradation of E. coli DNA after exposure to light in the presence and absence of riboflavin

is of importance because of the fact that mammalian systems do not normally contain enzymatic systems capable of repairing these types of lesions. This is in stark contrast to the predominant lesion (Thymine–Thymine dimers formed) upon exposure of nucleic acids or agents containing nucleic acids to light alone.[36–38] These studies identify the precise site of the lesions induced in nucleic acids treated with monochromatic 266, 308, or 355 nm light from exither an excimer or Yttrium Aluminum Garnet (YAG) laser in the presence and absence of riboflavin. Results from studies of light/riboflavin chemistry with plasmid DNA are depicted in Figure 6 and Table 1. The results demonstrate that in the presence of riboflavin, the predominant modifications occur to guanine bases, as evidenced by the formation of 8-oxodGuo. The extent of the oxidized guanines formed in the presence of riboflavin is far in excess of those observed upon exposure to light alone. These results are consistent with the literature reports of Cadet and co-workers[39] of the mechanism of action of riboflavin with regard to nucleic acid chemistry. The results were contrasted to those using UV light alone in the absence of riboflavin, and suggest that the addition of riboflavin to the system specifically enhances the damage to DNA induced by UV light alone.

5.4. Phage Reactivation

Virus reactivation is a phenomenon, which is known to occur through the use of host cell nucleic acid repair mechanisms. Viruses utilize normal cell repair processes that are in place to protect against nucleic acid modifications induced by oxidative, chemical, light induced, or replication errors. Several of these cell proteins for excision and repair can be utilized by viruses for repairing modifications, or damage induced in the viral agent through these same

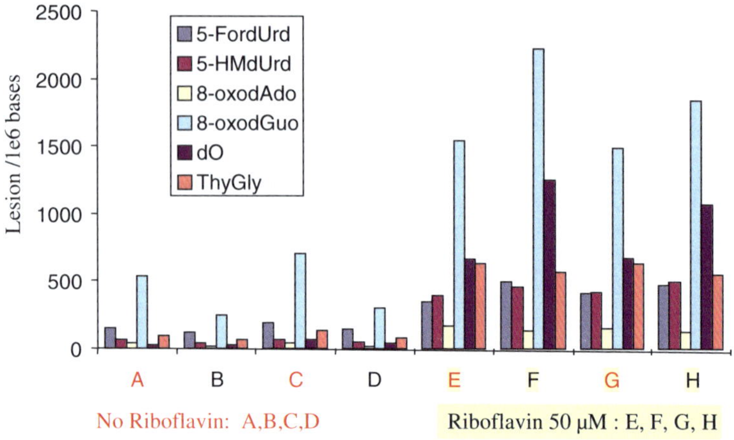

Figure 6. Types of DNA-base modifications induced in plasmid DNA following exposure to light in the presence and absence of riboflavin. The 5-formyl-2′deoxyuridine (5-FordUrd): methyl oxidation product of thymidine; 5-(hydroxylethyl)-2′-deoxyuridine (5-HMdUrd) another methyl oxidation product of thymidine; 8-oxo-7,8-dihydro-2′-deoxyadenosine, (8-OxodAdo) an oxidation product of 2′-deoxyadenosine; 8-oxo-7,8-dihydro-2′-deoxyguanosine, (8-OxodGuo) an oxidation product of 2′-deoxyguanosine; (2-deoxy-erythro-pentofuranosyl) oxaluric acid, (dO) another oxidation product of 2′-deoxyguanosine; 5,6-dihydroxy-5,6-dihydrothymidine, (ThyGly) oxidation product of thymidine that exist as diastereoisomers

Table 1. Formation of 8-oxodGuo/1000 dGuo in calf thymus DNA upon exposure to UV light alone and in combination with riboflavin. Levels of 8-oxodGuo measured in calf thymus DNA upon 266, 308, and 355 nm laser light treatment with and without riboflavin. The 8-oxodGuo formation was quantitatively measured by HPLC-MS/MS as described in the material and methods section of the publication (Kumar et al.)[34]

Wavelength (nm)	+Riboflavin	−Riboflavin	Change	Fold change +RB/−RB
No exposure	0.7 ± 0.4	0.9 ± 0.4	−0.2	0.8
266	109.8 ± 6.5	68.9 ± 2.9	40.9	1.6
308	3.5 ± 0.4	2.9 ± 0.5	1.5	1.2
355	4.5 ± 0.5	0.7 ± 0.1	3.8	6.4

mechanisms. Excision repair for example is well documented in the literature and arises through the actions of excinuclease enzymes in cells. In the context of virus inactivation, the desired end target for these treatments is the prevention of virus replication. It is also desirable, in this context, to prevent repair of damaged virus particles because such repairs may render non-infectious agents capable of transmitting the disease when re-infused. This may be accomplished by generating either an extent of damage that the host system cannot repair or a type of damage that the host system does not have the capability of repairing.

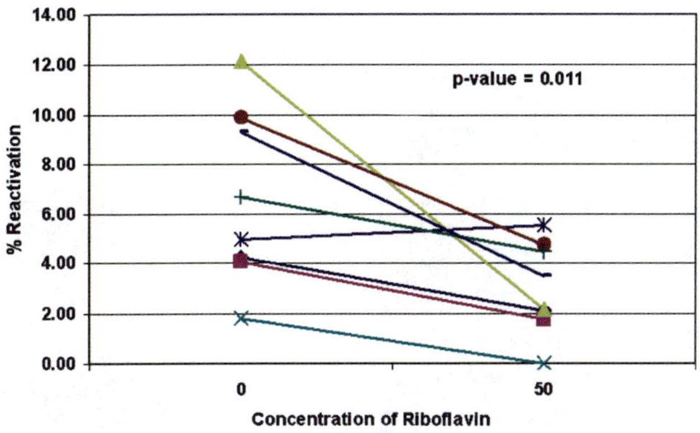

Figure 7. Reactivation of lambda phage in E. coli with UV light in the absence or presence of riboflavin. Percent reactivation of UV exposed lambda phage with and without riboflavin measured in irradiated and non-irradiated host cells. Phage and host cells were irradiated separately. Lines connect paired samples in order to show the relationship between treatment with and without riboflavin. This is not meant to imply a linear relationship in the response observed. In only 1 of the 8 cases presented was reactivation in the presence of riboflavin equal to that observed in the absence of riboflavin

Studies of the inability of bacteriophage to repair the lesions (Weigle reactivation) induced by riboflavin and light as contrasted to the observations with light exposure alone have been conducted (refer to Figures 7 and 8).[34] These studies confirm that the rescue of DNA damaged phage does not occur to the same extent when riboflavin is present in samples exposed to light.[40] These observations are consistent with the data suggesting that the presence of riboflavin and UV light selectively enhances damage to the guanine bases in DNA or RNA. These data also suggest that this type and extent of damage to nucleic acids of virus in the presence of riboflavin makes it less likely to be repaired by normal repair pathways available in host cells. This result is essential for a system intended to assure the highest and most complete levels of pathogen inactivation attainable.

In summary, the MIRASOL PRT process works through three independent mechanisms of action in rendering pathogens inactive. These include oxygen-dependent chemistry induced by the combination of riboflavin and light, electron transfer chemistry induced by the direct interaction of excited riboflavin molecules with nucleic acid base pairs (primarily guanine bases) leading to oxidation products, and effects that are due to the action of UV light alone. In essence, the presence of riboflavin in this system enhances the effects, that are due to UV light alone, creating a condition of greater sensitivity of the pathogen to the UV light to which the sample is exposed (photosensitization effect). The combination of these three modes affords broad and extended levels of pathogen inactivation with this process.

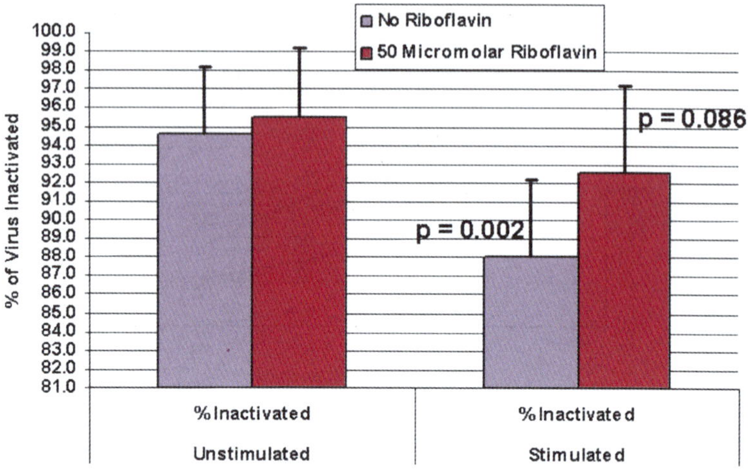

Figure 8. UV inactivation of lambda phage with and without riboflavin measured in irradiated and non-irradiated E. coli. Lambda-phage was titered before and after treatment in both treated and untreated E. coli. Lambda-phage samples were serially diluted in tryptone water and 100 μL of virus sample was added to 100 μL of either the treated or untreated E. coli. The samples were incubated for 10–20 min at 37°C. After incubation, 4 mL of LB top agar was added to each of the virus/bacteria mixtures, these solutions were vortexed and poured out onto TSA plates. The plates were then incubated overnight at 37°C. The numbers of plaques formed were determined for each serial dilution level. The ratio of the titers before treatment and after treatment was determined for each sample. This ratio represented the fraction of virus that survived treatment. This value was reported as the percentage of inactivated virus by subtracting from 100%. This measurement was reported in both E. coli strains that had been stimulated for activation of repair enzymes by exposure to UV light and in E. coli strains that were not stimulated. The difference in these values with and without stimulation represents the level of virus reactivation due to repair. This value was also reported for each test sample, both with and without riboflavin present during treatment

In an attempt to study the relative contributions of the oxygen-dependent and oxygen-independent mechanisms, we de-oxygenated samples using Argon and Nitrogen and added oxygen quenchers, such as ascorbic acid to samples of virus and riboflavin. Activation of the riboflavin chemistry was achieved using light at either 355 or 500 nm. The levels of inactivation observed under these conditions was determined and expressed as a percentage reduction in virus titer (Figure 9). In all cases, differences before and after physical (de-oxygenating) or quenching amounted to less than 1 log difference in virus reduction levels. This difference was more significant, however, when chemical quenching was achieved through the addition of ascorbic acid, suggesting that some portion of the inactivation achieved under the conditions utilized, was due to an oxygen-dependent mechanism of sensitizer action. Under the conditions utilized in this system, however, the contribution of this route to pathogen kill was not significant.

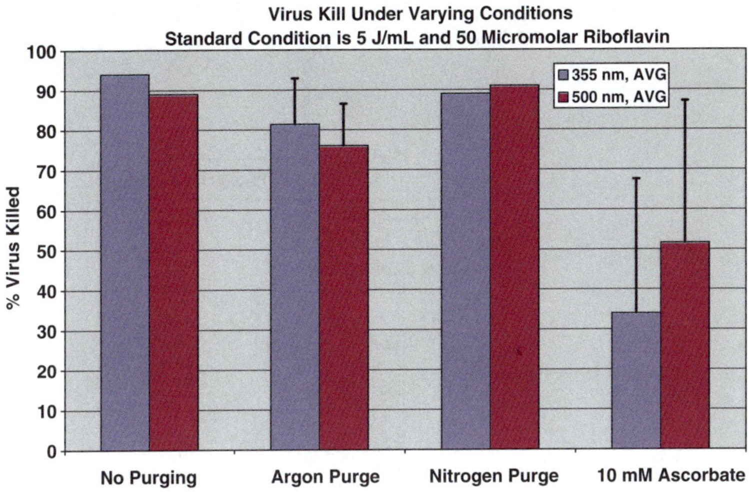

Figure 9. Inactivation of Lambda phage under standard conditions (no purging) and after purging with Argon, Nitrogen, or addition of 10 mM ascorbate to the solution. Light wavelengths of 355 nm and 500 nm were used in separate experiments to activate Riboflavin photochemistry

5.5. Treatment Procedure Overview

Transfer of this process to standard blood banking operations requires that the photochemical properties of riboflavin be utilized in containers and systems and under conditions that are routine for transfusion medicine practice. The following is a description of the treatment procedure for the MIRASOL PRT System for which NAVIGANT BIOTECHNOLOGIES is pursuing regulatory clearance. The process includes the steps of transferring the platelet product into an Illumination/Storage Bag, adding the Riboflavin Solution, illuminating the bag and placing the platelet product into storage.

5.5.1. Platelet Collection

Apheresis platelet products collected in Acid Citrate Dextrose Anticoagulant (ACDA) solutions using the Trima® Automated Blood Collection System (TRIMA) or manually collected buffy coat platelet products in Citrate Phosphate Dextrose (CPD) anticoagulant solutions have been evaluated and validated with this process. Buffy coat platelet products are platelets that are collected from individual whole blood donations and are separated from red cells and plasma by manual methods. Normally, 4–6 individual buffy coat platelet products from different donors are pooled in order to make one therapeutic dose of platelets for a patient. Apheresis platelets, because they are collected selectively with the return of red cells to the donor, can be collected in a volume

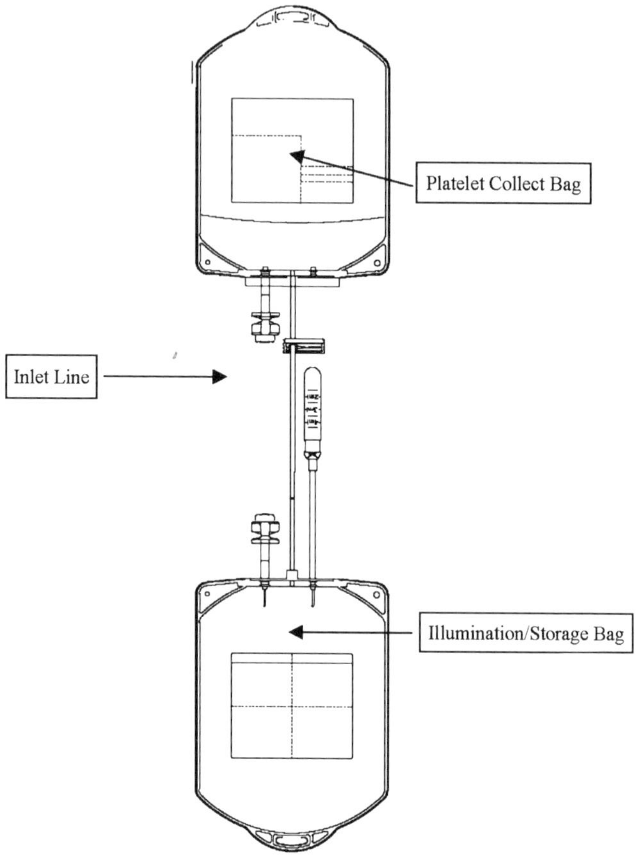

Figure 10. Platelet transfer to Illumination/Storage Bag

large enough to provide one therapeutic dose to a patient from a single donor. The bag containing the platelets is connected to the inlet line of the Illumination/Storage Bag using a sterile connection device (Figure 10). This device is commonly used in blood establishments to make such a connection. Following sterile connection, a platelet product within a specified volume and concentration range is transferred into the Illumination/Storage Bag. The transfer tubing is sealed using a standard laboratory hand-held tube sealer. Following the seal, the two bags are separated and the original collection bag is discarded.

5.5.2. Addition of Riboflavin Solution

The Riboflavin Solution bag is then connected to the Illumination/Storage Bag using a sterile connection device. The resulting disposable configuration is shown in Figure 11. The residual air in the Illumination/Storage Bag is

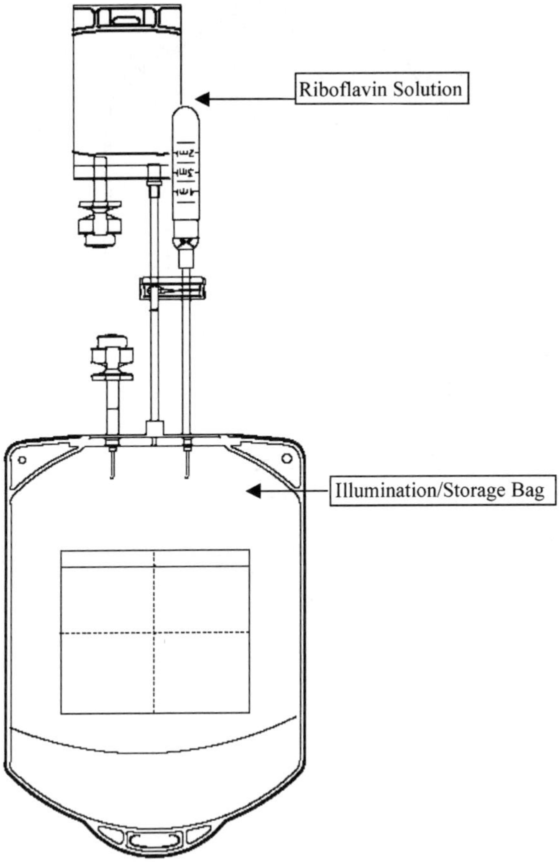

Figure 11. Adding Riboflavin Solution to Illumination/Storage Bag

expressed into the empty Riboflavin Solution bag. The transfer tubing is sealed using a hand-held tube sealer. Following the seal, the two bags are separated and the Riboflavin Solution bag is discarded. After the above process, the Illumination/Storage Bag will contain a mixture of platelets and riboflavin (Figure 12).

5.5.3. Illumination

The operator uses an Accessory Barcode Scanner or manual data entry screens provided on the MIRASOL™ Illuminator to enter the product identification number. The operator then places the Illumination/Storage Bag containing the product on an Accessory Scale or manually enters the weight of the product bag into the Illuminator so that the Illuminator can calculate the required light energy dose. The duration of the illumination step is established by monitoring

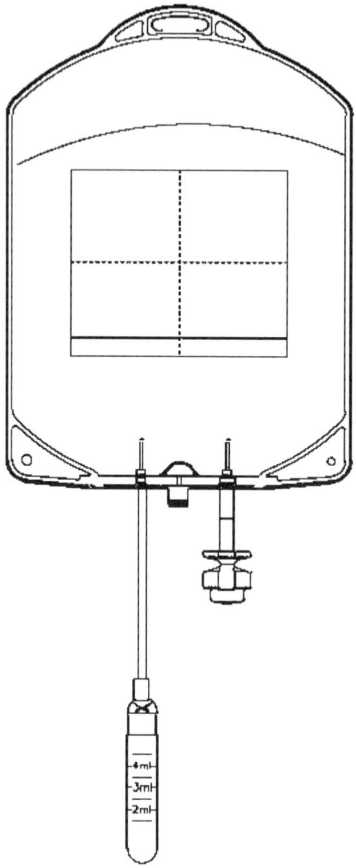

Figure 12. Adding Riboflavin Solution to Illumination/Storage Bag

the light output to achieve the required dose. The platelets are mixed using a back-and-forth motion at a rate of 120 cycles per minute during the illumination treatment procedure. Throughout the treatment procedure the Illuminator automatically monitors the temperature of the product and activates a cooling fan as needed to keep adequate product temperature. During illumination, the platelet product is exposed to light from 285 to 365 nm.

5.5.4. Platelet Storage

After the illumination treatment procedure is completed, the Illumination/Storage Bag is removed from the Illuminator and stored following standard blood establishment practices. This normally means storage for 5–7 days at 22°C, while being agitated on a horizontal mixing platform.

5.6. Performance

In vitro process, verification tests have been conducted to assess the performance of the MIRASOL PRT System in reducing a variety of viruses and bacteria in platelet products. *In vitro* studies have also been conducted to determine the effects of the MIRASOL PRT System process on platelet product quality following the pathogen reduction step. All *in vitro* pathogen reduction and platelet cell-quality studies were performed as Good Laboratory Practice (GLP) studies.[41]

5.6.1. Pathogen Reduction Studies

5.6.1.1. Viral Reduction Performance Studies

The MIRASOL PRT System pathogen reduction process has been evaluated for performance against five viral pathogens. These agents were selected owing to their representation of screened virus (HIV), non-screened virus (PPV and WNV), models for untested agents such as Hepatitis C (WNV), and models for other classes of virus recommended by the Committee for Proprietary Medicinal Products (CPMP).[42] These include a large RNA virus (VSV). In the case of HIV, the samples used for testing are representative of the various forms that this agent may take in contaminated blood products. The cell-free and cell-associated HIV (HIV *ca.*) samples are prepared using lymphocytes (H9 cells) that are infected with HIV. The cell-free and cell-associated samples contain virus in its various forms, including proviral, non-replicating forms, virus in the process of budding from cells, and extracellular virus that has been completely synthesized and released from infected cells. Intracellular HIV (HIV_i) is a form of HIV that has become integrated into the infected host cell's genome, known as the proviral form of HIV. The proviral-infected cells used are ACH-2 cells, which are added to platelet products. The HIV_i samples are representative of cases where blood products may be contaminated with lymphocytes containing non-replicating but infectious viral genomes. Table 2 lists the viral pathogens evaluated.

Log reduction values reported in the tables were calculated by determining the number of virus particles present in infectious form prior to treatment and the number of virus particles present after treatment. The level of log reduction is reported as the starting titer expressed in units of 10^x mL^{-1} minus the level after treatment expressed as the titer in 10^x mL^{-1}. Because volume was constant in the samples before and after treatment, the unit of volume cancels, resulting in a reported value of log reduction.

For example, a sample containing 1,000,000 infectious virus particles per mL would of course have 10^6 virus particles per mL. If after treatment, only 100 particles per mL were measured in tissue culture infectivity assays, this would correspond to 10^2 virus particles per mL. The log reduction reported for this system would be 10^4 or 4 logs. This corresponds to a reduction in virus level of 99.99%. Because values are reported in log units, 100% reduction is never achievable.

Table 2. Viral pathogens tested

Virus	Virus – full name	Model/family	Enveloped/non-enveloped	Average reduction factor (log)[b] N = 6
PPV	Porcine parvovirus	B-19 parvovirus/ parvoviridae	Non-enveloped	≥ 5.0[a]
HIV_{ca}	Cell-associated human immuno-deficiency virus	Extracellular and cell associated HIV/ retroviridae	Enveloped	5.9 ± 0.2
HIV_i	Intracellular human immuno-deficiency virus	HIV_i/retroviridae	Enveloped	4.5 ± 0.4
WNV	West Nile virus	HCV/flaviviridae	Enveloped	5.1 ± 0.5
VSV	Vesicular stomatitis Virus	Rabies, large RNA/ rhabdoviridae	Enveloped	≥ 6.3 ± 0.6[c]

[a]All test articles were inactivated to the limit of detection of the assay; therefore, results are reported as greater than or equal to the average value with no standard deviation.
[b]Values given in table are averages ± one standard deviation.
[c]Three of six test articles were inactivated to the limit of detection of the assay; therefore, results are reports as greater than or equal to the average.

The virus reduction studies were performed by spiking single donor platelet products collected by apheresis using the Trima apheresis device with sufficient virus to achieve a starting titer of 5–6 log $TCID_{50}$ mL^{-1}. Reduction factors were calculated in accordance with U.S. FDA[43] and CPMP[44] guidelines. Table 3 summarizes the viral reduction results. The data show reduction factors ranging from 4.0 to >6.3 log (99.99–99.99995% reduction) for each virus tested with the Mirasol PRT System. Based on contamination levels of agents that may be expected in blood products, the types of model agents studied, and the performance levels observed; these results suggest additional safety factors for these products on the order of up to >15,000-fold.[45–47]

5.6.1.2. Bacterial Reduction Efficacy Studies

Despite the fact that bacterial contamination of blood products poses one of the greatest risks for transfusion, there are currently no standards in place that establish a panel of species to test or a method to evaluate technology for pathogen reduction. Navigant Biotechnologies has, therefore, adopted a panel based on published hemovigilance studies incorporating those species responsible for the majority of morbidity and mortality in transfusion-associated reactions. NAVIGANT has also developed two complementary test methods, as described below, to measure bacterial reduction performance.

There are several possible causes of bacterial contamination of donor platelet products.

(i) Failure to adequately reduce the bacterial contamination of the donor's skin at the time of venipuncture

Table 3. Platelet transfusion-related sepsis events

Organism	Gram stain	Characteristics	Total observations	Fatalities	Frequency
S. epidermidis	Gram +	Frequent skin contaminant; implicated in transfusion related fatalities	35	1	61.4%
S. aureus	Gram +	Grows rapidly at body temperature, many sources; implicated in transfusion related fatalities	6	3	10.53%
E. coli	Gram −	Normal human flora; implicated in transfusion fatalities	1	1	1.75%
P. aeruginosa	Gram −	Implicated in transfusion fatalities	2	2	3.51%
B. cereus	Gram +	Frequent skin contaminant; implicated in transfusion related fatalities	3	0	5.26%
S. marcescens	Gram −	Opportunistic environmental contaminant; implicated in transfusion related sepsis	2	0	3.51%
All other species	—		8	0	14.04%
Total	—		57	7	100%

Source: Extracted from Ness[49] and Engelfriet[50].

(ii) Undetected donor bacteraemia
(iii) Environmental contamination during platelet collection and processing.

Of these potential causes of bacterial contamination, the overwhelming majority of cases are due to inadequate skin preparation.[48]

Ness[49] and Engelfriet[50] conducted retrospective surveillance studies to determine the number of platelet contamination-associated transfusion events at their institutions over a 7- and 12-year period, respectively. In these studies, two common skin bacteria, *Staphylococcus epidermidis* and *Staphylococcus aureus* accounted for over 70% of the reported events. In addition, although the number of reported contamination cases was low, three additional species, *E. coli, Pseudomonas aeruginosa*, and *Bacillus cereus* were associated with fatal transfusion events. Also, an additional species, *Serratia marcescens*, appears on many hemovigilance reports as a common platelet contaminant causing transfusion-associated sepsis. 3 summarizes the number of platelet transfusion-associated sepsis cases by bacterial species reported in the two studies.

Based on the clinical findings that these six bacteria species represent approximately 86% of the documented platelet transfusion-related sepsis events (and all of the fatalities) reported in the Ness and Engelfriet studies, NAVIGANT BIOTECHNOLOGIES deemed these to be an acceptable initial panel to demonstrate the performance of the MIRASOL PRT System against bacterial pathogens.

Results of the *in vitro* bacterial reduction studies for the six species are detailed below. These studies provide evidence that the MIRASOL PRT System is capable of reducing bacterial pathogens in donor platelet products. Future studies are planned to further assess the performance of the device against a broader spectrum of bacterial pathogens.

Two complementary test methods, known as "High Spike Bacterial Titer" and "Low Spike Bacterial Titer" tests, for bacterial reduction have been developed to measure the MIRASOL PRT System performance. Both methods involve inoculation of known titers of bacteria (a "spiking" study) into platelet products followed by PRT treatment and subsequent measurement of the presence of bacteria. The objective of the High Spike Bacterial Titer experiments is to determine the overall bacterial reduction ability of the MIRASOL PRT System against a severely contaminated platelet product. These studies may not, however, represent a clinically relevant finding in that viable bacteria remaining after treatment may grow to high titers through the storage period. The objective of the "Low Spike Bacterial Titer Experiments" is to spike a platelet product with a more clinically relevant bacterial titer, treat the product using the MIRASOL PRT System, and evaluate the platelet product using a standard culture system through a 5-day storage period to determine if it has remained culture negative, indicating that the platelet product meets release criteria for transfusion. The combined data from these studies demonstrates the bacterial reduction capability of the MIRASOL PRT System under conditions, that are still substantially higher challenges than may be anticipated in an actual clinical setting.

5.6.1.3. High Spike Bacterial Titer Experiments

In these experiments, single donor platelet products collected by apheresis using Trima were spiked with 5–6 log Colony Forming Units (CFU) mL^{-1} of bacteria. Published literature indicates that a freshly collected platelet product, if contaminated, will contain <100 CFU/product.[51,52] Hence, in the high-titer studies, the challenge levels used are on the order of 10,000–100,000-fold higher than challenges that may be expected in a typical clinical setting. Immediately

Table 4. Summary of bacterial reduction studies performed

Bacteria	High titer test average reduction factor (log) N = 6	Low spike titer bacterial reduction	
		Initial spike titer CFU/ product N = 10	*% Negative by day 5 of storage N = 10*
S. epidermidis, ATCC# 12228	≥ 4.15 (6 of 6 to LOD[a])	15,500	100
S. aureus, ATCC# 25923	3.56 ± 0.35[b] (2 of 6 to LOD[a])	100	80
S. aureus, ATCC# 700787	4.8 ± 0.8	—	—
B. cereus, ATCC# 7064	1.9 ± 0.3	—	—
B. cereus, Blood isolate	2.7 ± 0.6	150	100
E. coli, ATCC# 25922	≥ 4.38 (6 of 6 to LOD[a])	15,800	100
P. aeruginosa, ATCC# 43088	≥ 4.48 (6 of 6 to LOD[a])	21,400	90
S. marcescens, ATCC# 43862	4.0 ± 0.5	—	—

[a]LOD = Limit of detection (1.22 log CFU mL^{-1}). When all test articles were inactivated to the limit of detection of the assay, results are reported as greater than or equal to the average value with no standard deviation.
[b]Value is average ± one standard deviation.

after spiking, the platelet product was treated with the MIRASOL PRT System and the overall pathogen reduction performance was determined by calculating the reduction achieved between initial and final titers.

In these experiments, the titers were determined immediately before and immediately after illumination, with no storage period. Reduction factors were calculated in accordance with guidelines proposed by the U.S. FDA and the CPMP for virus-reduction performance quantification.[53,54]. Table 4 displays the results of the high-spike titer bacteria experiments for each bacterial species studied. These results demonstrate the MIRASOL PRT System reduced bacterial contamination by more than 3 logs for each test organism, with the exception of *B. cereus*. Average reduction factors for two different strains of the Gram positive, spore-forming bacteria, *B. cereus* were 1.9 and 2.7 logs, respectively. A second strain of *S. aureus* was also tested (ATCC #700787) to assess the performance against a methicillin resistant strain.

5.6.1.4. Low Spike Bacterial Titer Experiments

In these experiments, single-donor platelet products collected by apheresis using the Trima were inoculated with 0.5–2.0 log CFU mL^{-1} of *S. epidermidis*, *P. aeruginosa*, or *E. coli*, or with approximately 100 CFU/product of *S. aureus or B. cereus*. The levels of *S. epidermidis*, *P. aeruginosa*, and *E. coli* were greater than 155–2140 times that expected in freshly collected platelet products, while the *S. aureus and B. cereus* levels were equivalent to the highest levels of contamination expected. The platelets were then treated with the MIRASOL PRT System and stored under normal blood establishment conditions of 22–24°C with gentle agitation. After a minimum of 24 h of storage, the treated platelet products were sampled and cultured. Samples were monitored using an automated culturing system for blood products developed by BioMerieux. The system, known as the BacT/ALERT®, has been used clinically in Europe for several years to determine if platelet products remained culture negative after a minimum 5-day storage period. The device works by monitoring samples of platelet products in culture media for consumption of oxygen and production of carbon dioxide, which would be indicative of the presence of a microorganism in the sample. Because of concerns over bacteria transmission by blood products, the American Association of Blood Banks recommended implementation of bacteria screening for blood products in 2002. It is currently being implemented throughout blood centers in the United States.

Table 4 displays the results of the low-spike titer bacteria experiments for each bacterial species studied.[55] These results demonstrate the MIRASOL PRT System rendered platelet products initially contaminated with *S. epidermidis* and *E. coli*, at levels >150–1500 times higher than expected in freshly collected products, culture negative throughout storage. The results also demonstrate the MIRASOL PRT System rendered platelet products initially contaminated with *P. aeruginosa* at levels >214–2140 times higher than expected in freshly collected products, culture negative in 90% of test articles throughout storage. For platelet products initially contaminated with *B. cereus* at levels equal to the highest observed in freshly collected products, the results demonstrate treatment with the MIRASOL PRT System, which yields culture negative products throughout storage. For platelet products initially contaminated with *S. aureus* at levels equal to the highest observed in freshly collected products, the results demonstrate treatment with the MIRASOL PRT System yields 80% culture negative products throughout storage. The *S. marcescens* and the additional strains of *S. aureus* and *B. cereus* were not tested in the low-titer experiment.

A typical platelet product that is contaminated with bacteria may contain on the order of 10–100 organisms in total. The levels of pathogen reduction of bacteria that have been observed following MIRASOL PRT treatment suggests a very high margin for performance with a corresponding reduction in risk of bacteria transmission. If the results reported here reflect clinical performance, the application of MIRASOL PRT would be expected to result in a minimum of 1–2 log reduction in frequency of bacteria transmission *via* treated blood products. Given the frequency of bacteria contamination of platelet products of 1:2000–1:3000,[49,50] the potential benefit for patient morbidity and mortality in this setting is substantial.

5.6.2. In Vitro Cell Quality Studies

In transfusion medicine, platelets function to prevent bleeding or to stop bleeding in actively bleeding patients. The role of platelets in hemostasis is to migrate to sites of injury or damage, bind to the damaged surface, and release

Table 5. Platelet collection and treatment conditions. Values for volume and cell concentration are reported after the addition of 35 +/−5 mL of Riboflavin Solution to the collected products

Condition	Platelet source	Treatment concentration (cells nL^{-1})	Treatment volume (mL)
1	Apheresis	1000–1100	200–235
2	Apheresis	1700–1790	200–270
3	Apheresis	1000–1300	365–400
4	Buffy coat	1000–1100	200–235
5	Buffy coat	1700–1790	200–270
6	Buffy coat	1000–1300	365–400

agents that stimulate tissue repair and recruitment of other platelets to the site of injury. Through chemotactic responses, the recruitment of other platelets to the site of injury helps to form a plug that prevents blood or fluid loss. Platelets are essential for preventing spontaneous bleeding due to normal tissue wear and tear. When their numbers are reduced, as occurs in the case of immunosuppression or reduced production because of chemotherapy for malignancies, the patient is at risk of bleeding and must be transfused with platelets in order to prophylactically prevent bleeding.

Any secondary handling of platelets post collection, including PRT treatment, can induce activation or accelerate the platelet storage lesion. Although there is not one particular *in vitro* parameter that can predict the *in vivo* viability of platelets, many standard parameters can be monitored to create an overall platelet cell quality picture. Navigant Biotechnologies has utilized a broad panel of cell quality assays to assess platelet cell quality post Mirasol PRT treatment. This panel is based in part on recommendations of the Biomedical Excellence for Safer Transfusions (BEST) committee and other guidelines to industry on platelet testing.[56,57]

Navigant Biotechnologies has conducted *in vitro* cell-quality studies through 5 days of storage using both Trima apheresis and buffy coat platelet concentrates that cover the range of collection conditions described in Table 5. Assays included MPV (mean platelet volume), blood gases (pO_2 and pCO_2), pH, lactate and glucose concentrations, HSR (hypotonic shock response), P-selectin expression, platelet swirl, and ESC (Extent of Shape Change). All assays were conducted according to standard validated protocols. Table 6 provides the results for all assays performed at the end of a 5-day storage period.

5.6.2.1. Metabolic Results

In general, an increased metabolic rate was observed for the treated platelet products. Glycolysis, or the utilization of glucose by the cell to produce Adenosine Tri-Phosphate, is a way in which cells normally generate energy required for normal cell function and operation. When these rates are reduced, the cells are in jeopardy of not synthesizing sufficient ATP in order to maintain normal energy balance and metabolism. When these levels are increased, cells run the risk of increased lactic acid production, a by-product of glycolysis.

Table 6. Summary of in vitro results after 5 days of storage

Cell quality parameter	Units	Control apheresis (N = 20)	Mirasol PRT, apheresis (N = 12)	Mirasol PRT, buffy coat (N = 12)	p-value: day 5 control vs. mirasol PRT, apheresis	p-value: day 5 control vs. mirasol PRT, buffy coat
pH (22°C)	NA	7.48 ± 0.06^a	7.14 ± 0.09	7.06 ± 0.23	$<10^{-3}$	$<10^{-3}$
Lactate production rate	mmol 10^{-12} cells h^{-1}	0.032 ± 0.006	0.059 ± 0.012	0.075 ± 0.025	$<10^{-3}$	$<10^{-3}$
Glucose Consumption Rate	mmol 10^{-12} cells h^{-1}	0.019 ± 0.004	0.034 ± 0.005	0.042 ± 0.012	$<10^{-3}$	$<10^{-3}$
pO2	mm Hg	54 ± 15	48 ± 20	41 ± 22	NS	NS
pCO2	mm Hg	26 ± 3	28 ± 5	32 ± 8	NS	0.004
P-Selectin Expression	%	17.9 ± 7.0	57.8 ± 14.8	57.3 ± 9.6	$<10^{-3}$	$<10^{-3}$
Swirl	NA	3 ± 0^b	3 ± 0	3 ± 0	NS	NS
HSR	%	72.3 ± 10.9	67.0 ± 7.3	65.3 ± 9.3	NS	NS
ESC	%	24.7 ± 4.3^c	20.4 ± 4.8	17.8 ± 3.8	0.005	$<10^{-3}$
MPV	fL	6.4 ± 0.6	6.5 ± 0.6	8.1 ± 0.6	NS	$<10^{-3}$

[a] Values given in table are averages ± one standard deviation.
[b] $N = 17$.
[c] $N = 36$
NS = not significant ($p > 0.05$).

Increased lactic acid production can cause a decline in pH value, which poisons other cellular machinery, rendering treated products less viable. Despite the observed increase in lactic acid production observed in the treated cells, both test and control results met standard pH criteria through 5 days of storage, that are specified to assure normal platelet function *in vivo*.[57,58] Lactate production rates and glucose consumption rates were significantly higher for treated units compared to results from control units, but did not result in unacceptable generation of acid to cause a clinically relevant decline in pH.

The maintenance of pO_2 levels below ambient (120–130 mmHg in Lakewood, Colorado, USA) and above a minimum of 10 mmHg indicates that the platelets are still consuming oxygen. This observation suggests that treated platelets maintain the capacity to produce ATP through oxidative respiration, another metabolic source of ATP for platelets that is utilized in normal metabolic stasis.

5.6.2.2. Activation Results

P-selectin (GMP-140) is a measurement of expressed glycoprotein on the cell surface of activated platelets. An activated platelet has released granule contents to the surrounding media and, as a result, has granule markers, such as P-selectin present in the cell membrane. The P-selectin level is measured by flow cytometry with site-specific fluorescent marker labeled cells. The P-selectin data indicate that the Mirasol PRT platelets are more activated than controls. If the extent of this activation directly correlated with platelet granule release, the results would suggest that treated platelets would be less capable of providing necessary tissue growth stimulating and platelet recruiting chemicals at sites of vascular damage. Despite the elevated levels of activation seen in treated platelets, they remain within typical ranges seen routinely at the point of transfusion for a variety of clinically acceptable products in use today. Michelson *et al.*[59] has also found that even highly activated platelets, demonstrated by excessive P-selectin and GP IIb-IIIa expression, still circulate and function in three different animal coagulation models. These results, taken together, suggest that the alterations are not clinically relevant in the transfusion setting.

As platelets age and deteriorate, their shape changes from predominantly discoid to spherical, and the "swirl" effect in the bag correspondingly decreases. Swirl is scored on a scale of 0–3 with 3 being the best and 0 being no observable swirl. As indicated in Table 6, the swirl in all cases was scored as 3.

5.6.2.3. Extent of Shape Change

The ESC assay measures the change in optical density of a sample of platelets caused by the addition of Adenosine Di-Phosphate as based on principles outlined by Holme *et al.*[60] Ethylene Diamine Tetra-Acetic Acid is added to the sample to prevent aggregation. The ADP is then added, causing the platelets to change from a discoid shape to a spherical form, causing a decrease in the optical density of the sample. The extent of the decrease in light intensity correlates to the percentage of discoid platelets in a sample. Mirasol PRT platelets yield average ESC values around 15–20% after 5 days of storage.

Holme loosely correlated a 10–30% ESC with an *in vivo* recovery ranging from 40–70%. Therefore, correlation of this data to MIRASOL PRT platelets predicts *in vivo* recoveries within acceptable clinical performance ranges. Note that the control data is based on 36 replicates, rather than 20. These data are based on historical control tests conducted by NAVIGANT BIOTECHNOLOGIES in non-GLP studies.

5.6.2.4. Functional Results

HSR, another measurement of platelet functionality, is a spectrophotometric assay based on principles outlined by Holme *et al.*.[60] HSR measures the platelet's ability to recover its normal volume after swelling when placed in a hypotonic environment. On the fifth day of storage, there is very little difference in HSR between MIRASOL PRT and control platelets. Both MIRASOL PRT and control platelets yield average HSR values around 65–70% after 5 days of storage. Holme loosely correlated a 40–50% HSR with an *in vivo* recovery ranging from 40–70%. Therefore, correlation of this data to MIRASOL PRT platelets predicts *in vivo* recoveries within acceptable clinical performance ranges.

5.7. Conclusions

The MIRASOL PRT process is capable of inactivating significant levels of virus and bacteria in platelets and plasma. The levels of pathogens inactivated by this process would be expected to significantly reduce the possibility of disease transmission by these blood components. A technology with this capability (Solvent Detergent Plasma) has existed for some time now for plasma fractionation and plasma for transfusion. The consequence of its introduction has been a substantial reduction in disease transmission by these blood components and increased safety for the recipients of these products. The potential to extend this same level of safety to cellular components of blood, such as platelets and red cells appears to be achievable. Recently, a photochemical technology utilizing a synthetically produced psoralen compound and UV light has been approved in Europe for use with platelet concentrates.[61,62] The toxicology profile and photochemistry of riboflavin, vitamin B2, may afford the transfusion medicine community, another possibility in this area. In order to do so, pathogen reduction must be achievable without significant damage to the blood components being treated. This has been a significant challenge for technologies in this field.

The average *in vitro* cell quality of MIRASOL PRT treated platelets after 5 days of storage post-treatment are summarized in Table 6. A t-test analysis comparing MIRASOL PRT platelet products to control platelet products is also summarized in the table. For apheresis platelets, no significant differences were observed for %HSR, swirl, pO_2, and pCO_2 levels; for all other parameters, the difference between treated and control platelet products was significant. For buffy coat platelets, no significant differences were observed for %HSR, swirl, and pO_2 levels; for all other parameters, the difference between treated and control platelet products was significant. The MIRASOL PRT System induces an increase in platelet glycolysis as described in the preceding text. This is indicated by increases in glucose consumption and lactate production. However, the MIRASOL PRT System yields platelet products that still meet pH, P-selectin, ESC, and HSR response levels that have been observed for clinically transfused products and established by recent literature reports.[63,64]

The cell quality results obtained after 5 days of storage for MIRASOL PRT treated platelets in a US Clinical Study conducted in 2003–2004 were similar to those listed in Table 6. The results of the US Clinical Study in human subjects yielded values of mean recovery values for platelets of 51.4% and mean platelet survival times of 4.1 days.[63,64] These recovery and survival values are within the range of acceptable values for commercially available platelet products in use today. Although, in many cases, the *in vitro* cell quality parameters observed for MIRASOL treated platelets are different from control platelets, the recovery and survival results in human subjects have been clinically acceptable.[63,64]

Acceptable cell quality is thus achieved while simultaneously observing significant levels of pathogen inactivation. The levels of virus and bacteria reduction observed with this system would be expected to significantly reduce the risk of disease transmission by blood without increasing toxicological concerns or changing the performance of treated products clinically. Further studies with this technology in the clinical setting with patients requiring platelet and plasma transfusions will be ongoing in 2005. The introduction of this technology for the pathogen inactivation of blood products is expected in 2007. With that event, we believe that a new era in blood safety and approaches for safeguarding the blood supply against threats because of disease transmission will be initiated.

References

1. M. Busch, S. Kleinman and G. Nemo, Current and emerging infectious risks of blood transfusions, *J. Am. Med. Assoc.*, 2003, **289**, 959–962.
2. V.T. DeVita Jr., S. Hellman and S.A. Rosenberg, (eds), *AIDS: Etiology, Diagnosis, Treatment and Prevention*, Lippincott, Philadelphia, PA, 1992.
3. G. Schochetman, *AIDS Testing*, Springer, New York, NY, 1994.
4. P. Cummings, E. Wallace, J. Schorr and R. Dodd, Exposure of patients to human immunodeficiency virus through the transfusion of blood components that test antibody negative, *N. Engl. J. Med.*, 1989, **321**, 941–946.
5. L. Petersen, G. Satten, R. Dodd, M. Busch, S. Kleinman, A. Grindon and B. Lenes, Duration of time from onset of human immunodeficiency virus type I infectiousness to development of detectable antibody, *Transfusion*, 1994, **34**, 283–289.
6. L. Lin, D. Cook, G. Wiesehahn, R. Alfonso, B. Behrman, G. Cimino, L. Corton, P. Damonte, R. Dikeman, K. Dupuis, Y. Fang, C. Hanson, J. Hearst, C. Lin, H. Lunde, K. Metchette, A. Nerio, J. Pu, A. Reames, M. Rheinschmidt, J. Tessman, S. Isaacs, S. Wollowitz and L. Coresh, Photochemical inactivation of viruses and bacteria in platelet concentrates by use of a novel psoralen and long-wavelength ultraviolet light, *Transfusion*, 1997, **37**, 423–435.

7. E. Ben-Hur and B. Horowitz, Advances in photochemical approaches for blood sterilization, *Photochem. Photobiol.*, 1995, **62**, 383–388.
8. B. Lambrecht, H. Mohr, J. Knüver-Hopp and H. Schmitt, Photoinactivation of viruses in human fresh plasma by phenothiazine dyes in combination with visible light, *Vox Sang.*, 1991, **60**, 207–213.
9. S. Wagner, A. Skripchenko, D. Robinette, J. Foiley and L. Cincotta, Factors affecting virus photoinactivation by a series of phenothiazine dyes, *Photochem. Photobiol.*, 1998, **67**, 343–349.
10. J. Matthews, F. Sogandares-Bernal, M. Judy, A. Marengo-Rowe, J. Leveson, H. Skiles, J. Newman and T. Chanh, Preliminary studies of photoinactivation of human immunodeficiency virus in blood, *Transfusion*, 1991, **31**, 636–641.
11. J. Matthews, J. Newman, F. Sogandares-Bernal, M. Judy, H. Skiles, J. Leveson, A. Marengo-Rowe and T. Chanh, Photodynamic therapy of viral contaminants with potential for blood banking applications, *Transfusion*, 1988, **28**, 81–83.
12. J. Li, D. de Korte, M. Woolum, P. Ruane, S. Keil, O. Lockerbie, R. McLean and R. Goodrich, Pathogen reduction technology (PRT) for buffy coat platelet concentrates using riboflavin and light: comparison to PRT treated apheresis platelet products, *Vox Sanguinis*, 2004, **87**, 82–90.
13. R. Rivlin, Riboflavin Metabolism, *New Engl. J. Med.*, 1970, **293**, 463–472.
14. G. Treadwell and D. Jr. Metzler, Photoconversion of riboflavin of lumichrome in plant tissues, *Plant Physio.*, 1972, **49**, 991–993.
15. D. Metzler and W. Cairns, Photochemical degradation of flavins VI, *J. Am. Chem. Soc.*, 1971, **93**, 2772–2777.
16. G. Tapia and E. Silva, Photoinduced riboflavin binding to the tryptophan residues of bovine and human serum albumins, *Radiat. Environ. Biophys.*, 1991, **30**, 131–138.
17. E. Silva, M. Salim-Hanna, A. Edwards, M. Becker and A. De Ioannes, *A light-induced tryptophan-riboflavin binding: biological implications*, in *Nutritional and Toxicological Consequences of Food Processing*, N. Friedman (ed), Plenum, New York, NY, 1991, 33–48.
18. P. Joshi, Ultraviolet radiation-induced photodegradation and $^1O_2$, O_2 production by riboflavin, lumichrome and lumiflavin, *Ind. J. Biochem. Biophys.*, 1989, **26**, 186–189.
19. W. Speck, S. Rosenkranz and H. Rosenkranz, Further observations on the photooxidation of DNA in the presence of riboflavin, *Biochim. Biophys. Acta.*, 1976, **435**, 39–44.
20. M. Korycka-Dahl and T. Richardson, Photodegradation of DNA with fluorescent light in the presence of riboflavin and photoprotection by flavin triplet-state quenchers, *Biochim. Biophys. Acta.*, 1980, **610**, 229–230.
21. C. Martin, Ph.D. Thesis, The Ohio State University, 2004.
22. P. Burgstaller and M. Famulok, Flavin dependent photoclevage of RNA at G-U base Pairs, *J. Am. Chem. Soc.*, 1997, **119**, 1137–1138.
23. J. Cadet, S. Decarroz, Y. Wang and W. Midden, Mechanisms and products of photosensitized degradation and related model compounds, *Isr. J. Chem.*, 1983, **23**, 420–429.
24. P. Joshi, Comparison of the DNA damaging property of photosensitized riboflavin via singlet oxygen and superoxide mechanisms, *Toxicol. Letters*, 1985, **26**, 211–217.
25. H. Kasai, Z. Yamaizumi, M. Berger and J. Cadet, Photosensitized formation of 7, 8-dihydro-8-oxo-2'-deoxyguanosine (8-hudroxy-2'-deoxyguanosine) in DNA by riboflavin: a non singlet oxygen mediated reaction, *J. Am. Chem. Soc.*, 1992, **114**, 9692–9694.

26. F. Yamamoto, S. Nishimura and H. Kasai, Photosensitized reactions of 8-hydroxydeoxyguanosine in cellular DNA by riboflavin, *Biochem. Biophys. Res. Comm.*, 1992, **187**, 809–813.
27. J. Ennever, H. Carr and W. Speck, Potential for genetic damage from multivitamin solutions exposed to phototherapy illumination, *Pediatr. Res.*, 1983, **17**, 192–194.
28. T. Mori, K. Tano, K. Takimoto and H. Utsumi, Formation of 8-hydroxyguanine and 2,6-diamino-4-hydroxy-5-formamidopyrimidine in DNA by riboflavin mediated photosensitization, *Biochem. Biophys. Res. Comm.*, 1988, **242**, 98–101.
29. J. Ennever and W. Speck, Photochemical reactions of riboflavin: covalent binding to DNA and to poly (dA) poly (dT), *Pediatr. Res.*, 1983, **17**, 234–236.
30. T. Coohill, *Action spectroscopy ultraviolet radiation*, in *CRC Handbook of Organic Photochemistry and Photobiology*, W. Horspool (ed), CRC Press, Boca Raton, 1995, 1257–1266.
31. F. Gates, A study of the bactericidal action of UV light. III. The absorption of ultraviolet light by bacteria, *J. Gen. Physiol.*, 1930, **14**, 31–42.
32. P. Fan, A. Suri, R. Fiala, D. Live and D. Patel, Molecular recognition of the FMN-RNA aptamer complex, *J. Mol. Biol.*, 1996, **258**, 480–550.
33. J. Koziol and E. Knobloch, The solvent effect on the fluorescence and light absorption of riboflavin and lumiflavin, *Biochim. Biophys. Acta.*, 1965, **102**, 289–300.
34. V. Kumar, O. Lockerbie, S. Keil, P. Ruane, M. Platz, C. Martin, J. Ravanat, J. Cadet and R. Goodrich, Riboflavin and UV-light based pathogen reduction: extent and consequence of DNA damage at the molecular level, *Photochem. Photobiol.*, 2004, **80**, 15–21.
35. V. Kumar, R. McLean, S. Keil, D. Gilmour, G. Luzniak, O. Lockerbie, J. Rice, M. Janssens and R. Goodrich, Mirasol pathogen reduction technology for blood products using riboflavin and UV illumination: mode of action of riboflavin on pathogen nucleic acid chemistry, *Transfusion*, 2003, **43**, 79A.
36. T. Douki, M. Court, S. Sauvaigo, F. Odin and J. Cadet, Formation of the main UV-induced thymine dimeric lesions within isolated and cellular DNA as measured by high performance liquid chromatography-tandem mass spectrometry, *J. Biol. Chem.*, 2000, **275**, 11678–11685.
37. A. Gentil, F. LePage, A. Margot, C.W. Lawrence, A. Borden and A. Sarasin, Mutagenicity of a unique thymine-thymine dimer or thymine pyrimidine pyrimidone (6–4) photoproduct in mammalian cells, *Nucleic Acids Res.*, 1996, **24**, 1837–1840.
38. J. Butendandt, R. Epple, E. Udo Wallenborn, A.P.M. Eker, V. Gramlich and T. Carrell, Comparative repair study of thymine- and uracil-photodimers with model compounds and photolyase repair enzyme, *Chem. Eur. J.*, 2000, **6**, 62–72.
39. J. Cadet, C. Decarroz, S. Y. Wang and W.R. Midden, Mechanisms and products of photosensitized degradation of nucleic acids and related model compounds, *Isr. J. Chem.*, 1983, **23**, 420–429.
40. B. Lichtenberg and A. Yasai, Effects of recB and uvrA mutations on Weigle reactivation of lambda phages in Escherischia coli K12 treated with 8-methoxypsoralen or angelicin and 365 nm light, *Mutat. Res.*, 1983, **112**, 253–260.
41. Good Laboratory Practices, United States Code of Federal Regulations, 21 CFR Part 58.
42. Committee for Proprietary Medicinal Products (CPMP), CPMP/ICH/295/95 (Sections 4.0–4.3).
43. FDA Guidance for Industry, Q5A viral safety evaluation of biotechnology products derived from cell lines of human or animal origin, September 1998.

44. CPMP/ICH/295/95, Guidance for industry: Q5A viral safety evaluation of biotechnology products derived from cell lines of human or animal origin, March 4, 1997.
45. J. Weusten, H.A. van Drimmelen and P.N. Lelie, Mathematic modeling of the risk of HBV, HCV, and HIV transmission by window-phase donations not detected by NAT, *Transfusion*, 2002, **42**, 537–548.
46. A. Macedo de Oliveira, B.D. Beecham, S.P. Montgomery, R.S. Lanciotti, J.M. Linnen, C. Giachetti, L.A. Pietrelli, S.L. Stramer and T.J. Safranek, West Nile virus blood-transfusion related infection despite nucleic acid testing, *Transfusion*, 2004, **44**, 1695–1699.
47. Personal communication between R. Goodrich and L. Petersen.
48. S. Slichter, Extension of platelet storage: fact or fantasy, *Food and Drug Administration Blood Products Advisory Committee Meeting*, Gaithersburg, MD, March 14–15, 2002.
49. P. Ness, H. Braine, K. King, C. Barrassso, T. Kickler, A. Fuller and N. Blades, Single-donor platelets reduce the risk of septic platelet transfusion reactions, *Transfusion*, 2001, **41**, 857–861.
50. C.P. Engelfriet, H.W. Reesink, M.A. Blajchman, L. Muylle, J. Kragh-Kjeldsen, R. Kekomaki, R. Yomtovian, P. Hocker, H.G. Klein, K. Soldan, J. Barbara, A. Slopecki, A. Robinson and H. Seyfried, Bacterial contamination of blood components, *Vox Sang.*, 2000, **78**, 59–67.
51. M. Brecher, P.V. Holland, A.A. Pineda, G.E. Tegtmeier and R. Yomtovian, Growth of bacteria in inoculated platelets: implications for bacterial detection and the extension of platelet storage, *Transfusion*, 2000, **40**, 1308–1312.
52. Personal communication between R. Goodrich, R. Yomtovian and M. Brecher, July 2001.
53. Guidance for Industry For Platelet Testing and Evaluation of Platelet Substitute Products, US FDA (CBER), May 1999.
54. P. Ruane, R. Edrich, D. Gampp, S. Keil, L. Leonard and R. Goodrich, Photochemical inactivation of selected viruses and bacteria in platelet concentrates using riboflavin and light, *Transfusion*, 2004, **44**, 877–885.
55. R. Edrich, L. Benford, M. Urioste, O. Lockerbie, J. Rice, N. Hovenga, M. Janssens and R. Goodrich, Bacterial decontamination of apheresis platelets using a photochemical treatment process with riboflavin, *Transfusion*, 2003, **43**, 79A.
56. S. Murphy, P. Rebulla, F. Bertolini, S. Holme, G. Moroff, E. Snyder and R. Stromberg, *In vitro* assessment of the quality of stored platelet concentrates, BEST task force, *Trans. Med. Rev.*, 1994, **8**, 29–36.
57. V. Vengelen-Tyler (ed), *AABB Technical Manual*, American Association of Blood Banks, Bethesda, MD, 1999, 174.
58. Guide to the Preparation, Use and Quality Assurance of Blood Components, 10th edn, Council of Europe Publishing, Strasbourg, France, 2004.
59. A. Michelson, M.R. Barnard, H.B. Hechtman, H. MacGregor, R.J. Connolly, J. Loscalzo and C.R. Valeri, *In vivo* tracking of platelets: circulating degranulated platelets rapidly lose surface P-selection but continue to circulate and function, *Proc. Natl. Acad. Sci. USA*, 1996, **93**, 11877–11882.
60. S. Holme, G. Moroff and S. Murphy, A multi-laboratory evaluation of *in vitro* platelet assays: the test for extend of shape change and response to hypotonic shock, *Transfusion*, 1998, **38**, 31–40.
61. L. Lin, R. Dikeman, B. Molini, S.A. Lukehart, R. Lane, K. Dupuis, P. Metzel and L. Corash, Photochemical treatment of platelet concentrates with amotosalen and

long-wavelength ultraviolet light inactivates a broad spectrum of pathogenic bacteria, *Transfusion*, 2004, **44**, 1496–1504.
62. K. Janetzko, L. Lin, H. Eichler, V. Mayaudon, J. Flament and H. Kluter, Implementation of the INTERCEPT blood system for platlelets into routine blood bank manufacturing procedures: evaluation of apheresis platelets, *Vox Sanguinis*, 2004, **86**, 239–245.
63. J.P. AuBuchon, L. Herschel, J. Roger, H. Taylor, P. Whitley, J. Li, R. Edrich and R.P. Goodrich, Efficacy of apheresis platelets treated with riboflavin and ultraviolet light for pathogen reduction, *Transfusion*, 2004, **44**(Suppl), 16.
64. J.P. AuBuchon, L. Herschel, J. Roger, H. Taylor, P. Whitley, J. Li, R. Edrich and R.P. Goodrich, Efficacy of apheresis platelets treated with riboflavin and ultraviolet light for pathogen reduction, *Transfusion*, 2005, **45**, 1335–1341.

Chapter 6

Light-Induced Flavin Toxicity

ANA M. EDWARDS

Faculty of Chemistry, Biological Chemistry Laboratory, P.Universidad Catolica de Chile, Casilla 306, Correo 22, Santiago, Chile

Abstract

Riboflavin (RF), commonly known as vitamin B_2, is an essential component of living organisms, and it is also an efficient photosensitizer, widely used by chemists for modifying organic molecules that do not absorb visible light. Since the pioneer work by Wang (*Photochem. Photobiol.*, 1975, **21**, 373–375), who demonstrated that daylight fluorescence lamps have a lethal effect on human cells in culture media, many works have appeared concerning the toxic effect of RF photosensitization in biological systems; the effects on DNA, lipids and amino acids and proteins have been discussed. The mechanisms involved in light-induced cell toxicity of RF have not yet been elucidated. The possibility of inducing apoptosis in cells and tissues showing inhibition of this cell death mechanism (high proliferating tissues) has increased the interest in the research in this area.

6.1. Introduction

Riboflavin (RF), commonly known as vitamin B_2, is an essential component of living organisms. The major sources of RF are milk, eggs, cereals and grains, ice cream, some meats, and green vegetables. It is stable when heated but will leach into cooking water; however, RF is degraded by light. It is present, both in nature and in tissues, in free form or as flavin mononucleotide (FMN) or flavin-adenine-di-nucleotide (FAD). The function of the nucleotide forms is to act as prosthetic groups of important redox enzymes.[1] It is an important component of a healthy diet and is characterized by its yellow–orange colour. It is a safe colouring agent in most processed foods and pharmaceuticals. It is also permitted at quantum satisfied levels in most processed foods. It is included in multinutrients' supplements at levels up to 100 mg per daily dose. The minimal requirement for RF to prevent clinical signs of deficiency appears to be less than 0.35 mg per 1000 kcal, and in the UK the recommended amount is 1.1 mg per day for women and 1.3 mg per day for men.[2]

For many years, chemists have used RF as a photosensitizer for modifying organic molecules that do not absorb visible light. A photosensitizer (or photochemical sensitizer) is a compound that, besides showing a high visible (and/or UV) absorbance (compared to a dye), must have a high intersystem crossing quantum yield. Upon light absorption, it reaches the triplet excited state, whose lifetime must be long enough to allow interaction with oxygen to generate singlet oxygen (type II mechanism) or direct interaction with a suitable substrate to generate radical intermediates (type I mechanism).[3] RF is an efficient photosensitizer that can act both via singlet oxygen and via radical species generation (type II and I mechanism, respectively). The RF-sensitized photooxidation of free amino acids and/or proteins has been widely studied.[4–10] Nilsson *et al.*[5] reported that this process occurs at least in part via singlet oxygen. Silva *et al.*[9] demonstrated that, besides $^1O_2$ other reactive oxygen species such as $^{\bullet}OH$, H_2O_2, and $O_2^{\bullet-}$ are generated.

6.2. RF Toxicity

The first report of the potential toxic effect of RF was published in 1975 by Wang,[11] who demonstrated that daylight fluorescence lamps have a lethal effect on human cells in culture media. However, cell death does not occur when RF, tyrosine (Tyr) and tryptophan (Trp) are withdrawn from the medium prior to the irradiation.[12] These results were related with the photosensitizer properties of RF described above, and this resulted in a sharp increase in research on light-induced RF toxicity in biological systems. Many reports were dedicated to study the possible effect of RF photosensitization on biological macromolecules with special interest in DNA.

6.2.1. *Effects on DNA*

Coincidently with the paper of Wang[11] concerning the potential toxic effects of RF, it was reported that lumiflavin and lumichrome inhibits the growth of *Bacillus subtilus* cells and lumichrome represses the formation of RF precursors. Mutants resistant to both compounds were able to accumulate RF, FMN and FAD.[13] It is noteworthy that both lumiflavin and lumichrome are photoproducts from visible light (and UV) induced photodegradation of RF.[14]

A number of reports[15] related to carcinogenesis and other toxic effects of UV radiation have been published since the pioneer works by Verhoeff *et al.*[16] in 1916 about the pathological effects of radiant energy on the eye, by Bovie and Klein[17] in 1919 about heat sensitization due to exposure to light of short wavelengths, by Hausser and Vahle[18,19] in 1922 and 1927 about the dependence of light-induced erythema upon the frequency of the inducing radiation and by Duke-Elder[20] in 1926 about photophthalmia. However, it was in the 1960s and 1970s, when concomitantly with the explosive interest in all the processes related to the transmission of genetic information, the interest of the effects of UV radiation on DNA had a sharp increase. For an interesting review, see ref. 15 and the references herein. Most of the works deal with the direct effect of UV radiation on isolated or cellular DNA; however, there have appeared interesting reports about the effect of sensitizers in the toxicity of this radiation. Furocoumarins are natural products, and have been extensively studied[21–24] owing to the fact that they form additional photoproducts with DNA when irradiated (320–400 nm). The so-called "linear" furocoumarins such as psoralens have important applications in the treatment of skin diseases, particularly psoriasis. Other photosensitizers such as benzophenone and acetobenzophenone induce the formation of dimers and chain breaks in DNA;[25] toluidine and acridine yellow induce DNA lesions and mutations;[26] 1-cyclohexyl-3 (2-morpholinyl-4-ethyl) carbodiimide metho-*p*-toluene sulfonate (CMEC) binds to partially denatured region of DNA after irradiation at 254 nm.[27] Hutchinson[28] published a detailed review of the effect of 5-bromouracil on the toxicity of UV radiation on virus and cells. Wang *et al.*[29] reported in 1974, a work on the lethal effect of near UV irradiation on mammalian cells in culture, and the next year Wang[11] reported his work about the lethal effect of daylight's fluorescent light on human cells in tissue-culture medium. It is noteworthy that before the publication of this work[11] and that by Nixon and Wang[12] in 1977, it was not easy to find reports concerning the effect on DNA of visible (or UV) light irradiation in the presence of RF. The first reports came from groups that were active in the study of the effects of phototherapy in jaundiced infants. In 1977, Santella *et al.*,[30] studying the possibility that during phototherapy, physiological substances other than bilirubin may be photoactivated, develop a peroxidase technique for detecting changes in DNA of human cells (in culture) exposed to RF in the presence of light (450 nm). In 1979, Hoffmann and Meneghini[31] showed that when media containing RF and Trp were irradiated with visible light, hydrogen peroxide was formed; they related the action of this photoproduct with single-strand breaks in DNA from human fibroblast in culture. The same authors reported in 1980 that the strand breaks were not observed when purified DNA was exposed to H_2O_2, and they proposed that the damaging action of H_2O_2 on DNA of human fibroblast is mediated by a non-dialyzable compound.[32] The same year, Korycka-Dahl and Richardson[33] demonstrated, by electrophoresis in agarose gels, that RF-sensitized photoreaction causes single- and double-strand scissions in supercoiled DNA. They demonstrated that superoxide anion scavengers or quenchers have no effect in preventing the damage, but some flavin triplet quenchers protect DNA from photodegradation. In 1981, Ennever and Speck[34] demonstrated that singlet oxygen is not involved in the photodynamic reaction of RF and deoxyguanosine. In 1983, the same group[35] reported that the addition of a multivitamin concentrate to human cells under standard phototherapy illumination, enhances the generation of single-strand DNA breaks. They also demonstrated that the photoproduct responsible for the damage was hydrogen peroxide. The same year, these authors[36] described a new photochemical reaction between RF and purified DNA, in which an adduct is formed. The reaction is oxygen independent and involves deoxyadenosine and deoxythymidine.

In 1985, Joshi[37] reported a comparison of the DNA damage induced by RF photosensitization via singlet oxygen and superoxide radical mechanisms, concluding that singlet oxygen was largely responsible for the photodegradation of the guanine base of DNA and RNA when exposed to UV-A and UV-B radiation. The activated oxygen species were related to skin photosensitization, tumour promotion and carcinogenic properties.

Masayasu[38] reported that treatment with vitamin B_2 (RF) enhances the levels of breaks induced by chromate; however, the DNA–protein cross-links, the other major lesion induced by chromate, are not affected. In 1992, Kale *et al.*[39] evaluated the mutagenic potential of RF and its photodegradation product, lumiflavin. Both compounds by themselves are not mutagenic. After treatment with rat liver microsomal enzymes, lumiflavin acquires mutagenicity and RF remains unaffected. The same group[40] reported that both RF and lumiflavin, upon illumination show a mutagenic response, which is partially abolished by superoxide dismutase, while sodium azide does not have any effect. If the compounds were not illuminated, no mutagenicity could be observed. The results suggest the involvement of superoxide radicals in the light-induced mutagenecity of the compounds.

In 1992, a Japanese group in collaboration with a group of Grenoble, France, demonstrated the RF-photosensitized formation of 8-hydroxyguanine (7,8-dihydro-8-oxoguanine) in cellular DNA. They proposed a reaction mechanism involving the formation of guanine radical cation followed by hydration reaction. They confirmed this hypothesis by the incorporation of ^{18}O atom within guanine moiety in an isotopic experiments using $[^{18}O]H_2O$.[41] The same groups informed that the photoproduct 8-oxo-7,8-dihydro-2′-deoxyguanosine (8-oxodG) is predominantly generated in DNA in oxygen-free solutions. However, the photosensitized formation of 8-oxodG was very inefficient when 2′-deoxyguanosine is used as the substrate. The participation of reactive oxygen species such as OH radicals and $^1O_2$ was discarded, in accordance with the radical mechanism described in the previous communication.[42] The French group reported the isolation and characterization of the photoproducts of RF and other photosensitizers of d(TpG), and they obtained dinucleoside monophosphates containing the 4R and 4S diasteroisomers of the photoproducts.[43–44] They also demonstrated that photosensitization mediated by RF and other photosensitizers in aerated aqueous solutions of d(TpG) results in the occurrence of a covalent bonding between the pyrimidine ring on the 5′-OH terminal end and the imidazole ring on the 3′-OH terminal end through a methylene bridge. The initial event is the abstraction of a hydrogen atom from the Me group of the thymine base moiety of d(TpG), this is followed by the addition of the methyl-centred radical to the C-4 atom of the guanine ring, which after reaction with molecular oxygen gives rise to a novel vicinal lesion.[45] They also characterized the main photoproducts of 2′-deoxyguanosine arising from either the type I or the type II photosensitization mechanisms.[46]

The Japanese group[47] showed that exposure of double-strand DNA to 365 or 305 nm radiation in the presence of RF induces the sequence specific GG DNA cleavage, which is different from that caused by 302 or 254 nm irradiation in the absence of the photosensitizer. They suggested that photoexcited RF reacts with dGMP to produce RF anion radical and guanine cation radical, but not with other mononucleotides. These results were confirmed by Shulz *et al.*[48] in a study of the oxidative DNA-base damage induced by singlet oxygen and photosensitization, where they found the same specificity for RF photosensitization; and they classified RF as a type I photosensitizer. Herfeld *et al.*[49] reported the synthesis, DNA-binding properties and cytotoxic activity of a series of flavin derivatives, with the aim to search for sequence specific DNA-binding drugs. Seah and Burgoyne[50] found that type II mechanism is not important for RF photosensitization, and that free radicals appear to be involved in the mechanism, indicating that the predominant species in air appears to involve oxygen, but not exclusively or necessarily so. Similar results were reported by Riemschneider *et al.*[51] Okamoto *et al.*[52] found the same GG specificity in double-stranded DNA cleavage induced by RF; however, the specificity was lower for the effect on DNA/PNA duplex and DNA/$(PNA)_2$ triplex (where PNA are peptide nucleic acids) than that for DNA/DNA duplex. Kino and Sugiyama[53] reported that a new photoproduct, 2-aminoimidazolone (Iz) is formed in amounts larger than 8-oxo-G during the photoreaction of DNA with RF. They proposed that Iz could produce a stable base pair with C, thereby giving rise to a G:C to C:G transversion mutation. Jazzar and Nassem[54] studied the increased genotoxicity of photoilluminated RF in the presence of Cu(II). They suggested that a ternary complex of DNA–RF–Cu(II) is formed, and a redox reaction, involving RF and Cu(II) may then occur, with the formation of a DNA–oxidized RF–Cu(I) complex that probably acts as a catalyst for the oxidation of Cu(I) to Cu(II), generating a variety of active oxygen species.

In the last decade, it is observed that there is an increase in the study of apoptosis induction in cells that irradiated in the presence of photosensitizers. Since the publications by Agarwal *et al.*[55] and Zaidi *et al.*[56] on the apoptosis induction on tumoural cells after photodynamic therapy (PDT), many reports have appeared about the induction of apoptosis using different photosensitizers.[57–63] In a study on the visible light effects on tumour cells in culture media enriched with Trp and RF, Edwards *et al.*[61] in 1994 found (by transmission electron microscopy) morphological cell alterations similar to those described for apoptotic cells.[62] The same group[63] demonstrated typical apoptotic DNA fragmentation in human HL-60 and murine NSO/2 tumour cells by photoproducts of indole-3-acetic acid and RF.

All the results discussed above shed light on the complex mechanisms of RF-sensitized light effect on DNA. As an example of the difficult interpretation of the processes induced by RF photosensitization on cells, Heuser *et al.*[64] have recently proposed a cytotoxicity determination without photochemical artifacts by using "flavin-protecting" conditions: flavin-free culture media with foetal bovine serum as the only source of flavin. They proposed that flavin-mediated photosensitization should be strictly eliminated from *in vitro* experiments involving other cytotoxic drugs. Hirawaka *et al.*[65] have recently reported that the effect of a series of xanthone derivatives on the prevention of the DNA damage photosensitized by RF is based on the quenching of the triplet-excited state of RF through an electron-transfer mechanism. They proposed that the compounds would act as novel chemopreventive agents by quenching the excited photosensitizer.

6.2.2. *Effect on Lipids*

Most of the reports found in the literature about lipids and RF in biological systems concern metabolic problems induced by RF deficiency. They include, among many others, studies of the effect of the function and fluidity of rat erythrocyte membranes[66] on the possible mechanism of mitochondrial dysfunction during cerebral ischemia,[67] and recently, on the decreased oxidative folding and secretion of apolipoprotein B-100.[68]

In 1982, Suzuki *et al.*[69] published a work about RF-photosensitized haemolysis of rat erythrocytes, in the presence of serum. During illumination, a marked efflux of intracellular K^+ and an increase in osmotic fragility and lipid peroxidation occurred prior to haemolysis. However, in the absence of either oxygen or serum no haemolysis was observed. The authors suggested that photohaemolysis is initiated by oxidative cell-membrane damage.

Jernigan and Laranang[70] studied the effect of RF-sensitized oxidation on choline metabolism in cultured rat lenses. They concluded that the decrease in accumulation of choline in the lenses results from effects on the choline carrier. Santus and Reyftmann[71] reviewed the chemical mechanisms of the photosensitized peroxidation of unsaturated fatty acids and cholesterol. It also discussed the subsequent reaction of peroxide decomposition products with biological targets, and the importance of protein photooxidation and cross-linking in membrane function impairment. Lee and Tzeng[72] studied the photodynamic effect of methionine–RF mixture on membranous system of living cells. The strong lethal activity of this system was in great part due to the peroxidation of membrane lipids with substantial damages on the membrane function, with leakage of cellular proteins. Ali *et al.*[73] studied the damaging potential of photosensitized RF using erythrocyte membrane as model system. Considerable increase in lipid peroxidation, and significant inhibition of ATPases and other enzymes are caused by the exposure of the membranes to sunlight in the presence of RF. In 2001, Misra *et al.*[74] re-studied the photohaemolytic property of RF, previously described by Suzuki *et al.*[69] RF, unlike other photosensitizers, does not induce photohaemolyisis by itself; however, in the presence of serum or plasma, RF induces haemolysis when the system is illuminated. The authors used a known photohaemolytic agent, hematoporphyrin (HP), as a positive control, whose membrane damaging potential, when illuminated, was not affected by the presence of plasma in the reaction system. They studied the effect on blood and erythrocyte (RBCs) samples of different mammalian species such as hairless mice, rats and humans. They found a significant decrease in photohaemolitic potential of HP when human RBCs were incubated with RF; thus RF showed a preventive role in the absence of serum/plasma. They also observed that heat-inactivating plasma is as effective as native in carrying out RF-induced photohaemolysis, ruling out the involvement of complement. Human RBCs were found to be less sensitive to illuminated RF and HP than were those of hairless mice and rat. A possible explanation for the lower sensitivity of human RBCs may be their larger size, having a lower membrane surface area to volume ratio than that of the other cell types tested. Therefore, any photoproduct acting on the membrane might produce a greater effect on smaller cells. The authors concluded that during the process of photohaemolysis, the probable mechanism could be mediated by the long-lived triplet state of the RF–serum–proteins complex. Therefore, plasma helps in carrying out an interaction of photoexcited RF with the outer half of the bilayer of erythrocyte membrane, and increases its surface relative to the inner half, resulting in crenated erythrocytes. The uncoupling induced in the RBC's membrane is dependent on RF concentration and the sunlight dose. These membrane modifications leading to haemolysis may be a consequence of the generation of activated oxygen species.

Commercially available fat emulsions are widely used clinically as a component of parenteral nutrients, and they contain large amounts of polyunsaturated fatty acid. RF is present in the multivitamin infusate, which is added to the fat emulsion during a total parenteral nutrition.[75] The toxic effects of RF photosensitization of lipids described above, led to an increase in the interest of the study of photosensitized effects in parenteral nutrients. Silva *et al.*[76] studied the visible light-induced lipoperoxidation on parenteral nutrition fat emulsion sensitized by RF, FMN and FAD. They found that the efficiency of RF as photosensitizer is greater than that of FMN; however, FAD does not induce lipid peroxidation, probably because it remained excluded from the lipid emulsion, because of its polarity and molecular geometry. They demonstrated that ascorbic acid (also present in the multivitamin infusate) provides a significant protection against photoinduced lipoperoxidation, which was concomitant with an extensive photodecomposition of ascorbic acid. Silvers *et al.*[77] studied the lipid peroxidation and the formation of hydrogen peroxide in parenteral nutrition solutions containing multivitamins. They concluded that multivitamins protect against lipid peroxidation, but light-dependent hydrogen peroxide production and ascorbate loss also occurs, presumably because of light-dependent RF-catalysed reduction of oxygen by ascorbic acid. They proposed that any multivitamin solution requires protection of the delivery system from light. The same group[78] studied how to decrease light-induced lipid peroxidation and vitamin loss in infant parenteral nutrition. They found that parenteral lipids are susceptible to light-induced lipoperoxidation, particularly under photochemotherapy, and that multivitamins' preparations protect lipid emulsions against lipid peroxidation. They proposed that administration of multivitamins with lipids emulsions via dark delivery tubing, would provide a practical way of preventing both photosensitized effects, and that this procedure should be considered for routine use as well as together with phototherapy.

6.2.3. Effect on Amino Acids and Proteins

None of the 20 amino acids from which all the proteins in organisms are built, absorb UVA or visible radiation. Only Trp and Tyr, and in a lesser extent Phe, absorb UVB light, this being responsible for the characteristic absorbance by proteins at wavelengths between 270 and 280 nm. The photochemical alteration induced in proteins by sensitized photooxidation is circumscribed to the specific modification of the side chains of certain amino acid residues, with no alteration at the level of the peptide bond. Only the side chains of cysteine, histidine, methionine, Trp and Tyr residues are susceptible to modifications by sensitized photooxidation.[4,6,79]

There have been many reports concerning the mechanism of sensitized photooxidation of proteins and enzymes.[80,81] The kinetic analysis of the amino acid destruction in proteins, employing sensitized photooxidation as modification method, was used for many years with the aim to predict the degree of exposure of the photooxidable residues.[8,82,83] Lysozyme has been widely used[6–8] as a model compound because it was the first enzyme whose tridimensional structure was elucidated by X-ray crystallography.[84]

Silva *et al.*[6] reported that monochromatic irradiation of lysozyme in the presence of RF, using monochromatic visible light (448 nm) induces the selective destruction of His and Trp. However, if the enzyme is irradiated with light of 365 nm in the presence of RF, only Cys, Tyr and Met are destroyed. In 1977, Silva and Gaule[85] reported a light-induced binding of RF to lysozyme. The authors concluded that this result would explain the higher quantum yield observed when RF is used in comparison with other photosensitizers such as methylene blue. The same group[86] reported similar RF–protein binding for bovine and human serum albumins and for lens proteins.[87] They also demonstrated that the light-induced binding of RF was to Trp residues in proteins[88] and also to free Trp.[89]

The study of the effect of this photoinduced RF–Trp adduct in biological system led to the finding of toxicity in murine carcinoma cells and embryos.[90] These results motivated this group to study the mechanism of the RF-sensitized photoprocesses of Trp[9] and Tyr,[10] and also to explore the mechanisms of the cell toxicity. With this aim, they developed monoclonal antibodies against the RF–Trp adduct,[91,92] and also prepared and characterized, both photochemically[93] and spectroscopically,[94] hydrophobic esters of RF with the aim to obtain a higher cell uptake for the flavins. Martin *et al.*[95] by using density functional theory methods and lumiflavin as a model for RF, studied the reaction of triplet flavin with indole, and found that the most thermodynamically stable adduct is that being formed at C_4 of the isoalloxazine moiety. Vibrational spectra were also calculated for the transient species. Experimental time-resolved infrared spectroscopic data obtained using RF tetraacetate are in excellent agreement with the calculated spectra for the triplet flavin, the radical anion and the most stable hydroflavin radical.

6.2.4. Cell Toxicity and Apoptosis

Since the pioneer study of Wang[11] concerning the lethal effects of daylight on human cells in culture, many studies about RF-photosensitized cell toxicity have been published. Misra *et al.* have shown RF-induced phototoxicity in *Paramecium*[96] and *Tetrahymena thermophila.*[97] Donoso *et al.*[98] demonstrated that the Trp–RF photoinduced adduct described above,[89] induces hepatic dysfunction in rats when injected intraperitoneally for 12 days. Zigman *et al.*[99] studied the effects of Trp photoproducts and H_2O_2 on the growth of mouse embryonic fibroblasts. Sato *et al.* studied the cytotoxicity of RF activation by visible light[100] and also by UV radiation.[101] They found that in both conditions, the toxicity was dose dependent, and mediated by the long-lived reactive oxygen specie, H_2O_2. The same group[102] demonstrated that hypoxia potentiates UVA-induced RF phototoxicity on skin. This effect occurs with low UVA doses; however, at higher UVA doses the cytotoxicity and H_2O_2 production were similar to those observed under different oxygen conditions. Verrico and Moore[103] found that the expression of the collagen related heat-shock protein HSP41 in treated fibroblast was elevated when treated with hypothermia and with RF–PDT. The effect was not observed when other photosensitizers were used. Despite all these reports concerning the cytotoxicity of RF-sensitized irradiation of cells and tissues, Christensen[104] reported that RF can protect tissue from oxidative stress, and Mack *et al.*[105] studied a potential role of myocardial flavin reductase and RF in decreasing reoxygenation injury.

Edwards *et al.*[61] have demonstrated that the irradiation of tumour cells in culture with visible light in the presence of RF induces cell death. The phototoxic activity of RF is enhanced when the medium is enriched with Tyr[106] and becomes even higher when Trp is added.[61] The addition of photoproducts obtained by irradiation of a culture medium enriched with RF and Trp to non-irradiated cells induces morphological changes similar to those of apoptotic cells.[61] The effect was greater when the photoproducts were irradiated while bubbling up nitrogen.[61] They also studied the effect of RF photoproducts of Trp and indol-3-acetic acid (IAA),[63] on non-irradiated HL-60 human cells. Severe damage with a dose–response effect was observed for the IAA–RF photoproducts, being greater than that produced by Trp–RF photoproducts. On the basis of electron microscopy studies and flow cytometry analysis of DNA fragmentation, they demonstrated that RF photoproducts of IAA and Trp induce cell death by an apoptotic mechanism.[63] Edwards *et al.* have evaluated, light-mediated apoptosis induction by RF hydrophobic esters.[93,94]

They found morphological alterations, caspase-3 activation[107] and also DNA ladder,[108] all characteristic of apoptotic cells.

Cho *et al.*[109] reported that the reactive oxygen species produced by RF-sensitized irradiation induces apoptosis 30 min after irradiation, and necrosis after 2 h in bovine corneal endothelial cells. Wollesak *et al.*[110] evaluated the corneal endothelial cytotoxicity of RF/UVA treatment *in vitro*, and found a cytotoxic effect when the corneal thickness was less than 400 μm. They also found the induction of corneal collagen cross-linking using RF/UVA treatment[60] and they also detected keratocyte apoptosis in rabbit eyes enucleated 24 h postoperatively.[60] Interestingly, Marshall *et al.*[111] have recently purified and characterized a human apoptosis-inducing protein, AMID (apoptosis-inducing factor-homologous mitochondrion-associated inducer of death; also known as p53-responsive gene 3 (PRG3)) that is a human caspase-independent pro-apoptotic protein with some similarity to apoptosis-inducing factor. They have demonstrated that it is an oxidoreductase with a modified flavin cofactor (6-hidroxy-FAD) and possesses NAD(P)H oxidase activity, and catalyses NAD(P)H-dependent cytocrome *c* and other electron acceptors including molecular oxygen. AMID is a DNA-binding protein that apparently lacks DNA-sequence specificity. The authors establish a link between coenzyme and DNA binding that likely impacts on the physiological role of AMID in cellular apoptosis.[111]

The mechanisms involved in RF-photosensitized toxicity have not yet been elucidated, and the possibility of inducing apoptosis in cells and tissues that show an inhibition of this cell-death mechanism (high-proliferating tissues), have increased the interest in the research in this area.

Acknowledgements

The financial support received from Fondecyt (Chile) Grant N 1040667 is acknowledged.

References

1. A. Lehninger, In: *Principles of Biochemistry*, D.L. Nelson and M.M. Cox (eds), Worth Publishers, New York, 2000.
2. COMA, Committee on Medical Aspects of Food and Nutrition Policy, *Dietary Reference values for Food Energy and Nutrients for the UK*, HMSO, London, 1991.
3. L.I. Grossweiner, Molecular mechanisms in photodynamic action, *Photochem. Photobiol.*, 1969, **10**, 183–191.
4. E. Scoffone, G. Jori and G. Galiazzo, Selective photo-oxidation of amino acids in proteins, *Biochem. Soc. Symp.*, 1970, **31**, 163–170.
5. R. Nilsson, P.B. Merkel and D.R. Kearns, Unambiguous evidence for the participation of singlet oxygen in photodynamic oxidation of amino acids, *Photochem. Photobiol.*, 1972, **16**, 117–124.
6. E. Silva, S. Rissi and K. Dose, Photooxidation of lysozyme at different wavelengths, *Radiat. Environ. Biophys.*, 1974, **11**, 111–124.

7. E. Silva, Rate constants studies of the dye-sensitized photoinactivation of lysozyme, *Radiat. Environ. Biophys.*, 1979, **16**, 71–79.
8. A.M. Edwards and E. Silva, Photochemical reactivity of the homologous proteins α-lactalbumin and lysozyme, *Radiat. Environ. Biophys.*, 1985, **24**, 141–148.
9. E. Silva, R. Ugarte, A. Andrade and A.M. Edwards, Riboflavin-sensitized photoprocesses of tryptophan, *J. Photochem. Photobiol. B: Biol.*, 1994, **23**, 43–48.
10. E. Silva and J. Godoy, Riboflavin sensitized photooxidation of tyrosine, *Int. J. Vitam. Nutr. Res.*, 1994, **64**, 253–256.
11. R.J. Wang, Lethal effect of "daylight" fluorescent light on human cells in tissue-culture medium, *Photochem. Photobiol.*, 1975, **21**, 373–375.
12. B.T. Nixon and R.J. Wang, Formation of photoproducts lethal for human cells in culture by daylight fluorescent light and billirubin light, *Photochem. Photobiol.*, 1977, **26**, 589–593.
13. S.E. Bresler, D.A. Perumov, T.N. Shevchenko, E.A. Galzvnov and T.P. Chernik, Operon of riboflavin synthesis in bacillus subtilis. IX. Preparation and properties of lumiflavin- or lumichrome- resistant mutants, *Genetika (USRR)*, 1975, **11**, 129–138.
14. W.L. Cairns and D.E. Metzler, Photochemical degradation of flavins. VI. A new photoproduct and its use in studying the photolytic mechanism, *J. Amer. Chem. Soc.*, 1971, **13**, 2772–2777.
15. International Programme on Chemical safety, *Environmental Health Criteria 160. Ultraviolet Radiation*, Published under the joint sponsorship of the United Nations Environment Programme, the International Labour Organization, and the World Health Organization, Geneva, 1994.
16. F.H. Verhoeff, L. Bell and C.B. Walker, The pathological effects of radiant energy on the eye: an experimental investigation with a systematic review of the literature, *Proc. Am. Acad. Arts Sci.*, 1916, **51**, 630–818.
17. W.T. Bovie and A. Klein, Sensitization to heat due to exposure to light of short wavelengths, *J. Gen. Physiol.*, 1919, **1**, 331–336.
18. K.W. Hausser and W. Vahle, The dependence of light induced erythema and pigment formation upon the frequency (or wavelength) of the inducing radiation, *Strahlentherapie*, 1922, **13**, 41–71.
19. K.W. Hausser and W. Vahle, Sunburn and suntan, *Wiss Veröffent Siemens-Konzer*, 1927, **6**, 101–120.
20. W.S. Duke-Elder, The pathological action of light upon the eye. I. Action of the outer eye: photophthalmia, *Lancet*, 1926, **1**, 1137–1141.
21. P. Chandra, Photodynamic action: a valuable tool in molecular biology, *Res. Prog. Org. Biol. Med. Chem.*, 1972, **3**, 232–258.
22. M.A. Pathak, D.M. Kramer and T.B. Fitzpatrick, Photobiology and photochemistry of furocoumarins (psoralens), , in *Sunlight and man*, T.B. Fitzpatrick, M.A. Pathak, L.C. Harber, M. Seiji and A. Kukita (eds), University of Tokyo Press, Tokyo, 1974, 335–368.
23. L. Musajo, G. Rodighiero, G. Caporale, F. Dall'Acqua, S. Marciani, F. Bordin, F. Baccichetti and R. Bevilacqua, Photoreactions between skin-phosensitizing furocoumarins and nucleic acids, , in *Sunlight and Man*, T.B. Fitzpatrick, M.A. Pathak, L.C. Harber, M. Seiji and A. Kukita (eds), University of Tokyo Press, Tokyo, 1974, 369–387.
24. P. Chandra, G. Rodighiero, S. Palikcioglu and R.K. Biswas, Nucleic acid modification by furocoumarins and light: some biomedical implications, , in *Photochemotherapie*, E. Jung and F.K. Stuttgart (eds), Schattauer Verlag, 1976, 25–32.

25. M. Charlier and C. Helene, Photochemical reactions of aromatic ketones with nucleic acids and their components. I. Purine and pyrimidine bases and nucleosides, *Photochem. Photobiol.*, 1972, **15**, 71–87.
26. A.P. Harrison, Survival of bacteria. Harmful effects of light, with some comparisons with other adverse physical agents, *Annu. Rev. Microbiol.*, 1967, **21**, 143–156.
27. A.M. Rauth and M. Domon, Potentiation of ultraviolet light damage in mouse L cells by 1-cyclohexyl-3-(2-morpholinyl-4-ethyl)carbodiimide metho-*p*-toluene sulphonate (CMEC), *Int. J. Radiat. Biol.*, 1973, **24**, 189–198.
28. F. Hutchinson, The lesions produced by ultraviolet light in DNA containing 5-bromouracil, *Q. Rev. Biophys.*, 1973, **6**, 201–246.
29. R.J. Wang, Lethal effect of daylights fluorescent light on human cells in tissue-culture medium, *Photochem. Photobiol.*, 1975, **21**, 373–375.
30. R.M. Santella, H.S. Rosenkranz, S. Brem, B.W. Lubit, B.F. Erlanger and W.T. Speck, Peroxidase technique for the detection of photochemical lesions in intracellular deoxyribonucleic acid, *Pediatr. Res.*, 1977, **8**, 939–941.
31. M.E. Hoffmann and R. Meneghini, Action of hydrogen peroxide on human fibroblast in culture, *Photochem. Photobiol.*, 1979, **30**, 151–155.
32. R. Meneghini and M.E. Hoffmann, The damaging action of hydrogen peroxide on DNA of human fibroblast is mediated by a non-dialyzable compound, *Biochim. Biophys. Acta*, 1980, **608**, 167–173.
33. M. Korycka-Dahl and T. Richardson, Photodegradation of DNA with fluorescent light in the presence of riboflavin, and photoprotection by flavin triplet-state quenchers, *Biochim. Biophys. Acta*, 1980, **610**, 229–234.
34. J.F. Ennever and W.T. Speck, Photodynamic reaction of riboflavin and deoxyguanosine, *Pediatr. Res.*, 1981, **15**, 956–958.
35. J.F. Ennever, H.S. Carr and W.T. Sepck, Potential for genetic damage from multivitamin solutions exposed to phototherapy illumination, *Pediatr. Res.*, 1983, **17**, 192–194.
36. J.F. Ennever and W.T. Speck, Short communication. Photochemical reactions of riboflavin: covalent binding to DNA and to poly(dA), poly (dT), *Pediatr. Res.*, 1983, **17**, 234–236.
37. P.C. Joshi, Comparison of the DNA-damaging property of photosensitized riboflavin via singlet oxygen and superoxide radical mechanisms, *Toxicol. Lett.*, 1985, **26**, 211–217.
38. S. Masayasu, Effects of vitamin E vitamin B2 on chromate-induced DNA lesions, *Biol. Trace Element Res.*, 1989, **21**, 399–404.
39. H. Kale, P. Harikumar, P.M. Nair and M.S. Netrawali, Assessment of the genotoxic potential of riboflavin and lumiflavin. A. Effect of metabolic enzymes, *Mutation Res.*, 1992, **298**, 9–16.
40. H. Kale, P. Harikumar, S.B. Kulkarni and P.M. Netrawali, Assessment of the genotoxic potential of riboflavin and lumiflavin. B. Effect of light, *Mutat. Res.*, 1992, **298**, 17–23.
41. H. Kasai, Z. Yamaizumi, F. Yamamoto, T. Bessho, S. Nishimura, M. Berger and J. Cadet, Photosensitized formation of 8-hydroxyguanine (7,8-dihudro-8-oxoguanine) in DNA by riboflavin, *Nucl. Acids Symp. Ser.*, 1992, **27**, 181–182.
42. H. Kasai, Z. Yamizumi, M. Berger and J. Cadet, Photosensitized formation of 7,8-dihydro-8-oxo-2′-deoxyguanosine (8-hydroxy-2′-deoxyguanosine) in DNA by riboflavin: a nonsinglet oxygen-mediated reaction, *J. Am. Chem. Soc.*, 1992, **114**, 9692–9694.

43. G.W. Buchko, J. Cadet, M. Berger and J.L. Ravanat, Photooxidation of d(TpG) by phthalocyanines and riboflavin. Isolation and characterization of dinucleoside monophosphates containing the 4R* and 4S* diastereoisomers fo 4,8-dihydro-4-hydroxy-8-oxo-2′-deoxyguanosine, *Nucl. Acids Res.*, 1992, **20**, 4847–4851.
44. G.W. Buchko, J. Cadet, B. Morin and M. Weinfeld, Photooxidation of d(TpG) by riboflavin and methylene blue. Isolation and characterization of thymidylyl-(3′,5′)-2-amino-5-[(2-deoxy-β-D-erythro-pentofuranosyl)amino]-4H-imidazol-4-one and its primary decomposition product thymidylyl-(3′,5′)-2,2-diamino-4-[(2-deoxy-β-D-erythro-pentofuranosyl)amino]-5(2H)-oxazolone, *Nucl. Acids Res.*, 1995, **23**, 3954–3961.
45. T. Delatour, T- Douki, D. Gasparutto, M.C. Brochier and J. Cadet, A novel vicinal lesion obtained from the oxidative photosensitization of TpdG: characterization and mechanistic aspects, *Chem. Res. Toxicol.*, 1998, **11**, 1005–1013.
46. J.L. Ravanat, G. Remaud and J. Cadet, Measurement of the main photooxidation products of 2′-deoxyguanosine using chromatographic methods coupled to mass spectrometry, *Arch. Biochim. Biophys.*, 2000, **374**, 118–127.
47. K. Ito, S. Inoue, K. Yamamoto and S. Kawanishi, 8-Hydroxydeoxyguanosine formation at the 5′ site of 5′-GG-3′ sequences in double-stranded DNA by UV radiation with riboflavin, *J. Biol. Chem.*, 1993, **268**, 13221–13227.
48. I. Schulz, H.C. Mahler, S. Boiteux and B. Epe, Oxidative DNA base damage induced by singlet oxygen and photosensitization: recognition by repair endonucleases and mutagenicity, *Mutat. Res.*, 2000, **461**, 145–156.
49. P. Herfeld, P. Helissey, J. Nafziger and S. Giorgi-Renault, Synthesis, DNA-binding properties and cytotoxic activity of flavin-olilgopyrrolecarboxamide and flavin-oligoimidazolecarboxamide conjugates, *Anti-Cancer Drug Design*, 1998, **13**, 337–359.
50. L.H. Seah and L.A. Burgoyne, Photosensitizer initiated attacks on DNA under dry conditions and their inhibition: a DNA archiving issue, *J. Photochem. Photobiol. B*, 2001, **61**, 10–20.
51. S. Riemschneider, H.P. Podhaisky, T. Klapperstuck and W. Wohlrab, Relevance of reactive oxygen species in the induction of 8-oxo-2′-deoxyguanosine in HaCa T kertinocytes, *Acta Derm. Venereol.*, 2002, **82**, 325–328.
52. A. Okamoto, K. Tanabe, C. Dohno and I. Saito, Modulation of remote DNA oxidation by hybridization with peptide nucleic acids (PNA), *Bioorg. Med. Chem.*, 2002, **10**, 713–718.
53. K. Kino and H. Sugiyama, Molecular mechanism of GG-specific photooxidation of DNA, *J. Med. Dent. Sci.*, 1998, **45**, 161–169.
54. M.M. Jazzar and I. Naseem, Genotoxicity of photoilluminated riboflavin in the presence of Cu (II), *Free Radical Biol. Med.*, 1996, **21**, 7–14.
55. M.L. Agarwal, M.E. Clay, E.J. Harvey, H.H. Evans, A.R. Antúnez and N.L. Oleinick, Photodynamic therapy induced rapid cell death by apoptosis in L5178 Y mouse lymphoma cells, *Cancer Res.*, 1991, **51**, 5993–5996.
56. S.I.A. Zaidi, L. Oleinick and T.M. Zaim, Apoptosis during photodynamic therapy induced ablation of RIF-1 tumors in C_3H mice, *Photochem. Photobiol.*, 1993, **58**, 771–776.
57. X.Y. He, R.A. Sikes, S. Thomsen, L.W.K. Chung and S.L. Jacques, Photodynamic therapy with photofrin II induces programmed cell death in carcinoma cell lines, *Photochem. Photobiol.*, 1994, **59**, 468–473.
58. Y. Luo, C.K. Chang and D. Kessel, Rapid initiation of apoptosis by photodynamic therapy, *Photochem. Photobiol.*, 1996, **63**, 528–534.

59. J.Q. Li, R.C. Chen, K.X. Cai and Z.Y. Ye, Apoptosis of human gastric cancer cell induced by photochemical riboflavin, *Ai. Zheng*, 2003, **22**, 253–256.
60. G. Wollensak, E. Spoerl, M. Wilsch and T. Seiler, Keratocyte apoptosis after corneal collagen cross-linking using riboflavin/UVA treatment, *Cornea*, 2004, **23**, 43–49.
61. A.M. Edwards, E. Silva, B. Jofre, M.I. Becker and A.E. De Ioannes, Visible light effects on tumoral cells in a culture medium enriched with tryptophan and riboflavin, *J. Photochem. Photobiol. B: Biol.*, 1994, **24**, 179–186.
62. E.B. Thompson, Special topic: apoptosis, *Annu. Rev. Physiol.*, 1998, **60**, 525–532.
63. A.M. Edwards, F. Barredo, E. Silva, A.E. De Ioannes and M.I. Becker, Apoptosis induction in nonirradiated human HL-60 and murine NSO/2 tumor cells by photoproducts of indole-3-acetic acid and riboflavin, *Photochem. Photobiol.*, 1999, **70**, 645–649.
64. M. Heuser, M. Kopun, W. Rittgen and C. Granzow, Cytotoxicity determination without photochemical artifacts, *Cancer Lett.*, 2005, **223**, 57–66.
65. K. Hirakawa, M. Yoshida, A. Nagatsu A, H. Mizukami, V. Rana, M.S. Rawat, S. Oikawa and S. Kawanishi, Chemopreventive action of xanthone derivatives on photosensitized DNA damage, *Photochem. Photobiol.*, 2005, **81**, 314–319.
66. G. Levin, U. Cogan, Y. Levy and S. Mokady, Riboflavin deficiency and the function and fluidity of rat erythrocyte membranes, *J. Nutr.*, 1990, **120**, 857–861.
67. Y. Takeuchi, H. Morii, M. Tamura, O. Hayaishi and Y. Watanabe, A possible mechanism of mitochondrial dysfunction during cerebral ischemia: inhibition of mitochondrial respiration activity by arachidonic acid, *Arch. Biochem. Biophys.*, 1991, **289**, 33–38.
68. K.C. Manthey, Y.Ch. Chew and J. Zempleni, Riboflavin deficiency impairs oxidative folding and secretion of apolipoprotein B-100 in hepG2 cells, triggering stress response systems, *J. Nutr*, 2005, **135**, 978–982.
69. Y. Suzuki, T. Miura and T. Ogiso, Riboflavin photosnsitized hemolysis of rat erythrocytes in the presence of serum, *J. Pharmacobiodyn*, 1982, **5**, 568–575.
70. H.M. Jernigan Jr. and A.S. Laranang, Effects of riboflavin-sensitized photooxidation on choline metabolism in cultured rat lenses, *Curr. Eye Res.*, 1984, **3**, 121–126.
71. R. Santus and J.P. Reyftmann, Photosensitization of membrane components, *Biochimie*, 1986, **68**, 843–848.
72. M.H. Lee and D.D. Tzeng, Photodynamic effect of methionine-riboflavin mixture on rabbit red blood cells, *Zhonghua Min Guo Wei Sheng Wu Ji Mian Yi Xue Za Zhi*, 1990, **23**, 53–67.
73. N. Ali, R.K. Upreti, L.P. Srivastava, R.B. Misra, P.C. Joshi and A.M. Kidwai, Membrane damaging potential of photosensitized riboflavin, *Indian J. Exp. Bio.*, 1991, **29**, 818–822.
74. R.B. Misra, P.K. Bajpai, P.C. Joshi and R.K. Hans, An unusual photohaemolytic property of riboflavin, *Food Chem. Toxicol.*, 2001, **39**, 11–18.
75. J. Li and K.D. Caldwell, Structural studies of commercial fat emulsions used in parenteral nutrition, *J. Pharm. Sci.*, 1994, **83**, 1586–1592.
76. E. Silva, T. González, A.M. Edwards and F. Zuloaga, Visible light induced lipoperoxidation of a parenteral nutrition fat emulsion sensitized by flavins, *J. Nutr. Biochem.*, 1998, **9**, 150–154.
77. K.M. Silvers, B.A. Darlow and C.C. Winterbourn, Lipid peroxide and hydrogen peroxide formation in parenteral nutrition solutions containing multivitamins, *J. Parenteral Enteral Nutr.*, 2001, **25**, 14–17.

78. K.M. Silvers, K.B. Sluis, B.A. Darlow, F. McGill, R. Stocker and C.C. Winterbourn, Limiting light-induced lipid peroxidation and vitamin loss in infant parenteral nutrition by adding multivitamin preparations to intralipid, *Acta Pediatr.*, 2001, **90**, 242–249.
79. J.D. Spikes, Photochemotherapy: molecular and cellular processes involved, *SPIE*, 1988, **997**, 82–100.
80. C.A. Ghiron and J.D. Spikes, The flavin-sensitized photoinactivation of trypsin, *Photochem. Photobiol.*, 1965, **4**, 13–26.
81. J.D. Spikes and B.W. Glad, Photodynamic action, *Photochem. Photobiol.*, 1964, **3**, 471–487.
82. G. Jori, G. Galiazzo, A.M. Tamburro and E. Scoffone, Dye-sensitized photooxidation as a tool for determining the degree of exposure of amino acid residues in proteins, *J. Biol. Chem.*, 1970, **245**, 3375–3383.
83. I. Cozzani and G. Jori, Photo-oxidation of L-glutamate decarboxylase from *Escherichia coli* sensitized by the coenzyme pyridoxal phosphate and by proflavin, *Biochim. Biophys. Acta*, 1980, **623**, 84–88.
84. C.C.F. Blake, G.A. Mair, C.T. North, D.C. Phillips and V.R. Sarma, On the conformation of the hen egg-white lysozyme molecule, *Proc. R. Soc. London Ser. B*, 1967, **167**, 365–377.
85. E. Silva and J. Gaule, Light-induced binding of riboflavin to lysozyme, *Radiat. Environ. Biophys.*, 1977, **14**, 303–310.
86. G. Tapia and E. Silva, Photo-induced riboflavin binding to the tryptophan residues of bovine and human serum albumins, *Radiat. Environ. Biophys.*, 1991, **30**, 131–138.
87. R. Ugarte, A.M. Edwards, M. Diez, A. Valenzuela and E. Silva, Anaerobic riboflavin photosensitized modification of rat lens proteins. A correlation with aged-related changes, *J. Photochem. Photobiol. B: Biol.*, 1992, **13**, 161–168.
88. I. Ferrer and E. Silva, Study of a photo-induced lysozyme–riboflavin bond, *Radiat. Environ. Biophys.*, 1985, **24**, 63–70.
89. M. Salim-Hanna, A.M. Edwards and E. Silva, Obtention of a photo-induced adduct between a vitamin and an essential amino acid. Binding of riboflavin to tryptophan, *Int. J. Vitam. Nutr. Res.*, 1987, **57**, 155–159.
90. E. Silva, M. Salim-Hanna, M.I. Becker and A.E. de Ioannes, Toxic effect of a photoinduced tryptophan-riboflavin adduct on F-9 teratocarcinima cells and preimplantation mouse embryos, *Int. J. Vitam. Nutr. Res.*, 1988, **58**, 394–401.
91. M. Diaz, M.I. Becker, A.E. de Ioannes and E. Silva, Development of monoclonal antibodies against a riboflavin–tryptophan photoinduced adduct: reactivity to eye lens proteins, *Photochem. Photobiol.*, 1996, **63**, 762–767.
92. M. Mancini, A.M. Edwards, M.I. Becker, A.E. de Ioannes and E. Silva, Reactivity of monoclonal antibodies against a riboflavin–tryptophan adduct toward irradiated and non irradiated bovine eye lens protein fractions: an indicator of *in vivo* visible light mediated photo transformation, *J. Photochem. Photobiol. B: Biol.*, 2000, **55**, 9–15.
93. A.M. Edwards, C. Bueno, A. Saldano, E. Silva, K. Kassab, L. Polo and G. Jori, Photochemical and pharmacokinetic properties of selected flavins, *J. Photochem. Photobiol. B: Biol.*, 1999, **48**, 36–41.
94. A.M. Edwards, A. Saldano, C. Bueno, E. Silva and S. Alegria, Spectroscopic properties of hydrophobic flavin esters. A one and two-dimensional ^{1}H-NMR and ^{13}C-NMR study, *Bol. Soc. Chil. Quim.*, 2000, **45**, 423–431.
95. C.B. Martin, M.-L. Tsao, C.M. Hadad and M.S. Platz, The reaction of triplet flavin with indole. A study of the cascade of reactive intermediates using density

functional theory and time resolved infrared spectroscopy, *J. Am. Chem. Soc.*, 2002, **124**, 7226–7234.

96. R.B. Misra, V. Sundararaman and P.C. Joshi, Riboflavin induced phototoxicity to paramecium, *Ind. J. Exp. Biol.*, 1987, **25**, 194–201.
97. B. Misra, L.P. Srivastava and P.C. Joshi, Phototoxic effects of riboflavin in tetrahymena thermophila, *Ind. J. Exp. Biol.*, 1990, **28**, 858–861.
98. N. Donoso, A. Valenzuela and E. Silva, Tryptophan–riboflavin photoinduced adduct and hepatic dysfunction in rats, *Nutr. Rep. Int.*, 1988, **37**, 599–606.
99. S. Zigman, J.D. Hare, T. Yulo and D. Ennist, Differential effects of tryptophan photoproducts and H_2O_2 on the growth of mouse embryonic fibroblasts, *Photochem. Photobiol.*, 1978, **27**, 281–284.
100. K. Sato, H. Minami, H. Taguchi, T. Maeda, K. Yoshikawa and T. Tsuji, Activation of riboflavin by visible radiation, *Photomed. Photobiol.*, 1995, **17**, 125–126.
101. K. Sato, H. Taguchi, T. Maeda, H. Minami, Y. Asada, Y. Watanabe and K. Yoshikawa, The primary cytotoxicity in ultraviolet A-irradiated riboflavin solution is derived from hydrogen peroxide, *J. Invest. Dermatol.*, 1995, **105** 608–612.
102. H. Minami, K. Sato, T. Maeda, H. Taguchi, K. Yoshikawa, H. Kosaka, T. Shiga and T. Tsuji, Hypoxia potentiates ultraviolet A-induced riboflavin cytotoxicity, *J. Invest. Dermatol.*, 1999, **113**, 77–81.
103. A.K. Verrico and J.V. Moore, Expression of the collagen-related heat shock protein HSP47 in fibroblast treated with hyperthermia or photodynamic therapy, *Br. J. Cancer*, 1997, **76**, 719–724.
104. H.N. Christensen, Riboflavin can protect tissue from oxidative injury, *Nutr. Rev.*, 1993, **51**, 149–150.
105. C.P. Mack, D.E. Hultquist and M. Shlafer, Myocardial flavin reductase and riboflavin a potential role in decreasing reoxygenation injury, *Biochem. Biophys. Res. Commun.*, 1995, **212**, 35–40.
106. E. Silva, S. Furst, A.M. Edwards, M.I. Becker and A.E. de Ioannes, Visible light anaerobic photoconversion of tyrosine sensitized by riboflavin. Cytotoxicity on mouse tumoral cells, *Photochem. Photobiol.*, 1995, **62**, 1041–1045.
107. A.M. Edwards, A. Pacheco, A.E. de Ioannes and M.I. Becker, Apoptosis induction by flavin esters photosensitization on human tumoral cells, *Proceedings of the 5th international symposium on photodynamic diagnosis and therapy in clinical practice*, Bressanone, Italy, 2003, Abstract P10.
108. A. Pacheco, A.M. Edwards, E. Silva and M.I. Becker, Apoptotic effect on *in vitro* cultured human tumoral cells, photosensitized by hydrophobic flavins, in *VIII Encuentro Latinoamericano de Fotoquímica y Fotobiología (VIII ELAFOT)*, La Plata, Argentina, 2004, Abstract P96.
109. K.S. Cho, E.H. Lee, J.S. Choi and C.K. Joo, Reactive oxygen species-induced apoptosis and necrosis in bovine corneal endothelial cells, *Invest. Ophthalmol. Visual Sci.*, 1999, **40**, 911–919.
110. G. Wollensak, S. Eberhard, R. Friedmann, P. Lutz and F. Richard, Corneal endothelial cytotoxicity of riboflavin/UVA treatment *in vitro*, *Ophthalmic Res.*, 2003, **35**, 324–328.
111. K.R. Marshall, M. Gong, L. Wodke, J.H. Lamb, D.J. Jones, P.B. Farmer, N.S. Scrutton and A.W. Munro, The human apoptosis-inducing protein amid is an oxidoreductase with a modified flavin cofactor and DNA binding activity, *J. Biol. Chem.*, 2005, **280**, 30735–30740.

Chapter 7

Photoinduced Processes in the Eye Lens: Do Flavins Really Play a Role?

EDUARDO SILVA[a] AND FRANK H. QUINA[b]

[a] Laboratorio de Química Biológica, Facultad de Química, Pontificia Universidad Católica de Chile, Casilla 306, Correo 22, Santiago, Chile

[b] Instituto de Química, Universidade de São Paulo, São Paulo, Brazil

Abstract

In this chapter, the photochemical processes that can affect the eye lens as the result of the passage of the incident light required for vision are described, with special emphasis on those that involve sensitizers that are endogenous or that are age dependent. Riboflavin and the kynurenines are two of these endogenous photosensitizers present in the mammalian eye lens. Upon aging, the lens also accumulates brown fluorophores, derived mainly from the Maillard reaction between vitamin C oxidation products and lysine residues of the crystallins, which exhibit photochemical activity as sensitizers in the UVA and visible region. Although the riboflavin concentration in the lens is very low, the occurrence of *in vivo* riboflavin-mediated photoprocesses has been demonstrated. The present study quantifies the photosensitizing activity of advanced glycation end products (AGEs) at various oxygen pressures and compares this activity to that of lenticular riboflavin (RF), employing ascorbate and tryptophan (Trp) as the targets of the photochemical modifications. The comparison of the real effect of these two sensitizers in the eye lens takes into account (i) the solar irradiance spectrum; (ii) the absorption properties of AGEs and RF at the concentrations found in the eye lens; and (iii) the quantum yields for the AGEs- and RF-sensitized modification of Trp.

7.1. Introduction

Age- and environmental-related damage to eye lens proteins is a subject of permanent interest because the accumulation of modified proteins during aging is associated with lens opacification or senile cataract. Throughout life, the eye lens suffers constant exposure to light that is essential for vision and, in the rat, also required for optimal lens growth.[1] However, there is some evidence of a photochemical contribution to the development of cataract[2–9] and this subject is under permanent revision.[10–15] To understand and explain the photochemical processes that can occur in the eye lens, it is necessary to consider the principles governing this kind of reaction.

According to the first law of photochemistry, now known as the Grotthus–Draper law: "Only the light that is absorbed by a system can cause chemical change". The probability or rate of absorption is given by the Lambert–Beer Law, which states that the fraction of the incident radiation absorbed by a transparent medium is independent of the intensity of incident radiation and that each successive layer of the medium absorbs an equal fraction of the radiation incident on that layer. The analysis of the chemical constituents of the lens and their absorption properties allows an evaluation of the feasibility of the occurrence of photochemical processes in this tissue. The lens has the highest protein concentration of any tissue, with protein constituting approximately 35% of the wet weight and nearly all the dry weight of the lens. The main photochemical activity of the lens proteins should thus be due to the absorption capacity of the aromatic chromophores of the amino acids Phe ($\varepsilon_{259\ nm} = 2 \times 10^2\ M^{-1}\ cm^{-1}$), Trp ($\varepsilon_{279\ nm} = 5.2 \times 10^3\ M^{-1}\ cm^{-1}$) and Tyr ($\varepsilon_{278\ nm} = 1.1 \times 10^3\ M^{-1}\ cm^{-1}$) present in the lens proteins and only radiation in the UVB region could provoke direct photochemical modifications. Indeed, several authors have concluded, based on different irradiation protocols, that UVB radiation induces anterior lenticular opacities.[16–18] Under normal conditions, ozone and molecular oxygen block the electromagnetic radiation below 280 and 240 nm, respectively, and no UVC reaches the Earth's surface. UVB is received in varying amounts, depending on the location on Earth.[19] The cornea (100% absorption below 280 nm, 92% at 300 nm and 45% at 320 nm) and aqueous humor (6% absorption at 300 nm, 16% at 320 nm) filter most of the UVB that reaches the eye.[20–22]

The mechanism by which UVA radiation (315–400 nm) might cause damage to the crystallin is a subject of increasing interest. The occurrence of UVA-mediated photochemical processes requires the presence of chromophores that absorb in this region of the electromagnetic spectrum. The search for and study of this type of UVA-absorbing compound in the eye lens have been the subject of a number of studies.[23–34] In the human ocular lens, upon aging, the crystallin proteins become yellow colored and fluorescent, with a significant decrease in their solubility. During the development of senile nuclear cataract, the crystallins become brown and additional chromophores are formed.

Direct UVB absorption by the aromatic amino acid side chains, primarily the tryptophan (Trp) residues, induces the oxidation of this amino acid, producing *N*-formylkynurenine (NFK), kynurenine (Ky) and related products.

The absorption range of these oxidation products extends into the red, making them responsive to light in the UVA and even the visible range of the spectrum. The quantum yield of singlet oxygen generation upon irradiation of NFK is high ($\phi = 0.17$),[27] while Ky generates no detectable amount of singlet oxygen. The quantum yield of superoxide production by NFK ($\phi = 0.01$)[27] is low compared to that of singlet oxygen. The formation of oxidation products of Trp has been demonstrated in both the isolated intact eye lens and in crystallin solutions.[35–37] Nevertheless, the role of the Trp oxidation products as UVA sensitizers *in vivo* is unclear because amino acid analyses of lens proteins show no significant loss of Trp in either aged[38,39] lenses or in cataracts.[40,41] This is consistent with the suggestion that the colored products derived from the oxidation of the Trp residues of fiber cells act, together with other compounds, as UV filter compounds in primate lenses.[42–44] In a very recent publication, it has been demonstrated that 3-hydroxykynurenine modifies lens proteins independent of glycation to form products that may contribute to protein aggregation and browning during cataract formation.[45]

Another source of generation of eye lens chromophores that absorb in the UVA and, to a lower extent, in the visible region is the occurrence of a non-enzymatic browning reaction known as the Maillard reaction. This process is initiated by the condensation of an amino group with a reducing compound, usually a sugar. The long-lived crystallins, present at high concentrations in all mammalian lenses, are prone to modification by glycation of their N-terminal amino groups, of the amino groups in the side chain of their lysine residues, and/or the guanidine groups of their arginine residues. The non-enzymatic reaction of sugars with proteins occurs normally, but to an increased extent upon aging. Diabetic patients in particular are at higher risk for cataract formation, associated with a variety of complications.[46–48]

In diurnal animals and man, ascorbate concentrations in the lens and aqueous humor are 10–20 times higher than those in the plasma (up to 2 mM),[49] indicating active transport into the eye.[50] Furthermore, nocturnal animals have much lower concentrations of ascorbic acid in the lens than diurnal animals.[51] This suggests a protective role of ascorbic acid against photoinduced damage in eye lenses exposed to sunlight. A relationship has been shown[48] between decreased ascorbate levels and increasing age and cataract. Paradoxically, despite its antioxidant properties, vitamin C has also been implicated in the exacerbation of cataractogenesis. There is evidence that the oxidation products of ascorbate are 18 times more reactive than glucose in the glycation of lens proteins[52] and extensive crosslinking and formation of browning pigments and fluorophores accompany this process. The chromophores generated when the lens proteins are glycated *in vitro* in the presence of ascorbic acid have UVA sensitizer activity similar to that observed in human lenses with age.[53] The similarity of the yellow pigments isolated from human cataracts to those from ascorbic acid-modified lens proteins *in vitro* supports the hypothesis that the chromophores present in cataractous lenses are advanced glycation end products, probably owing to the glycation of proteins *in vivo* by ascorbic acid oxidation products.[54]

7.2. Riboflavin-Sensitized Photoprocesses

Among the endogenous sensitizers present at low concentration in the eye lens, the vitamin riboflavin (RF) deserves special mention.[55,56] In epithelial cells, it is a precursor of flavin mononucleotide (FMN) and flavin adenine dinucleotide (FAD), necessary cofactors for various metabolic pathways in the cells. $FADH_2$ also plays an important role in the glutathione redox cycle, which constitutes an important defensive barrier in the lens and is coupled with the high-ascorbate concentration present in this tissue. RF, or vitamin B2, has a very complex photochemistry. In the presence of visible light, RF can exhibit photosensitizing properties through a mixed Type I–Type II mechanism,[57,58] the former mechanism being favored at low oxygen concentrations.

Type I mechanism

$$RF + h\nu \rightarrow {}^{1}RF \rightarrow {}^{3}RF \quad (1)$$

$$^{3}RF + SH \rightarrow RF^{\cdot -} + SH^{\cdot +} \rightarrow RFH^{\cdot} + S^{\cdot} \quad (2)$$

$$RF^{\cdot -} + O_2 \rightarrow RF + O_2^{\cdot -} \quad (3)$$

$$2\ RFH^{\cdot} \rightarrow RF + RFH_2 \quad (4)$$

$$RFH_2 + O_2 \rightarrow RF + H_2O_2 \quad (5)$$

$$H_2O_2 + O_2^{\cdot -} \rightarrow {}^{-}OH + {}^{\cdot}OH + O_2 \quad (6)$$

$$S^{\cdot} \text{ and/or } SH^{\cdot +} + H_2O_2/O_2^{\cdot -}/{}^{\cdot}OH \rightarrow S_{Oxid} \quad (7)$$

Type II mechanism

$$RF + h\nu \rightarrow {}^{1}RF \rightarrow {}^{3}RF \quad (8)$$

$$^{3}RF + O_2 \rightarrow RF + {}^{1}O_2 \quad (9)$$

$$SH + {}^{1}O_2 \rightarrow S_{Oxid} \quad (10)$$

In these mechanistic formulations, RF, ^{1}RF and ^{3}RF represent RF in the ground state and in the excited singlet and triplet states, respectively; $RF^{\cdot -}$, $RFH^{\cdot}$ and RFH_2 are the radical anion, the radical and the reduced form of RF; SH corresponds to the reduced substrate and $SH^{\cdot +}$, $S^{\cdot}$ and S_{Oxid} represent the intermediate radical cation and radical and the oxidized form of the substrate, respectively.

Because the *in vivo* oxygen concentration of the lens is low ($<10^{-5}$ M),[54] both aerobic and anaerobic photoprocesses could contribute simultaneously to the photoinduced damage of lens proteins. Bose et al.[59] have presented evidence that the RF-sensitized conformational changes of α-crystallin differ significantly from those sensitized by either methylene blue or NFK. This behavior is consistent with previous results showing that the photooxidation of Trp and other amino acids present in lysozyme is exceptionally efficient when the vitamin RF is used as sensitizer.[60] The quantum yields with RF were significantly higher than those observed with other sensitizers, such as methylene blue or rose Bengal, known to involve a singlet oxygen-mediated Type II photooxidation mechanism, already suggesting that RF might act preferentially via a different mechanism. In this same context, it was subsequently reported that irradiation of lysozyme with visible light in the presence of RF and molecular oxygen resulted not only in photooxidation of amino acid residues of the enzyme, but also in covalent bonding of the sensitizer to the protein.[61] This RF–lysozyme photoadduct can also be obtained in an anaerobic atmosphere, thus avoiding photooxidative Type II processes. Adduct formation thus results from the Type I process, involving direct interaction between the triplet state of the flavin and the protein, and does not require the intervention of molecular oxygen.

In subsequent studies, it was demonstrated that a Trp residue of the enzyme was specifically involved in the covalent bonding between RF and lysozyme.[62,63] Upon irradiation of the amino acid Trp in its free form in solution in the presence of RF, the following species can be isolated and characterized spectrophotometrically: aggregate forms of RF, indolic products associated with flavins and indolic products of molecular weight higher than Trp as well as formylkynurenine and other Trp photodecomposition products.[57] Based on the infrared spectrum of a pure fraction of RF–Trp adduct, at least two modes of bonding between the indole and isoalloxazine rings were proposed.[64] More recently, the reaction of triplet flavin with indole has been studied theoretically by density functional theory and experimentally by time-resolved infrared spectroscopy.[65] This work provides important theoretical and experimental support for our previously proposed mechanism involving the triplet state of RF, which initiates a cascade of intermediates formed via electron-transfer, proton-transfer and radical–radical coupling reactions.[57] The occurrence of RF-photosensitized Trp–Trp[57] and Tyr–Tyr[66,67] coupling reactions has important biological implications,[68] given that they could result in intermolecular crosslinks between proteins. The concomitantly formed Trp–RF adduct is easily detected and could serve as a convenient indicator of the occurrence of such radical-initiated processes.

In an exploratory work, a photoinduced RF–protein bonding was observed for the water-insoluble protein fraction of lenses previously enriched with ^{14}C RF upon exposure to visible light.[69] The irradiation of rat lens homogenate (or its soluble protein fractions) in the presence of RF leads to a modification in the chromatographic elution pattern, with an increase in the high-molecular-weight fraction similar to that observed during normal aging of the lens.[70]

Because the covalent bonding of RF to Trp residues had been demonstrated in model studies[62,71] and with isolated fractions of eye lens[69] proteins, it was of interest to develop monoclonal antibodies against the RF–Trp adduct.[72] Such antibodies provide a very sensitive and specific tool for examining the possibility of *in vivo* formation of this adduct with lens proteins. ELISA reactivity of monoclonal antibodies to the RF–Trp adduct in soluble young and aged rat and human cataractous lens proteins clearly demonstrates the presence of photoinduced RF–Trp adducts in the ocular lens, with positive recognition by these specific monoclonal antibodies.[71] From these findings, it was also evident that the adduct is more prevalent in older lenses and that its presence becomes truly extreme in a human cataractous lens. The low molecular oxygen concentration present in this tissue[54] makes this process especially interesting because it does not require O_2. In addition, α-, β_H-, β_L- and γ-crystallins separated from bovine eye lens reacted by direct ELISA with the antiTrp–RF adduct monoclonal antibodies;[73] the recognition is enhanced when the isolated fractions were previously irradiated with visible light in the presence of RF. RF-sensitized irradiation of the protein fractions is accompanied by a concomitant decrease in Trp fluorescence and the appearance of new bands corresponding to compounds of higher molecular weight.

Immunofluorescent analysis of rat lens sections employing RF–Trp monoclonal antibodies indicated that the most differentiated and aged internal zone, which receives the greatest amount of light *in vivo*, is precisely the most immunoreactive to the antibodies. In the peripheral zone, where the younger cells that still maintain their subcellular corpuscles are located, the monoclonal antibodies show reactivity to nuclear structures.[71]

7.3. Photochemical Behavior of a System Composed of Ascorbate and Riboflavin

In apparent contrast to the above results, we demonstrated that the potent photosensitizer RF could not, at the very low concentration found in the human lens,[55] generate significant photodamage upon exposure to sunlight when ascorbate, the main barrier against oxygen radical species and protector of protein integrity, was present at near-physiological concentrations.[74] Nevertheless, it is important to consider in this analysis that the vitamin C concentration has been observed to decrease in human lenses during cataract progression.[48]

The photochemical behavior of a system consisting of ascorbate (AH^-) and RF at concentrations similar to those found in the human ocular lens has been a matter of study in our group.[73,75] In this study, we evaluated and compared the photosensitizing capacity of RF and advanced glycation end products (AGEs) on ascorbic acid and Trp. This amino acid is a particularly sensitive photooxidation target and the products derived from its reaction have been evidenced in cataractous lenses.[76,77] The photosensitized processes were examined as a function of oxygen pressure, including that (5% O_2) compatible with the lens physiology.[78] The efficiency of RF and AGEs induced decomposition of AH^- and Trp at the various O_2 levels was compared by calculating the respective initial photosensitization quantum yields (Table 1).

In the presence of O_2, quantum yields of RF-sensitized decomposition of Trp and AH^- were at least one order of magnitude higher than those in the presence of AGEs (absorbance of 0.2 at 365 nm), establishing that RF is a more efficient sensitizer.

A Type I mechanism implies a direct electron transfer between the substrate and the excited state of sensitizer, followed by the oxidation of the semioxidized substrate by molecular oxygen or by some other reactive species. On the contrary, in a Type II mechanism singlet oxygen is produced directly by energy transfer from the photosensitizer triplet-excited state to molecular oxygen and this singlet oxygen then oxidizes the substrate. Thus, by irradiating under increasing oxygen pressures, one can modulate the relative prevalence and efficiency of Type I or II mechanisms. In the absence of molecular oxygen, as expected and previously described,[73,74] there is negligible photodegradation of AH^- and Trp, independent of the sensitizer used. Upon increasing the oxygen concentration, quantum yields for Trp photooxidation initially increase and then decrease, being maximal under 5% O_2 for both sensitizers. This implies that, in the 5–20% O_2 concentration range, less oxygen results in more rapid photodegradation of Trp, compatible with a more effective Type I mechanism, which is not dependent on oxygen for initiation of the light-induced reactions. However, oxygen may be required in a later step to regenerate the sensitizer. These results confirm the predominance of a Type I mechanism for RF at low oxygen pressure.[69,73,74] A similar behavior of the Trp photodegradation quantum yield is found for AGEs, pointing to a similar mechanism for AGEs and RF. In contrast, at higher oxygen concentrations, the Type II mechanism is favored because there is increasing competition between oxygen and the substrate Trp for reaction with the excited photosensitizer.

The production of various oxygen-reactive species, including singlet oxygen, superoxide anion and H_2O_2, has been described in solutions containing ascorbylated lens proteins irradiated with UVA.[79–82] Under aerobic conditions, where a Type II mechanism is favored, singlet oxygen has been shown to be the major reactive oxygen species formed.[83] However, it must be pointed out that

Table 1. Initial quantum yields for ascorbate (AH^-) and Trp degradation during photosensitization by RF (RF 3.5 μM) or Maillard products (AGEs at OD_{365} of 0.2) in solutions irradiated at 365 nm and 37°C under various O_2 concentrations

	RF		AGEs	
Sensitizer (% O_2)	**$10^3 \times \phi$ (AH^-)**	**$10^3 \times \phi$ (Trp)**	**$10^3 \times \phi$ (AH^-)**	**$10^3 \times \phi$ (Trp)**
0	1.4 ± 0.6	0.9 ± 0.8	0.047 ± 0.027	0.016 ± 0.011
5	1670 ± 220	64.8 ± 3.7	46.1 ± 3.2	2.94 ± 0.14
20	1940 ± 200	23.0 ± 0.5	93.2 ± 9.5	1.04 ± 0.28

Note: Data are the mean ± SD of three independent experiments.

the aerobic condition (20% O_2) does not correspond to the real physiology of the lens (5% O_2), where a preponderant Type I mechanism may be expected. This hypothesis is also supported by the detection of Trp–RF photoadducts in the soluble protein fractions of the human eye lens[57,71] with specific monoclonal antibodies. Those lenticular photoadducts were shown to increase with age of the lens and cataract.[72] In contrast, in the case of AH^-, the AGEs-sensitized photooxidation quantum yields increased with increasing oxygen pressure, indicating that all oxygen-consuming reactions are favored. However, the absence of a linear relationship between ascorbate decomposition quantum yields and oxygen concentration suggests that the mechanism of photosensitization is not exclusively Type II (Table 1).

It should be noted that, in the presence of RF and O_2, quantum yields for decomposition of AH^- greater than one are observed, suggesting the existence of chain reactions[84] initiated by the ascorbyl radical, as follows:

$$^3RF + AH^- \rightarrow RF^{\cdot -} + AH^{\cdot} \tag{11}$$

$$AH^{\cdot} + O_2 \rightarrow A + HO_2^{\cdot} \tag{12}$$

$$HO_2^{\cdot} + AH^- \rightarrow H_2O_2 + AH^{\cdot} \tag{13}$$

Reaction (11), the generation of the ascorbyl radical $AH^{\cdot}$ by the reaction between triplet RF and ascorbate, is the only one that requires the absorption of photons (to form the excited triplet state of RF). The subsequent reactions are independent of the presence of light and contribute to further decomposition of AH^-. Moreover, the RF radical can also initiate an additional sequence of reactions, generating reactive oxygen species that can contribute to the decomposition of AH^- (reactions (4), (5) and (6)). Finally, if adventitious metals are present in trace amounts in the buffer, an active propagation of AH^- oxidation could also occur. Thus, transition-metal ions such as Fe^{2+} may catalyze the production of the hydroxyl radical ($HO^{\cdot}$) via the Fenton reaction in the presence of H_2O_2 (produced by oxidation of photoreduced RF, reaction (5), and/or via reaction (3)). The reduced form of the metal ion (*e.g.*, Fe^{2+}) can then be regenerated by metal-catalyzed oxidation of AH^-. This possibility is confirmed by the significant decrease in the AH^- photooxidation quantum yield in the presence of the metal chelator DTPA, which totally inactivates the trace amounts of copper and iron present in the buffer: the quantum yields for ascorbate-photosensitized oxidation decreased twofold at 20% O_2 and fourfold at 5% O_2 when DTPA was added to the incubation medium.

The occurrence of an *in vivo* Trp–RF photoadduct in the eye lens,[71] described above, has led to more detailed *in vitro* studies of the interaction between the triplet excited state of RF and the α-, β_H- and β_L-crystallins.[85] The transient absorption spectra of RF, determined by nanosecond laser flash photolysis, reveal an absorption band (625–675 nm) corresponding to the neutral RF triplet state ($\tau = 42$ μs), accompanied by a long-lived absorption ($\tau = 320$ μs) in the 500–600 nm region corresponding to the semireduced RF radical (Figure 1).

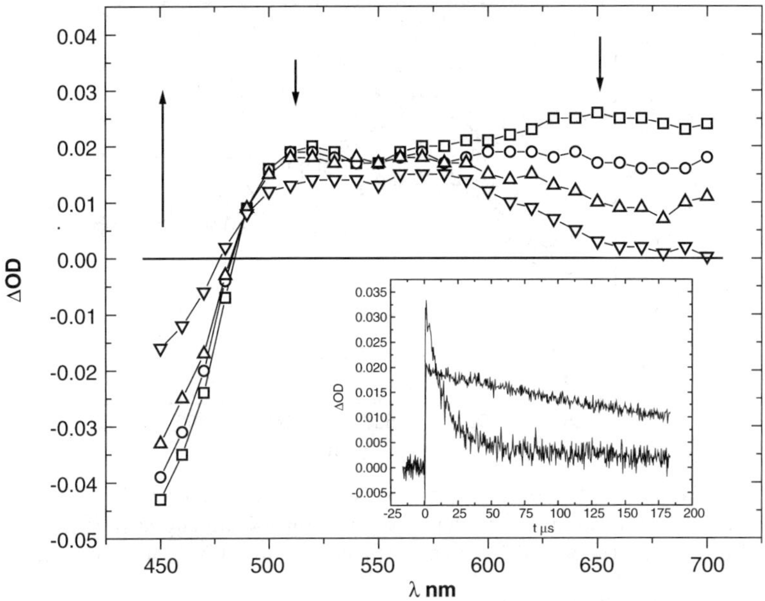

Figure 1. Transient absorption spectra of RF in phosphate buffer solution at 5 (□), 10 (o), 20 (Δ) and 100 (∇) μs after the laser pulse (Nd-YAG laser at 355 nm; 500 W xenon lamp monitoring lamp); the arrows indicate the evolution of the optical densities. The inset shows the transient absorption decay measured at 650 nm, $\tau = 42$ μs (lower line) and at 510 nm, $\tau \approx 320$ μs (upper line)

The RF triplet state is quenched by the α-, β_H- and β_L-crystallins proteins via a mechanism that involves electron transfer from the protein to the flavin, as shown by the decrease of the triplet RF band with a concomitant increase in the band of the semireduced form of RF (Figure 2).

From Stern–Volmer plots of the decrease in the ^{3}RF lifetime as a function of increasing protein concentration (expressed in milligrams of protein per milliliter), bimolecular quenching rate constants, k_q, of 3.31×10^4, 4.16×10^4 and 7.33×10^4 mL mg^{-1} s^{-1} were obtained for α-, β_H- and β_L-crystallins, respectively. The use of milligrams of protein per milliliter as the concentration unit permits the comparison of proteins with different molecular weights, independent of whether or not they exist as aggregates in solution. Trp loss upon RF-sensitized photooxidation of the crystallins was investigated by irradiation with monochromatic visible light (450 nm) in a 5%-oxygen atmosphere. A direct correlation was found between the rate constant for quenching of triplet RF by the different crystallin fractions and the corresponding rate constants for decomposition of the exposed and partially buried Trps in the proteins. RF-sensitized photooxidation of the crystallins is accompanied by a decrease in the low-molecular weight constituents that give rise to its multimeric forms. Thus, the initial rate of decrease of the low-molecular weight bands of the irradiated α-, β_H- and β_L-crystallins correlates with the rate constants for quenching of triplet

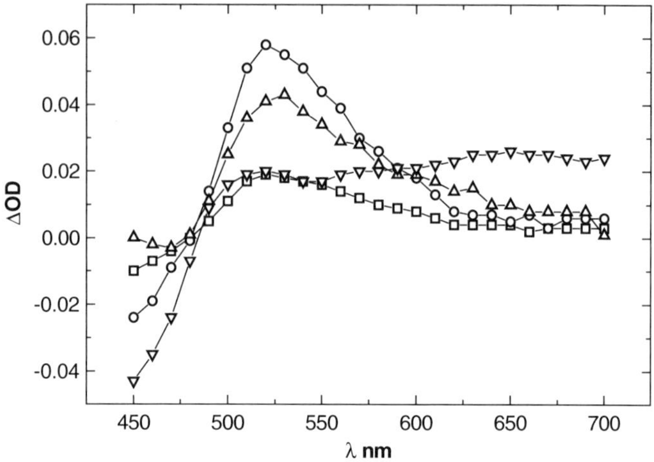

Figure 2. Transient absorption spectra of N_2-bubbled solutions of 5×10^{-5} M RF determined by nanosecond laser flash photolysis; alone (∇) and in the presence of 0.8 mg mL^{-1} of protein: (□) α-, (Δ) β_H- and (o) β_L-crystallins, at 5 μs after the laser pulse

RF by the different crystallins. It is noteworthy that β_L-crystallin shows both the highest efficiency of quenching of triplet RF and the highest rates of decomposition of the more exposed and the partially accessible Trp residues. The initial rate of decomposition of the more exposed Trp residues of β_L-crystallin (5.03×10^{-5} M min^{-1}) is only slightly smaller than that found for irradiation of a 3.5×10^{-4} M solution of free Trp under the same conditions (6.1×10^{-5} M min^{-1}). This result indicates a high degree of reactivity of this kind of residue in the protein. In general, there is a very good correlation between the photodecomposition rates of the Trp residues in each protein fraction and the ability of the respective protein fraction to quench triplet RF. This result indicates that, when the irradiation of the protein is performed at low oxygen concentrations, the quenching process involving triplet RF and the protein plays an important role in the photodecomposition of the exposed Trp residues. In addition, the number and degree of exposure of the Trp residues can also be important contributors to the net efficiency of quenching of the excited sensitizer.

Scheme 1 shows the different photophysical and photochemical processes involving RF, ^{3}RF, molecular oxygen, singlet oxygen, protein (P) and the Trp residues in the protein (P–Trp). The last of these is considered independently of the rest of the protein, because of the greater efficiency with which they quench the excited state of the flavin,[64,86] initiating a complex sequence of reactions leading to indole–indole and indole–flavin photoadducts.[57] The presence *in vivo* of a Trp–RF photoadduct in the eye lens[71] described above constitutes proof of the occurrence *in vivo* in this tissue of RF-sensitized photoprocesses with participation of the Trp residues of the proteins. The chemical interaction between triplet RF and Trp, which generates the radical forms of these compounds, leads to formation of the Trp–RF adduct and also to the

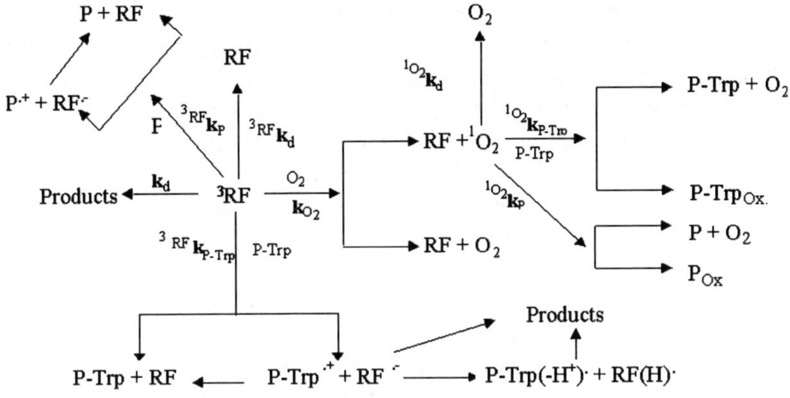

Scheme 1.

formation of adducts between two Trp residues.[57] The latter reaction is quantitatively more important[57] and can result in protein aggregation. These reactions can be especially important in a tissue like the eye lens with a relatively low oxygen concentration.

7.4. Advanced Glycation End Products and Riboflavin as Endogenous Sensitizers in the Eye Lens

A rigorous comparison of the photochemical potential of AGEs and RF as sensitizers, under conditions similar to those found in the eye, requires a consideration of the following factors:

(i) The intensity and the energy of the photons of the reflected solar radiation that actually reach the surface of the lens. Several studies have employed spectroradiometric and photometric methods to estimate the ocular exposure to solar UV and visible radiation under actual geometric-viewing conditions.[87–89] To obtain the true incident solar irradiance on the surface of the lens, it is necessary to correct these data by the fraction of reflected solar radiation that is filtered out by the cornea and the aqueous humor.

(ii) The light-absorption capacity of the endogenous sensitizers at the concentration found in the lens at each wavelength in the appropriately filtered reflected solar irradiance spectra.

(iii) The quantum yields for photosensitized decomposition of an appropriate target or indicator of the induced photodamage. Trp is an adequate photochemical sensor due to its elevated sensitivity to AGEs- and RF-sensitized photoprocesses.[74]

To perform this comparison, different AGEs concentrations were chosen, ranging from those at the beginning of the development of a cataract up to those found when this pathology has reached an advanced state. The amount of RF in the lens is so minute that quantitative determinations of its concentration have not yet been possible.[90] Consequently, although the lowest concentration (0.35 μM) employed in this study does not necessarily represent the true concentration in the lens, it represents the lowest concentration at which our experimental work could be reliably carried out.

The incident irradiance, expressed in Einsteins per second per square millimeter impinging on a sunlight-illuminated eye lens surface, is shown in Figure 3. The irradiance curve shown was obtained by transforming the solar irradiance entering the eye[86] into Einsteins per second per square millimeter and correcting for the fraction of the light absorbed by the cornea and the aqueous humor.[91] This figure also shows the absorbed irradiances for solutions containing various concentrations of AGEs and RF. The absorbed Einsteins per second per square millimeter shown in the plot were calculated by multiplying the incident solar irradiance by the fraction of the light absorbed at each wavelength, given by $(1–10^{-A})$, where 10^{-A} corresponds to the transmittance. The 10^{-A} values were obtained from the UV-visible absorption spectra at each concentration of AGEs or RF, taking the effective optical path length to be the thickness of the human lens in the central nuclear region (5 mm). The amount of reflected sunlight absorbed by the AGEs solutions is significant and much higher than that absorbed by RF at the two concentrations employed in this work.

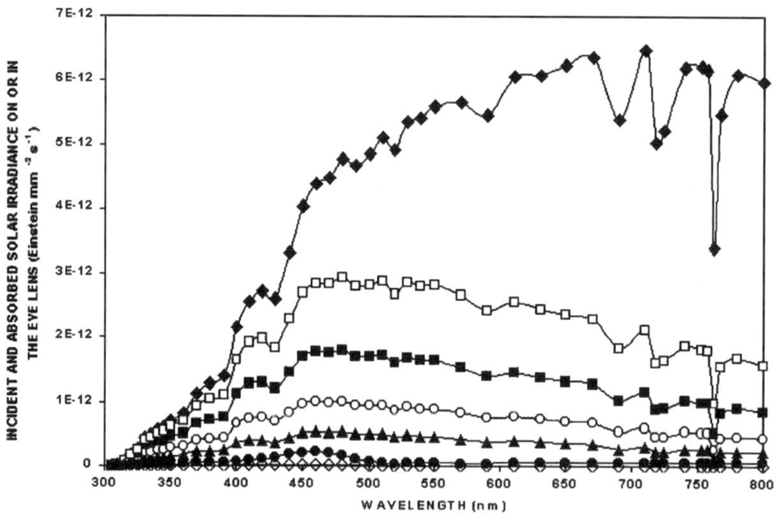

Figure 3. Incident solar spectral irradiance on the eye lens (◆) and absorbed irradiance (light path = 5 mm corresponding to the effective thickness of the central region of the human lens) by solutions of AGEs and RF at different concentrations. The AGEs solutions were prepared considering the absorbance at 367 nm ($A_{367\ nm}$) expressed in arbitrary units: $A_{367\ nm} = 0.8$ (□), $A_{367\ nm} = 0.4$ (■), $A_{367\ nm} = 0.2$ (○) and $A_{367\ nm} = 0.1$ (▲). RF: 3.5 μM (●) and 0.35 μM (◇)

Although the AGEs absorb light preferentially in the UVA region, they can also absorb in the visible region, though to a lesser extent. At these longer wavelengths, the solar spectrum shows the highest irradiance values and, therefore, the absolute amount of absorbed Einsteins per second per square millimeter can be more significant. Increased concentrations of AGEs in the eye lens are a typical consequence of aging and they reach the highest values in patients with cataract.[92,93] The decrease in the amount of light reaching the retina, because of the inner filter effect of the AGEs present in the lens, contributes to the diminished visual capacity observed in older and cataractous individuals.

The integrated area below the curve corresponding to the total incident solar irradiance spectrum (Figure 3) represents the total number of Einsteins emitted by the sun that arrive at each square millimeter of the illuminated surface of the eye lens per second. This value, together with the areas under the curves of irradiance absorbed by solutions of AGEs and RF, permits calculation of the respective percentage of solar irradiance absorbed at each concentration of AGEs or RF (Table 2). These percentages were calculated for two specific regions of the spectra, 305–400 nm and 400–800 nm. The former spectral region is representative of the contribution from the chromophores that are produced via the glycation processes. This process involves reaction of the ε-amino group of lysine of the eye lens proteins with the carbonyl group of glucose and/or of the sugars arising from ascorbate oxidation. These compounds are characterized by a fluorescence emission at 450–460 nm traditionally assigned to the AGEs.[94] Any absorption at lower energy than the fluorescence emission, *i.e.*, at

Table 2. Percentage of the total incident solar irradiance (Einsteins s^{-1} mm^2) absorbed by the AGEs and RF at different concentrations, considering the spectral regions comprised between 305–400 and 400–800 nm

Sensitizer	Percentage of absorbed solar irradiance (305-400 nm)	Mole/second square mm (305–400 nm)	Percentage of absorbed solar irradiance (400–800 nm)
AGEs: A = 0.1 (A.U.)	21.1	0.53×10^{-14}	8.1
A = 0.2 (A.U.)	36.8	0.94×10^{-14}	14.9
A = 0.4 (A.U.)	59.7	1.25×10^{-14}	27.2
A = 0.8 (A.U.)	83.0	2.12×10^{-14}	45.3
RF: 0.35 μM	0.5	0.14×10^{-14}	0.2
3.50 μM	4.1	1.42×10^{-14}	1.7

Note: For the determination of the absorbance and the respective transmittance values at the different wavelengths, a light path of 5 mm was considered, simulating the wide of the eye lens in the nuclear central part. The photodecomposed Trp mole per second and square millimeter of irradiated surface, when exposed to light (305–400 nm and 400–800 nm) in processes sensitized by AGEs or RF, were calculated multiplying the absorbed Einsteins per second, square millimeter and nanometer, by the quantum yield of the Trp photodecomposition. In the case of RF, two different values of ϕ were employed depending on the region of the spectra ($\phi = 37.7 \times 10^{-3}$ between 300 and 400 nm and $\phi = 24.1 \times 10^{-3}$ when $\lambda > 400$ nm).

wavelengths longer than 450 nm, must necessarily be attributed to other types of chromophoric compounds.

The quantum yields employed for the AGEs in this work were determined at an excitation wavelength of 367 nm and are taken to be representative of all photoproducts. The exact sensitization potential and efficiencies of the chromophores that absorb at longer wavelengths were not characterized. The determination of Trp decomposition quantum yields in the visible region would require the use of much higher concentrations of the AGEs; under these experimental conditions, the strong absorption of the AGEs in the UVA makes the spectrophotometric determination of Trp photoconversion impossible.

The number of moles of Trp photodecomposed per second per square millimeter of illuminated eye lens surface can be calculated by multiplying the percentage of solar irradiance absorbed in an optical path of 5 mm by the appropriate decomposition quantum yield for each photosensitizer. For RF as sensitizer, two different quantum yield values were employed, depending on the region of the spectra considered. Although the quantum yield for RF-sensitized Trp decomposition is higher than that for sensitization by the AGEs, the contribution of RF to Trp oxidation in the eye lens is smaller than that of the AGEs, reflecting the very low concentrations of RF in this organ.[89] Another factor that may also restrict the sensitizing capacity of RF is its non-homogeneous distribution *in vivo* in the eye lens[95] due to its transformation into FMN and FAD coenzymes, which are essential participants in many enzyme-catalyzed reactions. The AGEs should be more homogeneously distributed throughout the different crystallins and, in this regard, their sensitizing effect can be more deleterious. The highest AGEs concentrations used in this study correspond to those found in brunescent cataractous lenses and the lower concentrations to those found at the beginning of this process and in elderly lenses.

In summary, it can be concluded that RF is characterized by a high efficiency of photosensitized decomposition of ascorbate and Trp. In mixtures of the two, ascorbate is preferentially modified, *i.e.*, the amino acid only starts to be affected when ascorbate is almost completely depleted. Under conditions of oxidative stress, in which the ascorbate concentration is diminished, RF may contribute to the occurrence of photochemical processes in the eye lens, the products of which (RF–Trp adducts) have been detected in this tissue. At the same time, the participation of RF in the photochemical decomposition of vitamin C cannot be discarded. The sensitized photochemical reactions mediated by AGEs are characterized by a low quantum efficiency, but this is compensated by the importance of their contribution to the absorption of the incident radiation impinging on the lens.

Acknowledgments

The studies performed by our group and partner groups in this field were supported by FONDECYT (Grants 1980874, 1010722 and 1050965) and by ECOS/FONDECYT (Projects C97S01 and C01S04).

References

1. R.C. Augusteyn, The effect of light deprivation on the mouse lens, *Exp. Eye Res.*, 1988, **66**, 669–674.
2. S.D. Varma, D. Chand, Y.R. Sharma, J.F. Kuck Jr. and R.D. Richards, Oxidative stress on lens and cataract formation: role of light and oxygen, *Curr. Eye Res.*, 1984, **3**, 35–57.
3. A. Spector, G.-M. Wang and R.-R. Wang, Photochemically induced cataracts in rat lenses can be prevented by AL-3823 A, a glutathione peroxidase mimic, *Proc. Natl. Acad. Sci. USA*, 1993, **90**, 7485–7489.
4. A. Spector, G.-M. Wang and R.-R. Wang, The prevention of cataract caused by oxidative stress in cultured rat lenses. II. Early effects of photochemical stress and recovery, *Exp. Eye Res.*, 1993, **57**, 659–667.
5. A. Spector, G.-M. Wang, R.-R. Wang, W. Garner and M. Moll, The prevention of cataract caused by oxidative stress in cultured rat lenses. I. H_2O_2 and photochemical induced cataract, *Curr. Eye Res.*, 1993, **12**, 163–179.
6. E.L. Finley, M. Busman, J. Dillon, R.K. Crouch and K.L. Schey, Identification of photooxidation sites in bovine alpha-crystallin, *Photochem. Photobiol.*, 1997, **66**, 635–641.
7. G.B. Reddy and K.S. Bhat, UVB irradiation alters the activities and kinetic properties of the enzymes of energy metabolism in rat lens during aging, *J. Photochem. Photobiol. B: Biol.*, 1998, **42**, 40–46.
8. B.J. Ortwerth, A. Coots, H.L. James and M. Linetsky, UVA irradiation of human lens proteins produces residual oxidation of ascorbic acid even in the presence of high levels of glutathione, *Arch. Biochem. Biophys.*, 1998, **351**, 189–196.
9. M. Linetsky, N. Ranson and B.J. Ortweth, The aggregation in human lens protein blocks the scavenging of UVA-generated singlet oxygen by ascorbic acid and glutathione, *Arch. Biochem. Biophys.*, 1998, **351**, 180–188.
10. D. Balasubramanian, Photodynamic of cataract: an update on endogenous chromophores and antioxidants, *Photochem. Photobiol.*, 2005, **81**, 498–501.
11. N. Azzam, D. Levaron and A. Dovrat, Effects of UV-A irradiation on lens morphology and optics, *Exp. Geront.*, 2004, **39**, 139–146.
12. C.S. Rogers, L.M. Chan, Y.S. Sims, K.D. Byrd, H.L. Hinton and S.S. Twining, The effects of sub-solar levels of UV-A and UV-B on rabbit corneal and lens epithelial cells, *Exp. Eye Res.*, 2004, **78**, 1007–1014.
13. J.E. Roberts, E.L. Finley, S.A. Patat and K.L. Schey, Photooxidation of lens proteins with xanthurenic acid: a putative chromophore for cataractogenesis, *Photochem. Photobiol.*, 2001, **74**, 740–744.
14. J.E. Roberts, Ocular phototoxicity, *J. Photochem. Photobiol. B: Biol.*, 2001, **64**, 136–143.
15. M.J. Davies and R.J. Truscott, Photo-oxidation of proteins and its role in cataractogenesis, *J. Photochem. Photobiol. B: Biol.*, 2001, **63**, 114–125.
16. A. Bachem, Ophthalmic ultraviolet action spectra, *Am. J. Ophthalmol.*, 1956, **41**, 969–975.
17. D.G. Pitts, A.P. Cullen and P.D. Hacker, Ocular effects of ultraviolet radiation from 295 to 365 nm, *Investig. Ophthalmol. Vis. Sci.*, 1977, **16**, 932–939.
18. J.G. Jose and D.G. Pitts, Wavelength dependence of cataracts in albino mice following chronic exposure, *Exp. Eye Res.*, 1985, **41**, 545–563.

19. C.A. McCarty and H.R. Taylor, Recent developments in vision research: light damage in cataract, *Invest. Ophthalmol. Vis. Sci.*, 1996, **37**, 1720–1723.
20. D.H. Sliney and M.L. Wolbarsht, Safety with lasers and other optical sources. *A Comprehensive Handbook*, Plenum Press, New York, 1980.
21. E.A. Boettner and J.R. Wolter, Transmission of the ocular media, *Invest. Ophthalmol. Vis. Sci.*, 1962, **1**, 776–783.
22. D.H. Sliney, Physical factors in cataractogenesis: ambient ultraviolet radiation and temperature, *Invest. Ophthalmol. Vis. Sci.*, 1986, **27**, 781–790.
23. U.P. Andley, B. Clark and L.T. Chylak Jr., Photosensitized oxidation of lens crystallins: role of conformational changes in cataract, *Proc. SPIE-Int. Soc. Opt. Eng.*, 1988, **847**, 187–192.
24. J.S. Zigler Jr. and J.D. Goosey, Photosensitized oxidation in the ocular lens: evidence for photosensitizers endogenous to the human lens, *Photochem. Photobiol.*, 1981, **33**, 869–874.
25. A. Sen, N. Ueno and B. Chakrabarti, Studies on human lens. I. Origin and development of fluorescent pigments, *Photochem. Photobiol.*, 1992, **55**, 753–764.
26. J. Dillon, R.-H. Wang and S.J. Atherton, Photochemical and photophysical studies on human lens constituents, *Photochem. Photobiol.*, 1990, **52**, 849–852.
27. C.M. Krishna, S.U. Uppuluri, P. Riesz, J.S. Zigler Jr. and D. Balasubramanian, A study of the photodynamic efficiencies of some eye lens constituents, *Photochem. Photobiol.*, 1991, **54**, 51–58.
28. B.J. Ortwerth and P.R. Olesen, UVA photolysis using the protein-bound sensitizers present in human lens, *Photochem. Photobiol.*, 1994, **60**, 53–60.
29. B.J. Ortwerth, M. Prabhakaram, R.H. Nagaraj and M. Linetsky, The relative UV sensitizer activity of purified advanced glycation endproducts, *Photochem. Photobiol.*, 1997, **65**, 666–672.
30. E.R. Gaillard, L. Zheng, J.C. Merriam and J. Dillon, Age-related changes in the absorption characteristics of the primate lens, *Invest. Ophthalmol. Vis. Sci.*, 2000, **41**, 1454–1459.
31. S. Zigman, Lens UVA photobiology, *J. Ocul. Pharmacol. Therap.*, 2000, **16**, 161–165.
32. M.D. Arginova and W. Breipohl, Glycated proteins can enhance photooxidative stress in aged and diabetic lenses, *Free Rad. Res.*, 2000, **36**, 1251–1259.
33. R.Z. Cheng, B. Lin and B.J. Ortwerth, Separation of the yellow chromophores in individual brunescent cataracts, *Exp. Eye Res.*, 2003, **77**, 313–325.
34. D. Balasubramanian, Photodynamics of cataract: an update on endogenous chromophores and antioxidants, *Photochem. Photobiol.*, 2005, **3**, 498–501.
35. U.P. Andley, Photodamage to eye, *Photochem. Photobiol.*, 1987, **46**, 1057–1066.
36. Ch. Rao, D. Balasubramanian and B. Chakrabarti, Monitoring light induced changes in isolated intact eye lenses, *Photochem. Photobiol.*, 1987, **46**, 511–515.
37. U.P. Andley, P. Sutherland, J.N. Liang and B. Chakrabarti, Change in tertiary structure of calf-lens α-crystallin by near-UV irradiation: role of hydrogen peroxide, *Photochem. Photobiol.*, 1984, **40**, 343–349.
38. G.J.H. Bessems and H.J. Hoenders, Distribution of aromatic and fluorescent compounds within single human lenses, *Exp. Eye Res.*, 1987, **44**, 817–824.
39. M.H. Smeets, G.FR.J.M. Vrensen, K. Otto, G.J. Puppels and J. Greve, Local variations in protein structure in the human eye lens: a Raman spectroscopic study, *Biochim. Biophys. Acta*, 1993, **1164**, 236–242.
40. K.J. Dilley and A. Pirie, Changes to the proteins of the human lens nucleous in cataract, *Exp. Eye Res.*, 1974, **19**, 59–72.

41. J.S. Zigler Jr., J.B. Sidbury, B.S. Yamanasky and M. Wolbarssht, Studies on brunescent cataracts. I. Analysis of free and protein-bound amino acids, *Ophthal. Res.*, 1976, **8**, 379–387.
42. A.M. Wood and R.J.W. Truscott, UV filters in human lenses: tryptophan catabolism, *Exp. Eye Res.*, 1993, **56**, 317–325.
43. J.A. Aaquilina, J.A. Carver and R.J.W. Truscot, Oxidation products of 3-hydroxykynurenine bind to lens proteins: relevance for nuclear cataract, *Exp. Eye Res.*, 1997, **64**, 727–735.
44. B. Roll, Caratenoid and retinoid – two pigments in a gecko eye lens, *Comp. Biochem. Physiol. A-Molec. Intergr. Physiol.*, 2000, **125**, 105–112.
45. M.M. Staniszewska and R.H. Nagaraj, 3-Hydroxykynurenine-mediated modification of human lens proteins, *J. Biol. Chem.*, 2005, **280**, 22154–22164.
46. R.D. Lasker, Diabetes control and complication trial research group: the effect of intensive treatment of diabetes on the development and progression of long-term complications in insulin-dependent diabetes mellitus, *N. Engl. J. Med.*, 1993, **329**, 977–986.
47. M. Brownlee, Glycation products and the pathogenesis of diabetic complications, *Diabetes Care*, 1992, **15**, 1835–1843.
48. J.W. Baynes, Role of oxidative stress in development of complications in diabetes, *Diabetes*, 1991, **40**, 405–412.
49. F. Tessier, V. Moreaux, I. Birlouez-Aragon, P. Junes and H. Mondon, Decrease in vitamin C concentration in human lenses during cataract progression, *Int. J. Vit. Nutr. Res.*, 1997, **68**, 309–315.
50. N.P. Brown and A.J. Bron, Biology of the lens, *A Clinical Manual of Cataract Diagnosis*, Butterworth-Heinemann Ltd., Oxford, 1996.
51. V.N. Reddy, F.J. Giblin and L.-R.B. Chakrapani, The effect of aqueous humor ascorbate on ultraviolet-B-induced DNA damage in lens epithelium, *Inv. Ophthal. Vis. Sci.*, 1998, **39**, 344–350.
52. K.W. Lee, V. Mossine and B.J. Ortwerth, The relative ability of glucose and ascorbate to glycate and crosslink lens proteins *in vitro*, *Exp. Eye Res.*, 1998, **67**, 95–104.
53. B.J. Ortwerth, M. Linetsky and P.R. Olesen, Ascorbic acid glycation of lens proteins produces UVA sensitizers similar to those in human lens, *Photochem. Photobiol.*, 1995, **62**, 454–462.
54. R. Cheng, B. Lin, K.-W. Lee and B.J. Ortwerth, Similarity of the yellow chromophores isolated from human cataracts with those from ascorbic acid-modified calf lens proteins: evidence for ascorbic acid glycation during cataract formation, *Biochim. Biophys. Acta*, 2001, **1537**, 14–26.
55. J. Dillon and A. Spector, A comparison of aerobic and anaerobic photolysis of lens protein, *Exp. Eye Res.*, 1980, **31**, 591–599.
56. D.W. Batey and C.D. Eckhert, Analysis of flavins in ocular tissues of the rabbit, *Investig. Ophthalmol. Vis. Sci.*, 1991, **32**, 1981–1985.
57. A. Yoshimura and T. Ohno, Lumiflavin-sensitized photooxygenation of indole, *Photochem. Photobiol.*, 1988, **48**, 561–565.
58. E. Silva, R. Ugarte, A. Andrade and A.M. Edwards, Riboflavin-sensitized photoprocesses of tryptophan, *J. Photochem: Photobiol. B: Biol*, 1994, **23**, 43–48.
59. S.K. Bose, K. Mandal and B. Chakrabarti, Sensitizer-induced conformational changes in lens crystallin. II. Photodynamic action of riboflavin on bovine α-crystallin, *Photochem. Photobiol.*, 1986, **43**, 525–528.

60. E. Silva, S. Risi and K. Dose, Photooxidation of lysozime at different wavelenghts, *Radiat. Environ. Biophys.*, 1974, **11**, 111–124.
61. E. Silva and J. Gaule, Light-induced binding of riboflavin to lysozyme, *Radiat. Environ. Biophys.*, 1977, **14**, 303–310.
62. I. Ferrer and E. Silva, Isolation and photo-oxidation of lysozyme fragments, *Radiat. Environ. Biophys.*, 1981, **20**, 67–77.
63. I. Ferrer and E. Silva, Study of a photo-induced lysozyme–riboflavin bond, *Radiat. Environ. Biophys.*, 1985, **24**, 63–70.
64. M. Campos, R.G. Díaz, A.M. Edwards, S. Kennedy, E. Silva, J. Derouault and M. Rey-Lafon, Photo-induced generation of the riboflavin–tryptophan adduct and a vibrational interpretation of its structure, *Vibrat. Spectrosc.*, 1994, **6**, 173–183.
65. C.B. Martin, M.L. Tsao, C.M. Hadad and M.S. Platz, The reaction of triplet flavin with indole. A study of the cascade of reactive intermediates using density functional theory and time resolved infrared spectroscopy, *J. Am. Chem. Soc.*, 2002, **124**, 7226–7234.
66. E. Silva and J. Godoy, Riboflavin sensitized photooxidation of tyrosine, *Internat. J. Vit. Nutr. Res.*, 1994, **64**, 253–256.
67. E. Silva, S. Fürst, A.M. Edwards, M.I. Becker and A.E. De Ioannes, Visible light anaerobic photoconversion of tyrosine sensitized by riboflavin. Cytotoxicity on mouse tumoral cells, *Photochem. Photobiol.*, 1995, **62**, 1041–1045.
68. E. Silva, M. Salim-Hanna, A.M. Edwards, M.I. Becker and A. De Ioannes, A light-induced tryptophan–riboflavin binding: biological implications, *Adv. Exp. Med. Biol.*, 1990, **289**, 33–48.
69. M. Salim-Hanna, A. Valenzuela and E. Silva, Riboflavin status and photo-induced riboflavin binding to the proteins of the rat ocular lens, *Int. J. Vit. Nutr. Res.*, 1988, **58**, 61–65.
70. R. Ugarte, A.M. Edwards, M.S. Diez, A. Valenzuela and E. Silva, Riboflavin photosensitized anaerobic modification of rat lens proteins. A correlation with age-related changes, *J. Photochem. Photobiol. B: Biol.*, 1992, **13**, 161–168.
71. G. Tapia and E. Silva, Photoinduced riboflavin binding to the tryptophan residues of bovine and human serum albumins, *Radiat. Environ. Biophys.*, 1991, **30**, 131–138.
72. M. Diaz, M.I. Becker, A. De Ioannes and E. Silva, Development of monoclonal antibodies against a riboflavin–tryptophan photoinduced adduct: reactivity to eye lens proteins, *Photochem. Photobiol.*, 1996, **63**, 762–767.
73. M. Mancini, A.M. Edwards, M.I. Becker, A. de Ioannes and E. Silva, Reactivity of monoclonal antibodies against a tryptophan–riboflavin adduct toward irradiated and non-irradiated bovine-eye-lens protein fractions: an indicator of *in vivo* visible-light mediated phototransformations, *J. Photochem. Photobiol. B: Biol.*, 2000, **55**, 9–15.
74. A. de La Rochette, E. Silva, I. Birlouez-Aragon, M. Mancini, A.M. Edwards and P. Morliere, Riboflavin photodegradation and photosensitizing effects are highly dependent on oxygen and ascorbate concentrations, *Photochem. Photobiol.*, 2000, **72**, 815–820.
75. A. de La Rochette, I. Birlouez-Aragon, E. Silva and P. Morlière, Advanced glycation endproducts as UVA photosensitizers of tryptophan and ascorbic acid: consequences for the lens, *Biochim. Biophys. Acta*, 2003, **1621**, 235–241.
76. J.A. Aquilina, J.A. Carver and R.J. Truscott, Oxidation products of 3-hydroxykynurenine bind to lens proteins: relevance for nuclear cataract, *Exp. Eye Res.*, 1997, **64**, 727–735.

77. J.A. Aquilina, J.A. Carver and R.J. Truscott, Elucidation of a novel polypeptide cross-link involving 3-hydroxykynurenine, *Biochemistry*, 1999, **38**, 11455–11464.
78. M. Kwan, J. Niinikoski and T.K. Hunt, *In vivo* measurements of oxygen tension in the cornea, aqueous humor, and anterior lens of the open eye, *Invest. Ophthalmol.*, 1972, **11**, 108–114.
79. M. Linetsky, H.L. James and B.J. Ortwerth, Spontaneous generation of superoxide anion by human lens proteins and by calf lens proteins ascorbylated *in vitro*, *Exp. Eye Res.*, 1999, **69**, 239–248.
80. M. Linetsky and B.J. Ortwerth, The generation of hydrogen peroxide by the UVA irradiation of human lens proteins, *Photochem. Photobiol.*, 1995, **6**, 87–93.
81. B.J. Ortwerth, T.A. Casserly and P.R. Olesen, Singlet oxygen production correlates with His and Trp destruction in brunescent cataract water-insoluble proteins, *Exp. Eye Res.*, 1998, **67**, 377–380.
82. B.J. Ortwerth, H. James, G. Simpson and M. Linetsky, The generation of superoxide anions in glycation reactions with sugars, osones, and 3-deoxyosones, *Biochem. Biophys. Res. Commun.*, 1998, **245**, 161–165.
83. M. Linetsky and B.J. Ortwerth, Quantification of the singlet oxygen produced by UVA irradiation of human lens proteins, *Photochem. Photobiol.*, 1997, **65**, 522–529.
84. E.A. Lissi, M. Faure and N. Clavero, Effect of additives on the inactivation of lysozyme mediated by free radicals produced in the thermolysis of 2,2′-azo-bis-(2-amidinopropane), *Free Radic. Res. Commun.*, 1991, **14**, 373–384.
85. G. Viteri, A.M. Edwards, J. De la Fuente and E. Silva, Study of the interaction between triplet riboflavin and the α-, β_H- and β_L-crystallins of the eye lens, *Photochem. Photobiol.*, 2003, **77**, 535–540.
86. A. Yoshimura and T. Ohno, Lumiflavin-sensitized photo-oxygenation of indole, *Photochem. Photobiol.*, 1988, **48**, 561–565.
87. M. Hietanen, Ocular exposure to solar ultraviolet and visible radiation at high latitudes, *Scand. J. Environ. Health*, 1991, **17**, 398–403.
88. D.H. Sliney, UV radiation ocular exposure dosimetry, *J. Photochem. Photobiol. B: Biol.*, 1995, **31**, 69–77.
89. D.D. Duncan, B. Muñoz, K. Bandeen-Roche and S.K. West, Assessment of ocular exposure to ultravioletr-B for population studies, *Photoche. Photobiol.*, 1997, **66**, 701–709.
90. S. Ono and H. Hirano, Riboflavin metabolism in the single lens of rat, *Ophthalmic Res.*, 1983, **15**, 140–145.
91. E.A. Boettner and J.R. Wolter, Transmission of the ocular media, *Invest. Ophthalmol. Vis. Sci.*, 1962, **1**, 776–783.
92. S. Lerman, Lens fluorescence in aging and cataract formation, *Doc. Ophthalmol.*, 1976, **8**, 241–260.
93. B.J. Ortwerth, J.A. Speaker, M. Prabhakaram, M.G. Lopez, E.Y. Li and M.S. Feather, Ascorbic acid glycation: the reactions of L-threose in lens tissue, *Exp. Eye Res.*, 1994, **58**, 665–674.
94. S. Lerman, Lens fluorescence in aging and cataract formation, *Doc. Ophthalmol.*, 1976, **8**, 241–260.
95. K.S. Bhat and S. Nayak, Flavin nucleotides in human lens: regional distribution in brunescent cataracts, *Indian J. Ophthalmol.*, 1998, **46**, 233–237.

Chapter 8

Blue Light-Initiated DNA Repair by Photolyase

CHRISTOPHER W.M. KAY,[a] ADELBERT BACHER,[b] MARKUS FISCHER,[b] GERALD RICHTER,[c] ERIK SCHLEICHER[a] AND STEFAN WEBER[a]

[a] Fachbereich Physik, Freie Universität Berlin, Arnimallee 14, 14195 Berlin, Germany

[b] Lehrstuhl für Organische Chemie und Biochemie, Technische Universität München, Lichtenbergstr. 4, 85747 Garching, Germany

[c] School of Biological and Chemical Sciences, University of Exeter, Stocker Rd., Exeter EX4 4QD, UK

Abstract

In 1949 an enzyme was discovered that could repair UV-light-induced DNA lesions, by effectively reversing their formation with the aid of blue light. The class of enzymes has subsequently been named photolyases and the process, photoreactivation. All photolyases utilize a non-covalently bound cofactor, flavin adenine dinucleotide (FAD). With the exception of cryptochromes, with which photolyases share a high degree of structural homology in the *N*-terminal domain, photolyases show no similarity to other blue-light sensing flavoproteins, such as the phototropins or the recently discovered BLUF domains. Although, flavoproteins are ubiquitous, redox-active catalysts for one- and two-electron transfer reactions, it is the properties of the photoexcited states of the different redox states of FAD that are also important in photolyases. Two milestones mark the progress of photolyase research in recent years. The first was the elucidation of the three-dimensional structure of the enzyme in 1995 that revealed remarkable details, such as the FAD-cofactor arrangement in an unusual U-shaped configuration, and in late 2004 a structure was presented that included a bound fragment of repaired DNA. In the 10 years between, research on photolyases continued at a frenetic pace using all available biochemical and biophysical tools. This chapter focuses on recent biophysical studies and shows how different approaches have yielded details on the fundamental aspects of the modus operandi of this unique class of flavoenzymes.

8.1. UV Radiation Damage to DNA and its Repair by Photoreactivation

Ultraviolet (UV) radiation in the 100–320 nm spectral region has carcinogenic, mutagenic, and fatal effects on living organisms.[1] The most significant cellular target of UV is DNA. When DNA is exposed to radiation at wavelengths approaching its absorption maximum at around 260 nm, adjacent pyrimidines within the same DNA strand may become covalently linked by the formation of mainly (70–80%) four-membered ring structures referred to as cyclobutane pyrimidine dimers (CPDs)[†] and also (20–30%), of pyrimidine (6-4) pyrimidone photoproducts ((6-4)PPs)[‡1] (see Figure 1). The structure of the CPD was already elucidated by Blackburn and Davies 40 years ago.[2,3] The exact yields and the type of UV-induced damage depend on the sequence and structure, the curvature and the folding of DNA.

It has been known for a long time that UV damage to human skin increases the occurrence of certain types of cancer including melanoma. More recent life style changes have exacerbated the risk of skin cancer, which is at this time the most frequent cancer in North America. Moreover, UV exposure becomes a devastating risk in patients suffering from *Xeroderma pigmentosum*, a rare heritable disease affecting various proteins of the excision repair cascade; unless these patients are rigorously protected from UV exposure, they develop multiple cutaneous tumors at an early age. Whereas placental mammals repair

† Throughout the text the abbreviation Base(1)◇Base(2) is also used for CPDs (e.g., T◇T), where the "◇" represents the cyclobutane ring that links the two bases 1 and 2. As this review is mainly focused on the *cis-syn*-form, which is the main substrate in DNA photorepair, the qualifier "*cis-syn*" shall be dropped for convenience except for cases where the configuration of the CPD is relevant.
‡ The abbreviation Base(1)-64-Base(2) (e.g., T-64-C) will be used in cases where the base composition of the (6–4)PP is relevant. "-64-" represents the σ-bond linking C(6) on the 5'- base and C(4') on the 3' base.

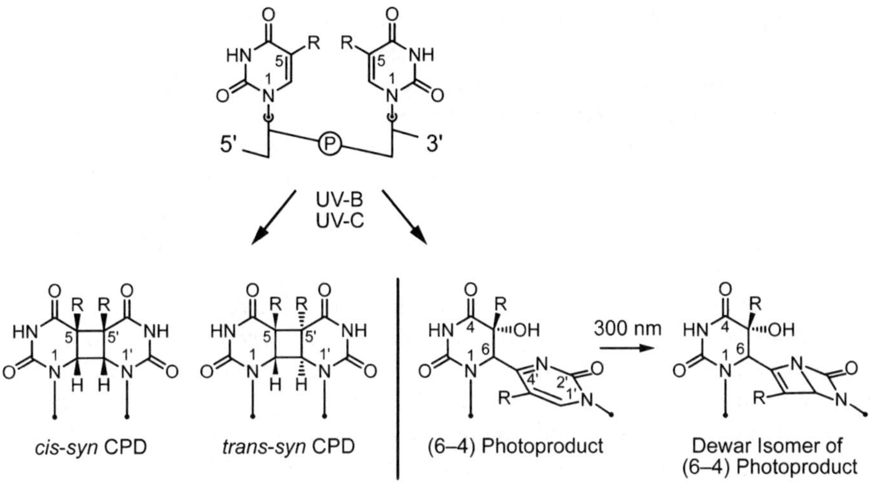

Figure 1. The structures of the most mutagenic UV-induced DNA lesions. R denotes either H in uracil (U) or CH_3 in thymine (T). The *cis-syn* CPD is a diastereoisomer of the *trans-syn* CPD

photodamaged DNA by excision repair, studies with recombinant mice expressing photolyase, initially called photoreactivating enzyme,[4–6] have confirmed that CPD lesions play a dominant role in the etiology of mammalian skin malignancies.

Photoreactivation by photolyase uses UV-A (320–400 nm) and blue light (400–500 nm) to monomerize pyrimidine dimers.[7–25] Photolyases have been reported in many vertebrates including fish, reptiles, and marsupials. The human genome has two CRY genes with similarity to photolyase sequences.[26–29] These, however, encode blue-light photoreceptors involved in setting of circadian rhythms[24,30–35] but not in repair of DNA damage. Additional homologues of DNA-repair photolyases in the human genome have not been detected.[36] Apart from some transgenic mice[37] it seems that placental mammals are generally devoid of DNA photolyases.[38–40]

8.2. Classification and Spectral Properties of Photolyases

All photolyases described to date are monomeric proteins with molecular masses in the 45–66 kDa range.[7] Photolyases are distinguished by their different substrate specificity: DNA photolyase (also called CPD photolyase) binds and repairs CPD lesions, whereas (6-4) photolyase reverses (6-4)PP-damages. DNA photolyases are further classified into class-I and class-II, depending on their amino-acid sequence similarity.[41–43] Class-I photolyases are found in many bacteria and archaea, but in eukaryotes, they are found only in fungi. Class-II photolyases are the most divergent group of the family and are additionally present in plants,[44–48] green algae,[49] dipterans,[50] and vertebrates.[51]

Interestingly, the class-II enzyme is also present in the genomes of several poxviruses from vertebrates and insects.[52] Sequence and phylogenetic analyses show that a photolyase gene from the microsporidian parasite *Antonospora locustae* encodes the enzyme of class-I type.[53]

All known photolyases contain stoichiometric amounts of non-covalently bound flavin adenine dinucleotide (FAD) as redox-active cofactor.[41,45,48,50,54–60] Nevertheless, there is no obvious amino-acid sequence homology to other classes of flavoproteins, perhaps because the enzymatic activity of photolyases depends on the photoexcited state of FAD in the two-electron reduced form, $FADH^-$ (see Figure 2), whereas the vast majority of flavoproteins utilize their ground states[61–63] or the excited states of the fully oxidized redox state of flavin, FAD^{ox}. $FADH^-$ in photolyases has weak absorbance at wavelengths below and vanishing absorbance above 400 nm, thus rendering the catalytically active enzyme nearly colorless to pale yellow. A second (also non-covalently bound) chromophore is also often present and functions as a light-harvesting cofactor to increase the absorption cross-section of the enzyme and to extend its absorbance to longer wavelengths.

The class-I photolyase is further categorized according to the type of light-harvesting cofactor present: either a folate-type (5,10-methenyltetrahydrofolyl-polyglutamate (MTHF)) or a deazaflavin-type (8-hydroxy-5-deazaflavin (8-HDF)). In solution, MTHF has maximum absorbance at 360 nm. Its absorption is red-shifted to 384 nm when bound to the photolyase enzyme.[64] The class of photolyases using folates, therefore, exhibits maximum catalytic activity at 377–410 nm. It includes enzymes from gram-negative bacteria,[64–66] yeast,[64,67] and the filamentous fungus *Neurospora crassa*.[58,68] The deazaflavin class has an action spectrum with a maximum in the range from 430 to 460 nm (its absorption maximum of 438 nm is also red-shifted compared to the peak absorbance at 420 nm of deazaflavin in solution) and includes photolyases from the cyanobacterium *Anacystis nidulans*,[57,69] the gram-positive bacterium

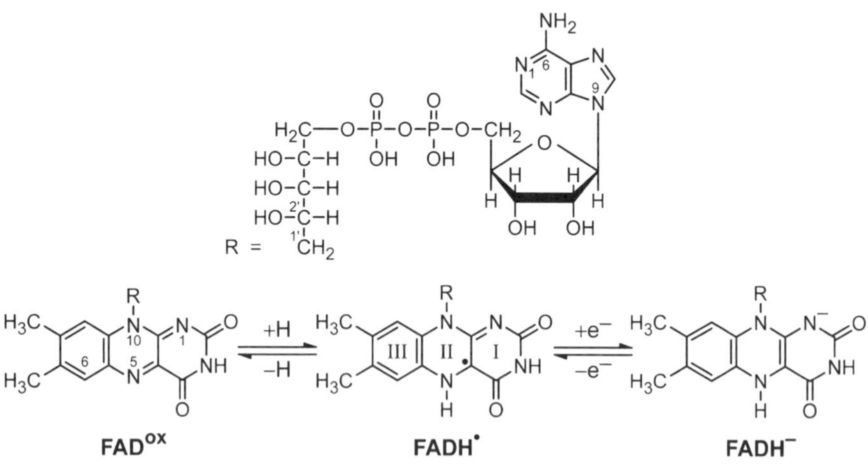

Figure 2. The redox states of the 7,8-dimethyl isoalloxazine moiety of FAD

(actinomycete) *Streptomyces griseus*,[70] the anoxygenic photosynthetic bacterium *Rhodobacter capsulatus*,[71] the green alga *Scenedesmus acutus*,[72] and the archae *Halobacterium halobium*.[73]

The amino-acid sequences of class-II photolyases share characteristics with both 8-HDF- and MTHF-type photolyases. Hence, from sequence comparison the nature of their second chromophore cannot be deduced.

(6-4) photolyases which also contain FAD as redox-active cofactor[59,74] have been found in *Drosophila melanogaster*,[26,75,76] *Xenopus laevis* (South African clawed toad),[59,77] *Crotalus atrox* (rattlesnake),[77] *Danio rerio* (zebrafish),[78] *Arabidopsis thaliana*,[46,79,80] and the marine hexactinellid sponge *Aphrocallistes vastus*.[81] In semi-purified fractions of *X. laevis* (6-4) photolyase, an absorption peak at 416 nm has been observed and assigned to the second chromophore, presumably MTHF, which, however, is lost upon further purification of the enzyme.[74] A fluorescent species with an excitation maximum at 360 nm and an emission maximum at 440 nm has been observed by fluorescence spectroscopy at pH 10 upon denaturing *D. melanogaster* (6-4) photolyase.[82] Taken together with the fact that neither *D. melanogaster* nor *Escherichia coli* (the organism in which the (6-4) photolyase enzyme was over-expressed) can synthesize deazaflavin,[83] it was concluded that (6-4) photolyases utilize MTHF as light-harvesting cofactor.

8.3. Structure of Photolyases

In a seminal contribution by Park and co-workers,[84] the three-dimensional structure of the class-I DNA photolyase from *E. coli* was determined by X-ray crystallography at 2.3 Å resolution. Subsequently, the crystal structures of the (class-I) 8-HDF-type DNA photolyase and the apophotolyase from *A. nidulans* were elucidated both at 1.8 Å resolution.[85,86] The *E. coli* and *A. nidulans* enzymes show similar backbone structures of their respective single polypeptide chain, consistent with their conserved function and the very homologous amino-acid sequences sharing 39% identical and 13% similar residues. The FAD cofactor is bound to a helical domain. An α/β-domain provides binding sites for the second chromophore.

The flavin chromophore is non-covalently bound by amino acids belonging to the helical domain. Both the location and conformation of the redox-active FAD are virtually identical in both photolyases. The molecule is bent in a unique U-shaped configuration so that the adenine ring is in close contact with the 7,8-dimethyl isoalloxazine ring.

The α/β-domains in the region which recognizes either 8-HDF or MTHF are quite similar in both enzymes. Nevertheless, the photolyase structures reveal completely different binding sites for the light-harvesting chromophores. 8-HDF is more deeply buried in the *A. nidulans* enzyme, whereas MTHF occupies a partially surface-exposed site in the *E. coli* photolyase. The average separation between 8-HDF and FAD (measured between the centers of their isoalloxazine moieties) in *A. nidulans* photolyase is 17.4 Å, while for MTHF and

FAD in *E. coli* photolyase the distance is 16.4 Å (measured between the center of the FAD's isoalloxazine moiety and *N*(5) of MTHF).

A three-dimensional structure determined to 2.1 Å resolution of the class-I DNA photolyase from the thermophilic bacterium *Thermus thermophilus*, has been reported by Komori and co-workers.[87] Although no significant electron density for the second cofactor was found, the crystal structure revealed a large cavity inside the cleft between the α/β-domain and the helical domains, which is large enough to accommodate an 8-HDF but is too small to accommodate an MTHF. In comparison with the 8-HDF-binding site in *A. nidulans* photolyase, however, the amino-acid residues interacting with 8-HDF are also found in the *T. thermophilus* enzyme. Hence, Komori and co-workers suggested that *T. thermophilus* photolyase is an 8-HDF-type.

In all three enzymes, the flavin cofactor is buried in the center of the helical domain with solvent access limited to a cavity leading from the surface to the edge of the adenine moiety of FAD. A band of positive electrostatic potential, suitable for contact with the negatively charged phosphate backbone of the damaged DNA strand, runs along the outside of the protein around the cavity entrance. One side of the cavity consists of hydrophobic amino-acid residues, and the other side has polar groups. This asymmetry is also a property of the CPD site of damaged DNA, in which the cyclobutane ring is hydrophobic and the opposite edges of the thymine bases have nitrogen and oxygen atoms capable of forming hydrogen bonds. The cavity opening is wide enough to accommodate a CPD provided that it is extruded from the duplex.[84] Such base-flipping out of double-helical DNA has been observed in many DNA-repair systems, in which the enzymes need to approach the DNA bases in order to perform a reaction on them.[88]

8.4. Substrate Binding

Photolyases are extremely efficient at distinguishing pyrimidine dimers in DNA from undamaged pyrimidine doublets as the number of photolyase molecules per cell is very small: yeast cells typically contain only about 75–300 molecules of DNA photolyase[89,90] and *E. coli* cells even less (about 16–17 molecules).[91] The flavin cofactor is essential for substrate binding,[92] although the redox state of the flavin is unimportant.[92,93] Photolyases are structure-specific DNA-binding proteins, whose specificity in recognizing pyrimidine dimers is determined by the configuration of the DNA-phosphodiester backbone at the damage site and/or the structure of the DNA lesion itself, in contrast to sequence-specific DNA-binding proteins which rely on hydrogen-bond donors and acceptors in the grooves of the duplex.[94–96]

CPD-substrate binding to DNA photolyase occurs with similar affinity to double-stranded and single-stranded DNA[96,97] with an association-rate constant in the 10^6-$M^{-1}\ s^{-1}$ range.[98] Studies on the *E. coli* and *S. cerevisiae* DNA photolyase enzymes have shown that the equilibrium association constant for *cis-syn* CPDs in DNA (specific binding) is $2.6 \times 10^8\ M^{-1}$ to $2.2 \times 10^9\ M^{-1}$,[95,99] and 3.7×10^9 M^{-1},[100] respectively, whereas the association constant for the undamaged DNA (non-specific binding) is more than four orders of magnitude lower.

The affinity of DNA photolyase to the *trans-syn* configured CPD damage is very low and identical or very close to the enzyme's affinity to undamaged DNA (association constant in the 10^4 to 10^5 M^{-1}-range). Once bound, however, the *trans-syn* CPD is repaired with photochemical efficiency comparable to that of the *cis-syn* isomer.[101] The enzyme furthermore binds to UV-dimerized thymine–uracil (T◇U) and cytosine–cytosine (C◇C) bases, as well as to uracil–uracil (U◇U) bases in DNA and RNA.[102]

X. laevis and *D. melanogaster* (6-4) photolyases specifically bind to (6-4)PP-damaged double-stranded DNA with association constants of 2.1×10^8 M^{-1},[74] and 2×10^9 M^{-1},[82] respectively. (6-4) photolyase also binds to single-stranded (6-4)PP-damaged DNA substrates and to DNA containing the Dewar isomer of the (6-4)PP. However, the affinities for these substrates are somewhat lower than that of (6-4)PP in double-stranded DNA.[74,82]

Extensive biochemical and biophysical studies were carried out to unravel the mode of substrate docking, binding sites, and docking geometries because obtaining a crystal structure of photolyase bound to a CPD-containing substrate proved rather difficult. Although such a structure has now been determined, see ref. 103, the methods used to approach this problem are well worth discussing.

Van Noort and co-workers[104] used atomic force microscopy to visualize the conformation of DNA when *A. nidulans* DNA photolyase is bound to intact and damaged DNA. In the former, DNA was found to have its usual straight geometry, whereas in the latter average DNA bending angles of about 36° were observed.

Following previous studies of the contact sites of UV-damaged DNA at various class-I and class-II photolyases,[94,105,106] nuclear-magnetic resonance (NMR) methods have been recently applied by Torizawa and co-workers[107] to identify the respective amino-acid residues in *T. thermophilus* photolyase. Detailed analyses showed that the bases of the CPD lesion are bound in a cavity of the enzyme making close contact to a tryptophan residue W247.[107] The equivalent residues W277 and W387 in *E. coli* and yeast photolyase, respectively, lie at the rim of the cavity. They contribute to substrate specificity by either van der Waals or π-stacking interactions as has been demonstrated previously by site-directed mutagenesis.[99]

The amino-acid residues at the enzyme surface of yeast photolyase that contribute to high-affinity substrate binding and binding specificity have also been identified by chemical modification and site-directed mutagenesis.[100] K517 was found to play an important role in binding of photolyase to DNA. In all microbial photolyases sequenced so far, the residue at the equivalent position is either lysine or arginine, suggesting a functional requirement for a positive charge at this location. Alanine substitutions of the conserved amino acids R507, K463, and W387 reduced both the overall affinity for substrate and discrimination for different substrates. The equivalent residues of K517 and

R507 in *E. coli* photolyase, K407 and R397, are located within the band of positive electrostatic potential that runs across the surface of the enzyme near the putative substrate-binding site.[84]

Alanine-substitution mutations in yeast photolyase at conserved sites K330, E384, and F494 within the cavity also slightly reduced substrate binding and discrimination.[108] Because these residues lie too deeply buried inside the cavity to interact with normal B-DNA, the only explanation for these results is that the substrate undergoes a dramatic structural modification, such as CPD-base flipping, that allows the CPD to have a direct interaction with these amino acids. These findings together with recent ^{1}H-^{15}N-HSQC-NMR and ^{1}H-NMR studies[107] have therefore been interpreted to support the dinucleotide-flip model.

Evidence for CPD base flipping by photolyase was also presented by Christine and co-workers[109] from an examination of fluorescence yields in DNA substrates in which the adenine base opposing the 3′-thymine of a thymine doublet has been replaced by a fluorescent reporter of helical structure, the adenine analog 2-aminopurine. Due to base stacking facilitated by favorable Watson–Crick base pairs in annealed double-stranded DNA, the fluorescence of 2-aminopurine is significantly quenched, regardless of whether 2-aminopurine is hydrogen-bonded to a non-dimerized or to a CPD-dimerized thymine in the complementary strand. The fluorescence of 2-aminopurine is almost completely restored to its value in single-stranded DNA when the CPD-containing duplex is bound to photolyase. Such a large change in fluorescence yield indicates that by interaction of UV-damaged DNA with photolyase a severe distortion of the local helical structure around the 2-aminopurine base has occurred. This is consistent with base flipping of the lesion into the protein-binding cavity with a concomitant de-stacking of the opposing complementary 2-aminopurine nucleotide.

The modes of substrate-binding have also been modeled by several groups using molecular-dynamics simulations. Sanders and Wiest[110] examined various models for *E. coli* DNA photolyase enzyme-substrate complexes. Distances between the CPD and the isoalloxazine ring of FAD of 15 and 12 Å were predicted for the non-flipped and the flipped-out configurations of a CPD-containing double-stranded dodecamer with and without CPD flipping, respectively. Even for a simple dinucleotide CPD, a minimum distance between *C*(5) on the 3′-base of the CPD and the ribose chain of FAD of larger than 6 Å was calculated.

Predictions of large CPD-to-FAD cofactor distances based on molecular dynamics simulations were also presented by Hahn and co-workers.[111] They have examined the docking of photolyase to "bare" (with both pyrimidine bases linked only by the cyclobutane ring) and "dressed" (with the pyrimidine bases also linked by the deoxyribose phosphate backbone) pyrimidine dimers.

In contrast to the aforementioned model calculations, Antony and co-workers[112] have calculated a very short CPD-to-FAD distance of less than 3 Å in a third molecular dynamics study. In the average configuration, the docked thymine dimer is sitting deeply buried in the enzyme cavity, approaching the isoalloxazine ring of FAD from the "open" side between the ribityl chain and the methyl group at *C*(8). At a distance of ≈2.5 Å between the hydrogen atoms on the FAD's *C*(8α) methyl group and those on the methyl group of the 3′-thymine in the CPD, a significant deformation of the damaged strand in the DNA is expected.

Van de Berg and Sancar[108] examined the docking of a CPD to *S. cerevisiae* photolyase. Beginning with the structure of the putative DNA binding domain of the *E. coli* enzyme as a model, the structure of the equivalent region in the yeast enzyme was calculated but neither coordinates nor distances were presented. However, shorter rather than longer CPD-to-FAD distances were preferred.

Kay and co-workers[113] recently used electron paramagnetic resonance (EPR) and electron-nuclear double resonance (ENDOR) to probe docking of a CPD to *E. coli* photolyase utilizing the FAD cofactor in its neutral radical form as a naturally occurring electron-spin probe. Small shifts of the flavin ^{1}H hyperfine couplings were observed when UV-damaged DNA was bound to the enzyme in comparison to substrate-free enzyme. The changes of the hyperfine couplings upon binding of substrate are too small to be consistent with a distance of less than 6 Å between the CPD and the location of maximum unpaired electron-spin density in the FAD's isoalloxazine ring. In a separate contribution, using density functional theory calculations, Weber and co-workers[114] showed that the orbital in which the unpaired electron spin is located has maximum density at the *C*(4a) and *N*(5) positions. Hence, the 6-Å distance determined from ENDOR should be applied from this position rather than from the *C*(8α) methyl group. The observed subtle shifts of the isotropic ^{1}H hyperfine-coupling constants of the isoalloxazine moiety could be rationalized in terms of changing polarity of the cofactor environment once the substrate is docked to the enzyme. The trends observed are consistent with the substrate-binding pocket becoming less polar because of the displacement of water molecules upon substrate binding.

MacFarlane VI. and Stanley[115] used an electrochromic-shift model to yield information on the CPD-to-FAD distance in photolyase. They described a spectral blue shift of the optical absorption bands of the flavin cofactor (in the fully oxidized redox state, FAD^{ox}) observed upon binding of the enzyme to a polynucleotide single-stranded CPD substrate.[116] No band shift was observed for a dinucleotide CPD, suggesting that there are significant differences in the binding geometry of dinucleotide versus polynucleotide dimer lesions (see also ref. 117). Under the assumption that the band shifts arise from a strong dipolar electric field in the enzyme cavity due to the permanent dipole moment of the two coplanar thymine bases in the CPD, the authors obtained the binding distance from the CPD to the FAD^{ox} in photolyase, however, as a function of the effective dielectric constant of the protein. Assuming a realistic range of dielectric constants of $\varepsilon = 2.6$–10, the analysis yielded a distance range between 5.5 and 8 Å, implying that substrate and flavin cofactor are not in van der Waals contact. Using transient absorption spectroscopy it was furthermore demonstrated that substrate binding does not significantly alter the structure of the protein around the FAD binding site. This observation is consistent with findings from EPR and ENDOR that also revealed only minor changes of the electronic structure of the flavin upon docking of the enzyme to photodamaged DNA.[113] Recently, Gindt and co-workers showed that the redox potential of the $FADH^-/FADH^{\bullet}$ couple increased by 65 mV following substrate binding. The modulation in reduction potential has only a minor effect on the electron-transfer processes, and the authors concluded that the physiological relevance appeared to be an increase in stability of the catalytic form of the enzyme with respect to oxidation.[118]

An electrochromic shift of the flavin cofactor's optical absorption bands due to the presence of the dipolar thymidine-dimer substrate, previously reported for FAD^{ox},[115,116] has recently also been observed by Schelvis and co-workers[119] in the absorption spectrum of the flavin-cofactor in the neutral radical redox state, $FADH^{\bullet}$. However, because the difference dipole moments for transitions from the $FADH^{\bullet}$ ground state to excited doublet states are unknown, CPD-to-FAD distance and orientation parameters could not be extracted from these experiments.

Using resonance Raman spectroscopy, the interaction between photolyase and UV-damaged DNA has been further investigated.[119] Changes in Raman intensities could be largely explained by changes in the Raman excitation profiles due to the electrochromic shift. The observed subtle frequency shifts of some $FADH^{\bullet}$ vibrations on binding of substrate to photolyase were proposed to arise primarily from small changes in the hydrogen-bonding interaction between $FADH^{\bullet}$ and its protein environment. Interestingly, when the resonance Raman spectrum of *E. coli* DNA photolyase containing the light-harvesting MTHF chromophore was compared with the one obtained previously by Murgida and co-workers[120,121] from a photolyase mutant (E109A) lacking the ability to bind MTHF, slight differences were revealed that suggest that the second chromophore may play a structural role in stabilizing the protein environment of the FAD cofactor.

By analyzing the paramagnetic relaxation enhancement in NMR resonances induced by the flavin cofactor in the radical semiquinone form, Ueda and co-workers[122] calculated a distance between the FAD and the CPD of (16 ± 3) Å in *T. thermophilus* photolyase, thus also favoring a relatively large separation of the redox-active center of the enzyme and its substrate. In their proposed model the side chains of the tryptophans W247 and W353 bridge the gap between the CPD and the FAD's adenine moiety thus supporting the idea that electron transfer in DNA repair may proceed through these aromatic residues.[110,114,122]

Very recently, the long-awaited X-ray crystal structure of *A. nidulans* DNA photolyase bound to a CPD-like DNA lesion was presented by Mees and co-workers,[103] finally showing that the CPD lesion is indeed flipped out of the DNA helix into the active site (see Figure 3). As has been predicted from biophysical studies (see above), aromatic interactions with tryptophan residues W286 and W392 (equivalent to W277 and W386 in *E. coli*, and W247 and W353 in *T. thermophilus* DNA photolyase, respectively) seem to play an important role in binding of both CPD thymines. W286 and W392 form an

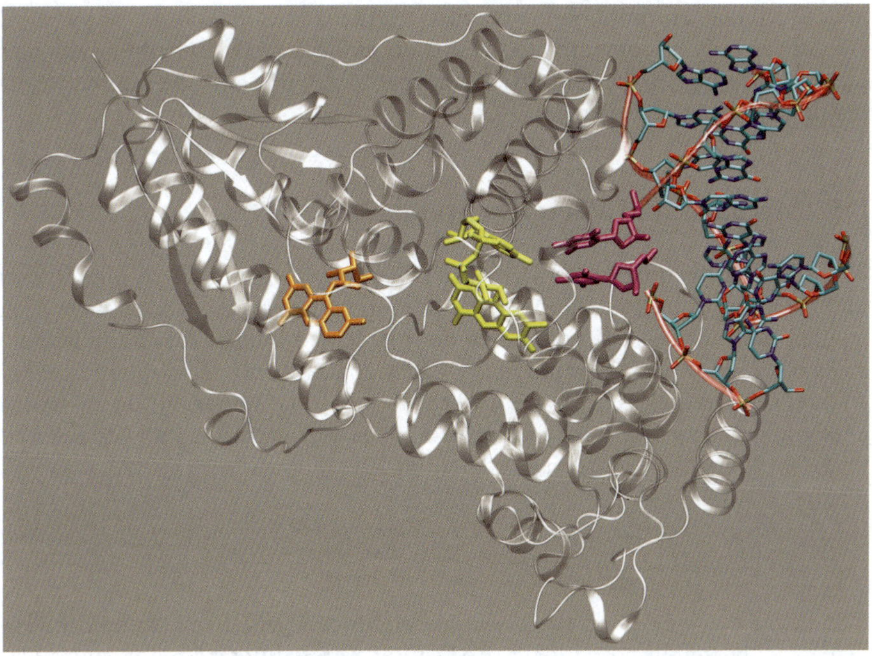

Figure 3. The arrangement of the CPD-like DNA substrate with respect to the flavin chromophore in *A. nidulans* DNA photolyase (PDB-entry: 1TEZ[103]). The repaired thymine bases in DNA, the flavin cofactor, and the light-harvesting second chromophore 8-HDF are drawn in purple, yellow, and orange, respectively

L-shaped wedge making van der Waals interactions with the ring plane of the 5′-thymine and the edge of the thymine dinucleotide. The *C*(4) carbonyl groups of both thymines form hydrogen bonds with the amino group at *C*(6) of the adenine moiety of FAD. Distances of 3.1 and 3.2 Å have been determined between the amino-group nitrogen and the *O*(4) oxygens of the *C*(4)=*O*(4) carbonyl groups of 5′- and 3′-thymine, respectively. Thus, the separation between the CPD-like DNA lesion and the flavin is closer than commonly expected. At a distance of only 4.3 Å between the *C*(8) methyl group of the FAD isoalloxazine moiety and the 3′-thymine, and about 7 Å between the center of the electron-donating $FADH^-$ ring system to the CPD-like lesion, a direct electron-transfer pathway for CPD repair cannot be excluded. From the co-crystal structure it appears likely that the flavin's adenine moiety is involved in forward and back electron transfer (see below).

It is noteworthy, however, that the *C*(5)–*C*(5) and *C*(6)–*C*(6) bonds of the synthesized CPD lesion used for the elucidation of the three-dimensional structure of the photolyase-substrate complex were found to be broken as a result of the prolonged exposure of the sample to synchrotron radiation. Thus, the active site of photolyase as revealed by X-ray crystallography harbors a cleaved thymine dinucleotide as the product of CPD repair by radiolytically produced electrons.[103] Consequently, the distances between the DNA substrate and the flavin chromophore obtained by the above-mentioned biophysical studies can only be compared to those from the crystal structure if one assumes that the binding of the photolyase enzyme to a CPD and the binding to its repaired product are identical.

8.5. Mechanism of Photoreactivation

The photoreactivation of UV-damaged DNA which is catalyzed by photolyases and the role of the FAD and second chromophore in this process have been studied quite extensively. The commonly accepted model, for the individual steps of the enzymatic reaction in the CPD-repairing protein, is summarized in Figure 4. Nevertheless, it should be kept in mind that important details in the catalytic pathway of DNA repair have not yet been experimentally established and therefore deserve further examination.[24]

8.5.1. Energy Transfer from the Light-Harvesting Chromophore to the Flavin Cofactor

The flavin cofactor plays the essential role in catalysis. Even though the enzyme can be stabilized with the FAD in all three different redox states (see Figure 2), only the fully reduced $FADH^-$ is catalytically active both *in vitro*[93,123,124] and *in vivo*.[125] DNA repair does not require the presence of the second chromophore.[126,127] However, it enhances the DNA-repair efficiency by gathering light in the regions of near-UV and visible wavelengths where $FADH^-$ absorbs only weakly. The excitation energy of the second chromophore is then transferred to $FADH^-$ to yield its photoexcited singlet state, $*FADH^-$.[96,116,126,128] Both excitations are singlet π–π^* transitions from the highest occupied molecular orbital (HOMO) to the lowest unoccupied molecular orbital (LUMO). The mechanism of energy transfer between the cofactors is of Förster type via long-range dipole–dipole interaction from either MTHF or 8-HDF to $FADH^-$. Under physiological conditions, the excitation energy is transferred over a distance of 16.8 Å from the excited *MTHF to $FADH^-$ in *E. coli* DNA photolyase within about 290 ps, the energy-transfer efficiency being only 55%.[129]

Recent resonance Raman[120] and Fourier transform infrared data[121,130] suggest that the glutamine residue E109 stabilizes the ground state of the antenna cofactor but stabilizes the excited state to an even greater extent. The importance of E109 for MTHF binding has also been shown with the complete loss of MTHF upon replacement of this residue by other amino acids such as alanine or asparagine.

8.5.2. Electron Transfer from the Flavin Cofactor to the CPD Lesion

The excited singlet state of the fully reduced flavin, $*FADH^-$, transfers an electron to the CPD lesion, generating an $FADH^{\cdot}$ radical and a CPD anion radical. The latter subsequently undergoes cycloreversion. Evidence for photoinduced electron transfer in the catalytic pathway has been obtained by Okamura and co-workers[131] by observation of an intermediate radical in a picosecond laser-flash photolysis experiment on MTHF-deficient *E. coli* DNA photolyase. The broad absorption in the 500–900 nm region, that has been assigned to the first excited singlet state of $FADH^-$, decays much faster in the presence of U◇U-containing substrate than it does without substrate (160 ps versus 1.4 ns). This is consistent with previous reports by Jordan and Jorns who showed selective $FADH^-$-fluorescence quenching in an enzyme-substrate complex.[132] From the $*FADH^-$ lifetime, the rate of forward electron transfer from $FADH^-$ to the CPD (in either a dimeric or a 15-mer oligothymidine substrate), $k_{FET} = 5.5 \times 10^9\ s^{-1}$, and the quantum yield, $\Phi_{FET} = 0.88$, have been determined by using time-correlated single-photon counting.[133] Experiments on the *A. nidulans* enzyme have yielded $k_{FET} = 6.5 \times 10^9\ s^{-1}$ and $\Phi_{FET} = 0.92$.[134] Subsequent UV transient-absorption experiments at a probe wavelength of 265 nm with the enzyme bound to a pentameric oligothymidine substrate, however, revealed a considerably higher electron-transfer rate ($k_{FET} = 3 \times 10^{10}\ s^{-1}$)[135] which is consistent with the shorter than expected FAD-to-CPD distance observed recently by X-ray crystallography on the enzyme-substrate complex (see above).[103]

Okamura and co-workers[131] have investigated an MTHF-deficient *E. coli* DNA photolyase bound to a U◇U substrate with transient absorption spectroscopy. They observed a new band close to 400 nm that appeared with increasing delay time after the excitation pulse. It was suggested that this transient represents an essential intermediate in the reaction pathway leading to CPD splitting. However, owing to the lack of reference spectra, it was not possible to assign the transient to either a flavin radical or a U◇U radical. Subsequently, Kim and co-workers[136] re-examined these findings by using different photodimer substrates (namely U◇T and T◇T). The 400-nm-species previously observed with U◇U was also observed with U◇T (with peak-absorbance at 420 nm) but not with T◇T. The decay time of the transient was close to the spectrometer response time but was nevertheless estimated to be in the 0.5 to 2-ns-range. The lack of 400–420-nm-signal with T◇T was assumed to be due to an even shorter lifetime of this transient or due to a shift of the absorbance of the transient outside the spectral region probed. Based on their observations, Kim and co-workers concluded that the transient absorbance arises from the substrate rather than the flavin.

Langenbacher and co-workers[137] have used transient absorption spectroscopy with picosecond time resolution to study the lifetime of the first excited singlet state of $FADH^-$ at different temperatures and the acceleration of the $*FADH^-$ decay by electron transfer in various enzyme-substrate complexes. In MTHF-deficient *E. coli* photolyase without substrate, $*FADH^-$ decays with a time constant of about 1.8 ns at 275 K and 7.5 ns at 90 K. In the presence of substrate, the decay is more than 10 times faster (*e.g.*, 110 ps in an enzyme-T◇T-complex at 275 K) depending on the nature of the substrate. This substrate dependence is even more pronounced at low temperatures.

For example, at 90 K, $*FADH^-$ decays faster when photolyase is complexed with U◇U ((170 ± 80) ps) than with T◇T ((350 ± 60) ps). The temperature dependence of the quantum yield of dimer repair was analyzed under the assumption that back electron transfer is temperature-independent and that dimer splitting is a thermally activated process. An activation energy of (450 ± 200) meV for dimer splitting of U◇U was obtained. The whole photoreactivation process is considered to be "photon powered" as the activation energy is easily overcome at ambient temperatures.[11]

In a theoretical examination of electron transfer between $FADH^-$ and the CPD, Antony and co-workers[112] calculated the electron-transfer matrix element for the coupling between the LUMOs[§] of electron donor and acceptor in a computer-modeled enzyme-substrate complex. The root-mean-square transfer matrix element along the dynamic trajectory of a molecular dynamics simulation was found to be 6 cm^{-1} with a variance of 5 cm^{-1} for a T◇T substrate. The calculations predicted that despite the short distance of less than 3 Å assumed between $FADH^-$ and the CPD, the electron-transfer mechanism is not direct, but indirect with the adenine moiety of $FADH^-$ acting as an intermediate. The proposed superexchange-type mechanism utilizes the unusual conformation of the flavin cofactor specific for photolyases, in which the isoalloxazine ring and the adenine ring are in close proximity[84,85,87,103] leading to an overlap of their π-systems. The strong coupling of the donor and acceptor states with the same intermediate electronic states of the adenine causes an effective coupling between the CPD and $FADH^-$ without the necessity for direct orbital overlap. The pathway of electron transfer between these moieties has been calculated by Medvedev and Stuchebrukhov[138] using the method of interatomic tunneling currents.[138] Initially, the electron (probability) density is predicted to be delocalized mostly on the rings I and II (see Figure 2) of the FAD's isoalloxazine moiety. The main tunneling path leads via $C(1')$, $H(1')$ in the ribityl side chain to $N(1)$, $C(6)$, and 6-NH_2 on the adenine ring, and from there to the $C(4){=}O(4)$ carbonyl groups as the main acceptor sites on both thymine bases of the CPD.

Results from EPR and ENDOR experiments by which the extent of delocalization of the unpaired electron spin over the $FADH^{\cdot}$ cofactor's isoalloxazine moiety is probed support the superexchange model of photolyase-mediated CPD repair.[114,139] The singly occupied HOMO (= SOMO) of $FADH^{\cdot}$ constitutes the electron acceptor site in back electron transfer from either $T◇T^{\cdot -}$ or the repaired $TT^{\cdot -}$. Proton hyperfine couplings as determined by ENDOR spectroscopy provide an indirect probe of the electron-density distribution in this frontier orbital, as spin density at the protons usually arises from spin polarization and/or hyperconjugation with the main π-electron system situated on the carbons and nitrogens. Compared to other flavoenzymes that catalyze one-electron transfer reactions,[63] the unpaired electron-spin density in $FADH^{\cdot}$ of *E. coli* DNA photolyase is more confined to the rings I and II (see Figure 2) of the isoalloxazine moiety whereas the electron-spin density on

[§]The LUMO is taken as representative of the electronic distribution of the donor as an electron of the HOMO is promoted to the LUMO before being transferred.

the outer ring III is very low. This is manifested by the smallest detected isotropic proton hyperfine coupling of an 8-methyl group in a neutral flavin radical in flavoenzymes so far.[139]

The immediate protein environment plays an important role in fine-tuning the electronic structure of the flavin for highly optimized electron transfer.[114,140] The fact that the unpaired electron-spin density and also the electron density on the flavin's benzene ring are small, even though this part of the FAD's isoalloxazine moiety is in the direction of the CPD lesion, is a clear indication that the electron transfer between the isoalloxazine ring and the dimer is not direct. The large amplitudes of the SOMO on the central pyrazine and the pyrimidine rings increase orbital overlap between the π-systems of the isoalloxazine and the adenine moieties of FAD (see Figure 3). It was therefore assumed that the adenine bridges the gap between FAD and the CPD. Whether the electron transfer between the CPD and the adenine of FAD is direct or sequential through further aromatic residues such as tryptophan as intermediates is a question that deserves further experimental and theoretical studies.

8.5.3. Splitting of the CPD

Mechanistic details of enzymatic cyclobutane-ring opening in photodamaged DNA bound to photolyase are still rudimentary. The little information available is derived form either model studies, such as pulse-radiolysis experiments in the presence of photosensitizers (see, *e.g.*, refs. 12 and 141), or from quantum-chemical calculations at various levels of theory.[142–146] However, most of the computational studies were performed for CPDs in the gas phase, and their applicability to reactions in solution or with the enzyme is not clear. The consensus to date is that in a CPD radical anion, such as $T\diamond T^{\cdot -}$ or $U\diamond U^{\cdot -}$, the *C*(5)–*C*(5′) σ-bond breaks first with small activation energy, leading to the formation of an intermediate (see Figure 4).[142–145] This is followed by the cleavage of the *C*(6)–*C*(6′) σ-bond.

MacFarlane IV and Stanley have used sub-picosecond UV transient absorption spectroscopy to follow the repair kinetics of *A. nidulans* photolyase bound to either dimeric or pentameric oligothymidine substrate.[135] In contrast to previous flash-photolysis experiments they have monitored the UV spectral region at 265 nm because the breaking of the CPD's cyclobutane ring leads to the re-formation of the double bonds *C*(5)=*C*(6) and *C*(5′)=*C*(6′) of the adjacent thymines, which both strongly absorb in this range. The results suggest that the cyclobutane-ring opening proceeds sequentially rather than in a concerted fashion. The breaking of the first σ-bond was found to be almost activation-less and takes place within approximately 60 ps. A slower increase of the 265-nm absorption up to 1.5 ns was assigned to the re-formation of the second double bond. This second step has a significant energy barrier.

Interestingly, from their optical experiments MacFarlane IV and Stanley concluded that the flavin cofactor remains in the semiquinone state ($FADH^{\cdot}$)

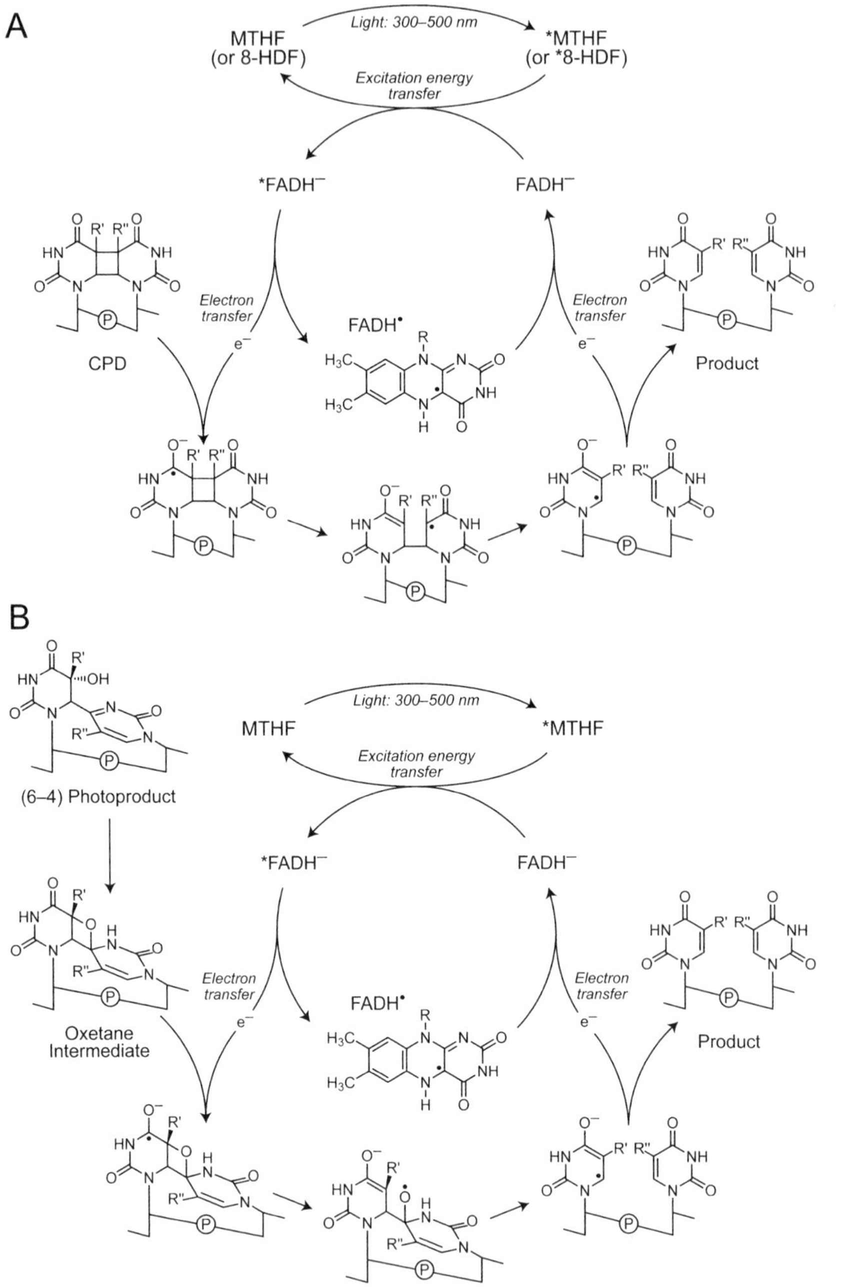

Figure 4. The putative reaction mechanism of CPDs by DNA photolyase (A) and (6-4)PPs by (6-4) photolyase (B). R is defined in Figure 2. R′ and R″ are either CH_3 in thymine or H in uracil

after repair rather than being re-reduced to $FADH^-$ by back electron transfer from $TT^{\cdot -}$.[135] On the other hand, no $FADH^{\cdot}$ buildup was observed in the absence of substrate. Hence, if these preliminary findings prove to be correct, then the excess electron on the thymine bases would have to be donated to electron acceptors different from $FADH^{\cdot}$, which are abundant in a cell. In this case, the $FADH^{\cdot} \rightarrow FADH^-$ reactivation would not occur within the repair process but would require an additional photoinduced electron-transfer process in order to re-reduce the flavin cofactor (see below) and allow further enzymatic activity. Accumulation of $FADH^{\cdot}$ in CPD repair has been reported previously, however, under high excitation intensities (>100 μJ mm^{-2}) regardless of whether substrate was present or not.[137] That $FADH^{\cdot}$ buildup was not observed in other repair studies was explained with the longer laser pulses utilized in previous experiments. If the excitation pulses were sufficiently long and intense so that two sequential electron-transfer processes (reductive T◇T-dimer cleavage and subsequent $FADH^{\cdot}$ photoreduction) are initiated with one single pulse, a photoreduction step leading to fully reduced $FADH^-$ could in principle immediately follow the repair reaction to complete the catalytic cycle. In this case, however, the overall quantum yield of CPD repair may not exceed 0.5. Hence, the results from time-resolved experiments as reported by MacFarlane IV and Stanley[135] are at odds with steady-state quantum-yield studies from other laboratories that gave repair quantum yields that range from 0.7 to 0.9.[93,96,147] Recent data on the *in vivo* and *in vitro* photoreactivation kinetics also strongly indicate that the photocycle does not generate $FADH^{\cdot}$ as an end product and that photolyase repairs CPDs in a cyclic electron-transfer reaction.[24,91]

Kim and Rose have observed that in isotropic solution reductive CPD splitting dramatically depends on the dielectric constant of the solvent.[148,149] In non-polar solvents the cleavage reaction is much more efficient than in polar solvents, reaching a maximum for the least polar solvent investigated. It was suggested that this observation may be rationalized by the fact that back electron transfer from the dimer to the photosensitizer is slowed in non-polar solvents as it pushes the reaction into the Marcus-inverted region, thus increasing the efficiency of the cleavage reaction. Likewise, in enzymatic DNA photorepair it may be that photolyases use their low-polarity active site to slow down back electron transfer from $T◇T^{\cdot -}$ to $FADH^{\cdot}$ and thereby to enhance the competitive splitting reaction.

8.5.4. Photoreactivation of (6-4)PP

The mechanism of photorepair of (6-4)PPs by the enzyme (6-4) photolyase still remains to be clarified, and at the time of writing no three-dimensional structure is available. In the past, spectroscopic studies on the enzyme were often hampered by the limited quantities of enzyme available. Furthermore, enzyme preparations frequently contained sub-stoichiometric amounts of the two chromophores and were, therefore, not very well defined.[76] For these reasons, many conclusions on the catalytic cycle of (6-4)PP photoreactivation have been drawn based only on the protein sequence homology between (6-4) photolyase and DNA photolyase and therefore remain preliminary.[26]

Although the (6-4) photolyase enzyme binds to both DNA containing only (6-4)PP and DNA containing the Dewar isomer, its repair activity on the latter form is very low.[76,82] The enzyme's efficiency in repairing *cis-syn* CPDs is also negligible.[82,150] Even the overall quantum yield of repair of (6-4)PPs, $\Phi = 0.11$, is significantly lower under *in vitro* conditions than that of CPD repair by DNA photolyase under comparable experimental conditions ($\Phi = 0.42$ for *E. coli* DNA photolyase).[74] Nevertheless, the enzyme faithfully restores various (6-4)PPs, namely T-64-T and T-64-C, to the original bases despite the difference in the functional group at the *C*(4) of the 3′-base.[74,151] These results suggest that the transfer of the amino or hydroxyl group of the (6-4)PP is an intramolecular reaction within the substrate.

The absolute action spectrum of enzyme deficient of second chromophore and with the flavin cofactor in its fully reduced form closely matches the absorption spectrum of $FADH^-$ in the 350–600 nm region.[74,82] This suggests that the catalytically active enzyme contains the fully reduced flavin. This similarity to DNA photolyase together with the finding that the oxetane species that is known to be an intermediate in the formation of a (6-4)PP is unstable at temperatures above −80°C, led Kim and co-workers[76] to propose that the equilibrium between the (6-4)PP and oxetane is shifted toward the latter upon enzyme binding. Subsequent electron transfer from $FADH^-$ to the oxetane could thus lead to oxetane splitting by a mechanism analogous to that proposed for CPD photorepair (see Figure 3B) to give a neutral base and a base anion radical. The latter restores its electron to the $FADH^{\bullet}$, resulting in the fully repaired bases.

An implicit assumption in the mechanism depicted in Figure 4B is that the binding energy of the (6-4) photolyase enzyme to the (6-4)PP is sufficient to perturb the (6-4)PP $\rightleftharpoons$ oxetane equilibrium in favor of the latter. Interaction of critically placed amino-acid side chains might stabilize the oxetane intermediate relative to the (6-4)PP. Specifically, interactions between the highly conserved amino-acid residues E304, D397, and D399 of the *D. melanogaster* (6-4) photolyase have been proposed by Zhao and co-workers.[82] The respective carboxylates (E279, D372, and D374) of *E. coli* DNA photolyase are indeed found at the substrate-binding site.

Hitomi and co-workers[152] used a different approach to find the amino-acid residues that are involved in (6-4)PP repair. Four amino-acid residues that are specific to (6-4) photolyases, Q288, H354, L355, and H358, and two conserved tryptophans, W291 and W398 (equivalent to W277 and W384 in *E. coli* DNA photolyase), were substituted with alanine. Only the L355A mutant had a lower affinity for the substrate, which suggested a hydrophobic interaction with the (6-4)PP. Both the H354A and H358A mutants resulted in an almost complete loss of repair activity, while W291A and W398A mutants retained some activity. In a structural model of the DNA-binding domain of *X. laevis* (6-4) photolyase that was constructed using its similarity to *E. coli* photolyase, H354, L355, and H358 are in close contact to the (6-4)PP. It was therefore proposed

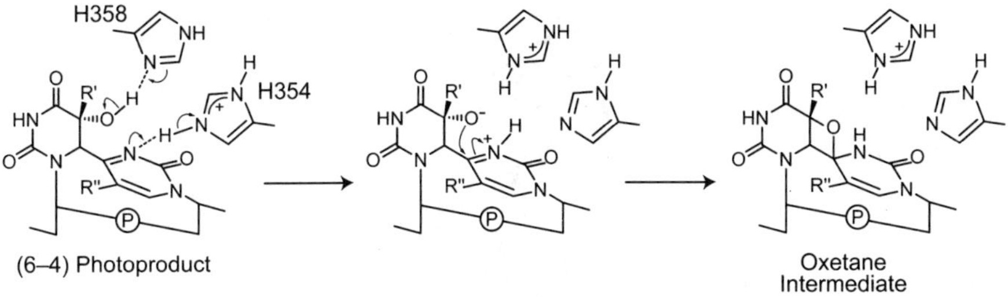

Figure 5. The proposed mechanism for oxetane intermediate formation in (6-4) photolyase.[152] It should be noted, however, that the chemical reactivity of each of the histidines has not yet been determined experimentally

that both H354 and H358 are crucial for the activity and catalyze the oxetane intermediate formation by a mechanism outlined in Figure 5. The observation of a marked pH dependence of the (6-4)PP photorepair and an isotope effect upon H → D exchange are in accord with such a mechanism.

8.6. Photoreduction of the Flavin Cofactor

A second photoreaction has also been observed in photolyases that is characteristic of inactive enzyme with the flavin cofactor in the neutral radical form, $FADH^{\cdot}$, or the fully oxidized redox state, FAD^{ox}. By absorption of light in the visible range, both $FADH^{\cdot}$ and FAD^{ox} undergo a photoreduction in the presence of reducing agents to yield the catalytically active $FADH^{-}$-form. This process has therefore been named photoactivation (not to confuse with photoreactivation of DNA).

8.6.1. DNA Photolyase Flavin Photoreduction

Photoactivation in photolyases has been studied quite extensively. Using nanosecond flash photolysis, an absorption peak at 510 nm at 4 μs after the excitation flash has been observed in DNA photolyase with the flavin initially in the $FADH^{\cdot}$-form. Based on this finding it was proposed that the immediate electron donor in flavin photoactivation was a tryptophan residue which, following its oxidation, was reduced by reducing agents in the medium.[153] By individual replacement of each one of the 15 tryptophan residues in *E. coli* DNA photolyase with phenylalanine, W306 was identified as the terminal electron donor.[154] Kim and co-workers[155] have reviewed the early work on the involvement of tryptophans in photoactivation and photoreactivation. After the three-dimensional structures of the *E. coli*[84] and *A. nidulans*[85] photolyases were solved, biophysical examinations of flavin-cofactor photoactivation have been intensified. W306 in *E. coli* photolyase is located near the protein surface, however, more than 14 Å away from the flavin. Such a large distance excludes a direct or superexchange-mediated electron transfer on a picosecond

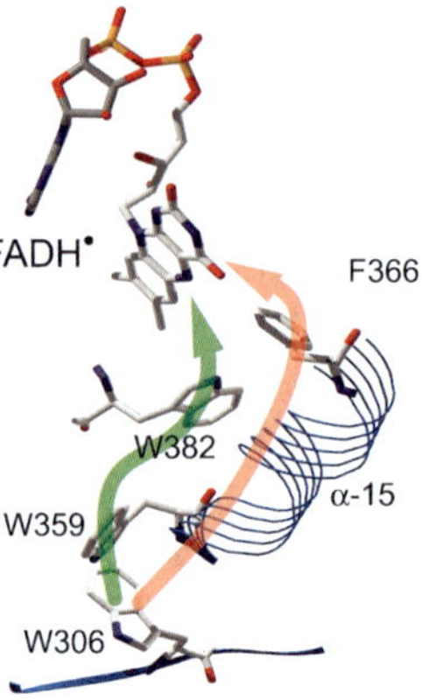

Figure 6. The proposed electron-transfer pathways from W306 to FADH˙ of flavin-cofactor photoreduction in *E. coli* DNA photolyase[84]

timescale.[156] Therefore, two hypothetic pathways for the electron transfer between W306 and FADH˙ have been proposed.[84] One is via two other tryptophans, W382 and W359, which are located between W306 and FADH˙ (see Figure 6). The other putative pathway involves the α-helix (α-15) between residues D358 and F366, and the phenyl ring of F366. Subsequent studies have mainly been focused on the W382 ← W359 ← W306-pathway, because these three residues are highly conserved throughout the photolyase family, including DNA photolyases of class I and class II as well as (6-4) photolyases (see below) and cryptochromes.

W382 has been identified as one primary electron donor in FADH˙ photoactivation of *E. coli* DNA photolyase.[157,158] The electron-transfer rate constant for the primary electron-transfer step is $k = 2.6 \times 10^{10}\ s^{-1}$, corresponding to a time constant of 38 ps. The two subsequent electron-transfer steps from W359 to $W382^{\cdot+}$ (to yield $W359^{\cdot+}$ and W382), and from W306 to $W359^{\cdot+}$ to yield W359 and the radical cation $W306^{\cdot+}$ could not be resolved but were estimated to be completed within less than 10 ns.[159] Within 300 ns, the surface-exposed $W306^{\cdot+}$ releases a proton to buffer molecules or water in the solvent rather than to a proton-accepting residue within the enzyme to yield a neutral W306˙ radical.

Recent steady-state photoreduction experiments with a W382F mutant, however, showed an increase in the photoreduction quantum yield by a factor of 2, which was interpreted to be inconsistent with the above model of exclusive electron transfer via W382.[24] Byrdin and co-workers[157] on the other hand, observed almost complete photoreduction inactivity for the W382F mutant. Using femtosecond transient absorption spectroscopy with a pump wavelength of 580 nm (to exclusively excite the FADH˙ state of the flavin), a bi-exponential decay has been observed at the probe wavelength of 690 nm, revealing two time constants for the decay of the *FADH˙ absorbance, 11 and 42 ps, with amplitudes of 36 and 64%, respectively.[129,140] Saxena and co-workers assigned the fast component to *FADH˙ decay due to electron transfer from W382, and the longer component as arising to photoreduction via F366. Both electron donors, W382 and F366, are found at an edge-to-edge distance of about 4 Å next to the flavin. Electron transfer is thought to be faster from W382 because of the favored redox potentials ($W/W^{\cdot +}$ versus $F/F^{\cdot +}$), although its distance to the flavin is slightly larger. The electron transfer steps following initial oxidation of F366 may either proceed via tunneling along the α-15 helix or hopping through the W382 ← W359 ← W306 triad again.[129,160] Clearly, further studies are necessary to remove the above-mentioned inconsistencies regarding the photoreduction pathway.

Flavin cofactor photoactivation has also been studied in *A. nidulans* photolyase.[161] In contrast to the *E. coli* enzyme, a tyrosine residue was observed as the final electron donor, reducing a tryptophan intermediate with a time constant of about 50 μs. In this enzyme, however, the photooxidized $W^{\cdot}$ is only partially (approximately 40%) reduced by a tyrosine, whereas the majority (approximately 60%) of $W^{\cdot}$ is re-reduced by back electron transfer from the flavin. It is not clear to date whether this non-uniform behavior is caused by sample heterogeneity or results from competing reactions where the reduction of $W^{\cdot +}$ by tyrosine is energetically much less favorable than back electron transfer from the flavin. The tyrosyl radical can also recombine with $FADH^{-}$ but on a much slower time scale than $W^{\cdot}$ (76 ms versus 1 ms).

Aubert and co-workers[162] detected $Y^{\cdot}$ in *A. nidulans* photolyase also by continuous-wave EPR spectroscopy with a 10-ms time resolution. From the *g*-value of the tyrosine resonance, $g \approx 2.005$, it was concluded that the tyrosine residue is deprotonated. To date, however, it is unclear which tyrosine residue participates in flavin photoactivation of *A. nidulans* photolyase. If one assumes that W314 is the terminal tryptophan electron donor (W314 corresponds to W306 in *E. coli* photolyase) then the tyrosine closest to this residue is Y468 at an edge-to-edge distance of about 8.6 Å. There is, however, a further tryptophan, W309, bridging the gap between W314 and Y468. Y468 is at the surface of the enzyme and might therefore mediate electron transfer to reducing agents in the solvent.[163]

Gindt and co-workers re-examined earlier EPR observations[155,164,165] of a transient spin-polarized radical-pair signal detected upon photoreduction of *E. coli* DNA photolyase.[166] By measurement of the transient EPR signal amplitude as a function of the excitation wavelength and depending on the amplitude of the stationary (dark) EPR signal of $FADH^{\cdot}$ in different enzyme preparations, they concluded that the spin-polarized EPR spectra observed previously did not originate from photoactivation initiated by flavin-semiquinone photochemistry, but rather by photoactivation starting from the fully oxidized redox state of the flavin cofactor. They proposed that electron abstraction from a nearby tryptophan residue occurs via the triplet state of FAD^{ox} generated by intersystem crossing from an excited singlet state of FAD^{ox}.

$$FAD^{ox}\ F366/W382\ W359\ W306 \xrightarrow{h\nu} {}^{*}FAD^{ox}\ F366/W382\ W359\ W306 \xrightarrow{ISC}$$
$${}^{3}FAD^{ox}\ F366/W382\ W359\ W306 \longrightarrow {}^{3}(FAD^{\cdot -}\ F366^{\cdot +}/W382^{\cdot +})\ W359\ W306$$
$$\rightarrow \ldots\ FADH^{\cdot}\ F366/W382\ W359\ W306^{\cdot}.$$

Evidence for the involvement of a flavin triplet state came from the disappearance of the spin-polarized EPR signal in the presence of potassium iodide, an effective triplet quencher for flavins.

Regardless of whether the reduction of the flavin cofactor starts out from $FADH^{\cdot}$ or FAD^{ox}, Kavakli and Sancar have recently concluded that photoreduction by intraprotein electron transfer is not part of each individual photolyase photocycle under physiological conditions as a W306F mutant was capable of carrying out at least 25 rounds of photorepair at the same rate as the wild type.[91] However, flavin photoreduction might still be involved in maintaining the enzyme's catalytic activity under oxidative conditions.

8.6.2. (6-4) Photolyase Flavin Photoreduction

The photoactivation of the flavin cofactor has recently been studied in *X. laevis* (6-4) photolyase by time-resolved EPR spectroscopy with a high time resolution in the 50-ns-range.[167] Isolation and purification typically renders the FAD cofactor in this enzyme fully oxidized. Upon photoexcitation with pulsed laser operating at 440 nm, a characteristic electron spin-polarization pattern of a radical-pair was observed after 100 ns. The assignment of the signal contributions of the individual radical-pair halves was accomplished from their characteristic resonance positions in the EPR spectrum, which is determined by their individual g-values. Precise measurements of the principal values of the **g**-matrix of $FADH^{\cdot}$ in *E. coli* DNA photolyase were obtained from high-frequency/high-field EPR/ENDOR at 94 GHz/3.4 T[139,168] and 360 GHz/12.8 T,[169] $g_X = 2.00429(2)$, $g_Y = 2.00359(2)$, and $g_Z = 2.00218(2)$, giving an average value, $g = 2.00336(5)$. Broad spectral wings in the transient EPR spectrum of the radical-pair species in *X. laevis* (6-4) photolyase centered at $g = 2.0034(3)$ have therefore been assigned the flavin-semiquinone cofactor in the neutral radical form, $FADH^{\cdot}$. A narrow emissive/absorptive polarized feature of the signal centered at $g = 2.0048(3)$ was assigned to a neutral (deprotonated) tyrosine radical, for which average g-values of $2.0045 \leq g \leq 2.0050$ are typically expected.[170] Further studies are required to identify the tyrosine residue in the amino-acid sequence that contributes to photoreduction in (6-4) photolyase. Intermediate radicals of other aromatic residues in (6-4) photolyase have not been detected. Their lifetimes must therefore either be less than 100 ns or an electron-transfer pathway different to that proposed in DNA photolyases must be in effect in (6-4) photolyases.

8.7. Concluding Remarks

To summarize, a wealth of information on photolyase-mediated DNA repair has been obtained in the last 20 years from the combined efforts of biologists, biochemists, molecular biologists, and biophysicists, both in experiment and theory. Nevertheless, many key aspects of this light-driven repair reaction remain hypothetic. The determination of a three-dimensional structure of a complex comprising functionally competent enzyme bound to a CPD-like lesion has finally given a framework with which to explore electron-transfer pathways between the flavin cofactor and the DNA lesion. On the other hand, a structure of the related (6-4) photolyase enzyme is still missing. This is important to compare both enzymes due to their different DNA-repair mechanisms. Other important questions that remain to be answered are: How does photolyase differ in recognizing CPDs and (6-4)PPs? What are the time constants and exact modes of splitting of CPDs and (6-4)PPs in the enzyme? Modern biophysical characterization methods should make substantial contributions to unraveling the remaining secrets of light-induced enzymatic DNA repair, and thus, are expected to fundamentally enhance our understanding of flavin-mediated photochemistry and photobiology.

Acknowledgments

This work was supported by the Deutsche Forschungsgemeinschaft (Sfb-498 (TP A2 and B7), and Sfb-533 (TP A5)) and the Volkswagen Stiftung (grant I/77100).

References

1. E.C. Friedberg, G.C. Walker and W. Siede, *DNA Repair and Mutagenesis*, ASM Press, Washington, DC, 1995.
2. G.M. Blackburn and R.J.H. Davies, Structure of DNA-derived thymine dimer, *Biochem. Biophys. Res. Commun.*, 1966, **22**, 704–706.
3. G.M. Blackburn and R.J.H. Davies, Structure of thymine photo-dimer, *Chem. Commun.*, 1965, 215–216.
4. C.S. Rupert, Photoreactivation of transforming DNA by an enzyme from baker's yeast, *J. Gen. Physiol.*, 1960, **43**, 573–595.
5. C.S. Rupert, Photoenzymatic repair of ultraviolet damage in DNA. I. Kinetics of the reaction, *J. Gen. Physiol.*, 1962, **45**, 703–724.
6. S. Minato and H. Werbin, Excitation and fluorescence spectra of the chromophore associated with the DNA-photoreactivating enzyme from the blue-green alga *Anacystis nidulans*, *Photochem. Photobiol.*, 1972, **15**, 97–100.
7. G.B. Sancar, DNA photolyases: physical properties, action mechanism, and roles in dark repair, *Mutat. Res.*, 1990, **236**, 147–160.
8. A. Sancar, *Photolyase: DNA repair by photoinduced electron transfer*, in: *Advances in Electron Transfer Chemistry*, P.E. Mariano (ed), JAI Press, London, 1992, 215–272.
9. P.F. Heelis, S.-T. Kim, T. Okamura and A. Sancar, The photo repair of pyrimidine dimers by DNA photolyase and model systems, *J. Photochem. Photobiol. B Biol.*, 1993, **17**, 219–228.
10. S.-T. Kim and A. Sancar, Photochemistry, photophysics, and mechanism of pyrimidine dimer repair by DNA photolyase, *Photochem. Photobiol.*, 1993, **57**, 895–904.

11. A. Sancar, Structure and function of DNA photolyase, *Biochemistry*, 1994, **33**, 2–9.
12. P.F. Heelis, R.F. Hartman and S.D. Rose, Photoenzymic repair of UV-damaged DNA: a chemist's perspective, *Chem. Soc. Rev.*, 1995, **24**, 289–297.
13. T. Carell, Sunlight-damaged DNA repaired with sunlight, *Angew. Chem., Int. Ed. Engl.*, 1995, **34**, 2491–2494.
14. P.F. Heelis, R.F. Hartman and S.D. Rose, Energy and electron transfer processes in flavoprotein-mediated DNA repair, *J. Photochem. Photobiol. A Chem.*, 1996, **95**, 89–98.
15. A. Sancar, No "End of History" for photolyases, *Science*, 1996, **272**, 48–49.
16. X. Zhao and D. Mu, (6-4) photolyase: light-dependent repair of DNA damage, *Histol. Histopathol.*, 1998, **13**, 1179–1182.
17. F. Thoma, Light and dark in chromatin repair: repair of UV-induced DNA lesions by photolyase and nucleotide excision repair, *EMBO J.*, 1999, **18**, 6585–6598.
18. T. Todo, Functional diversity of the DNA photolyase/blue light receptor family, *Mutat. Res.*, 1999, **434**, 89–97.
19. G.B. Sancar, Enzymatic photoreactivation: 50 years and counting, *Mutat. Res.*, 2000, **451**, 25–37.
20. J. Deisenhofer, DNA photolyases and cryptochromes, *Mutat. Res.*, 2000, **460**, 143–149.
21. T. Carell, L.T. Burgdorf, L.M. Kundu and M. Cichon, The mechanism of action of DNA photolyases, *Curr. Opinion Chem. Biol.*, 2001, **5**, 491–498.
22. R.P. Sinha and D.-P. Häder, UV-induced DNA damage and repair: a review, *Photochem. Photobiol. Sci.*, 2002, **1**, 225–236.
23. C.L. Thompson and A. Sancar, Photolyase/cryptochrome blue-light photoreceptors use photon energy to repair DNA and reset the circadian clock, *Oncogene*, 2002, **21**, 9043–9056.
24. A. Sancar, Structure and function of DNA photolyase and cryptochrome blue-light photoreceptors, *Chem. Rev.*, 2003, **103**, 2203–2237.
25. S. Weber, Light-driven enzymatic catalysis of DNA repair: a review of recent biophysical studies on photolyase, *Biochim. Biophys. Acta*, 2005, **1707**, 1–23.
26. T. Todo, H. Ryo, K. Yamamoto, H. Toh, T. Inui, H. Ayaki, T. Nomura and M. Ikenaga, Similarity among the *Drosophila* (6-4) photolyase, a human photolyase homolog, and the DNA photolyase–blue-light photoreceptor family, *Science*, 1996, **272**, 109–112.
27. D.S. Hsu, X. Zhao, S. Zhao, A. Kazantsev, R.-P. Wang, T. Todo, Y.-F. Wei and A. Sancar, Putative human blue-light photoreceptors hCRY1 and hCRY2 are flavoproteins, *Biochemistry*, 1996, **35**, 13871–13877.
28. P.J. van der Spek, K. Kobayashi, D. Bootsma, M. Takao, A.P.M. Eker and A. Yasui, Cloning, tissue expression, and mapping of a human photolyase homolog with similarity to plant blue-light receptors, *Genomics*, 1996, **37**, 177–182.
29. T. Todo, H. Tsuji, E. Otoshi, K. Hitomi, S.-T. Kim and M. Ikenaga, Characterization of a human homolog of (6-4) photolyase, *Mutat. Res.*, 1997, **384**, 195–204.
30. P. Emery, W.V. So, M. Kaneko, J.C. Hall and M. Rosbash, CRY, a *Drosophila* clock and light-regulated cryptochrome, is a major contributor to circadian rhythm resetting and photosensitivity, *Cell*, 1998, **95**, 669–679.
31. Y. Miyamoto and A. Sancar, Vitamin B_2-based blue-light photoreceptors in the retinohypothalamic tract as the photoactive pigments for setting the circadian clock in mammals, *Proc. Natl. Acad. Sci. USA*, 1998, **95**, 6097–6102.
32. P.F. Devlin and S.A. Kay, Cryptochromes – bringing the blues to circadian rhythms, *Trends Cell Biol.*, 1999, **9**, 295–298.

33. A. Sancar, Cryptochrome: the second photoactive pigment in the eye and its role in circadian photoreception, *Annu. Rev. Biochem.*, 2000, **69**, 31–67.
34. P.F. Devlin and S.A. Kay, Circadian photoperception, *Annu. Rev. Physiol.*, 2001, **63**, 677–694.
35. A.R. Cashmore, Cryptochromes: enabling plants and animals to determine circadian time, *Cell*, 2003, **114**, 537–543.
36. R.D. Wood, M. Mitchell, J. Sgouros and T. Lindahl, Human DNA repair genes, *Science*, 2001, **291**, 1284–1289.
37. W. Schul, J. Jans, Y.M.A. Rijksen, K.H.M. Klemann, A.P.M. Eker, J. de Wit, O. Nikaido, S. Nakajima, A. Yasui, J.H.J. Hoeijmakers and G.T.J. van der Horst, Enhanced repair of cyclobutane pyrimidine dimers and improved UV resistance in photolyase transgenic mice, *EMBO J.*, 2002, **21**, 4719–4729.
38. C.C.-K. Chao, Lack of DNA enzymatic photoreactivation in HeLa cell-free extracts, *FEBS Lett.*, 1993, **336**, 411–416.
39. Y.F. Li, S.-T. Kim and A. Sancar, Evidence for lack of DNA photoreactivating enzyme in humans, *Proc. Natl. Acad. Sci. USA*, 1993, **90**, 4389–4393.
40. R.D. Ley, Photoreactivation in humans, *Proc. Natl. Acad. Sci. USA*, 1993, **90**, 4337.
41. A. Yasui, A.P.M. Eker, S. Yasuhira, H. Yajima, T. Kobayashi, M. Takao and A. Oikawa, A new class of DNA photolyases present in various organisms including aplacental mammals, *EMBO J.*, 1994, **13**, 6143–6151.
42. S. Kanai, R. Kikuno, H. Toh, H. Ryo and T. Todo, Molecular evolution of the photolyase–blue-light photoreceptor family, *J. Mol. Evol.*, 1997, **45**, 535–548.
43. K. Hitomi, K. Okamoto, H. Daiyasu, H. Miyashita, S. Iwai, H. Toh, M. Ishiura and T. Todo, Bacterial cryptochrome and photolyase: characterization of two photolyase-like genes of *Synechocystis* sp. PCC6803, *Nucl. Acids Res.*, 2000, **28**, 2353–2362.
44. M. Ahmad, J.A. Jarillo, L.J. Klimczak, L.G. Landry, T. Peng, R.L. Last and A.R. Cashmore, An enzyme similar to animal type II photolyases mediates photoreactivation in *Arabidopsis*, *Plant Cell*, 1997, **9**, 199–207.
45. O. Kleiner, J. Butenandt, T. Carell and A. Batschauer, Class II DNA photolyase from *Arabidopsis thaliana* contains FAD as a cofactor, *Eur. J. Biochem.*, 1999, **264**, 161–167.
46. C.-Z. Jiang, J. Yee, D.L. Mitchell and A.B. Britt, Photorepair mutants of *Arabidopsis*, *Proc. Natl. Acad. Sci. USA*, 1997, **94**, 7441–7445.
47. S. Takahashi, N. Nakajima, H. Saji and N. Kondo, Diurnal change of cucumber CPD photolyase gene (*CsPHR*) expression and its physiological role in growth under UV-B irradiation, *Plant Cell Physiol.*, 2002, **43**, 342–349.
48. T. Hirouchi, S. Nakajima, T. Najrana, M. Tanaka, T. Matsunaga, J. Hidema, M. Teranishi, T. Fujino, T. Kumagai and K. Yamamoto, A gene for a class II DNA photolyase from *Oryza sativa*: cloning of the cDNA by dilution-amplification, *Mol. Genet. Genom.*, 2003, **269**, 508–516.
49. J.L. Petersen, D.W. Lang and G.D. Small, Cloning and characterization of a class II DNA photolyase from *Chlamydomonas*, *Plant Mol. Biol.*, 1999, **40**, 1063–1071.
50. S.-T. Kim, K. Malhotra, H. Ryo, A. Sancar and T. Todo, Purification and characterization of *Drosophila melanogaster* photolyase, *Mutat. Res.*, 1996, **363**, 97–104.
51. T. Kato, Jr., T. Todo, H. Ayaki, K. Ishizaki, T. Morita, S. Mitra and M. Ikenaga, Cloning of a marsupial DNA photolyase gene and the lack of related nucleotide sequences in placental mammals, *Nucl. Acids Res.*, 1994, **22**, 4119–4124.

52. V. Srinivasan, W.M. Schnitzlein and D.N. Tripathy, Fowlpox virus encodes a novel DNA repair enzyme, CPD-photolyase, that restores infectivity of UV light-damaged virus, *J.Virol.*, 2001, **75**, 1681–1688.
53. C.H. Slamovits and P.J. Keeling, Class II photolyase in a microsporidian intracellular parasite, *J. Mol. Biol.*, 2004, **341**, 713–721.
54. N. Iwatsuki, C.O. Joe and H. Werbin, Evidence that deoxyribonucleic acid photolyase from baker's yeast is a flavoprotein, *Biochemistry*, 1980, **19**, 1172–1176.
55. A. Sancar and G.B. Sancar, *Escherichia coli* DNA photolyase is a flavoprotein, *J. Mol. Biol.*, 1984, **172**, 223–227.
56. M.S. Jorns, G.B. Sancar and A. Sancar, Identification of a neutral flavin radical and characterization of a second chromophore in *Escherichia coli* DNA photolyase, *Biochemistry*, 1984, **23**, 2673–2679.
57. A.P.M. Eker, P. Kooiman, J.K.C. Hessels and A. Yasui, DNA photoreactivating enzyme from the cyanobacterium *Anacystis nidulans*, *J. Biol. Chem.*, 1990, **265**, 8009–8015.
58. A.P.M. Eker, H. Yajima and A. Yasui, DNA photolyase from the fungus *Neurospora crassa*. Purification characterization and comparison with other photolyases, *Photochem. Photobiol.*, 1994, **60**, 125–133.
59. T. Todo, S.-T. Kim, K. Hitomi, E. Otoshi, T. Inui, H. Morioka, H. Kobayashi, E. Ohtsuka, H. Toh and M. Ikenaga, Flavin adenine dinucleotide as a chromophore of the *Xenopus* (6-4) photolyase, *Nucl. Acids Res.*, 1997, **25**, 764–768.
60. R. Kato, K. Hasegawa, Y. Hidaka, S. Kuramitsu and T. Hoshino, Characterization of a thermostable DNA photolyase from an extremely thermophilic bacterium, *Thermus thermophilus* HB27, *J. Bacteriol.*, 1997, **179**, 6499–6503.
61. V. Massey, Introduction: flavoprotein structure and mechanism, *FASEB J.*, 1995, **9**, 473–475.
62. M.W. Fraaije and A. Mattevi, Flavoenzymes: diverse catalysts with recurrent features, *Trends Biochem. Sci.*, 2000, **25**, 126–132.
63. C.W.M. Kay and S. Weber, *EPR of radical intermediates in flavoenzymes*, in: *Electron Paramagnetic Resonance*, B.C. Gilbert, M.J. Davies and D.M. Murphy (eds), Royal Society of Chemistry, Cambridge, UK, 2002, pp. 222–253.
64. J.L. Johnson, S. Hamm-Alvarez, G. Payne, G.B. Sancar, K.V. Rajagopalan and A. Sancar, Identification of the second chromophore of *Escherichia coli* and yeast DNA photolyases as 5,10-methenyltetrahydrofolate, *Proc. Natl. Acad. Sci. USA*, 1988, **85**, 2046–2050.
65. Y.F. Li and A. Sancar, Cloning, sequencing, expression and characterization of DNA photolyase from *Salmonella typhimurium*, *Nucl. Acids Res.*, 1991, **19**, 4885–4890.
66. E.N. Worthington, I.H. Kavakli, G. Berrocal-Tito, B.E. Bondo and A. Sancar, Purification and characterization of three members of the photolyase/cryptochrome family blue-light photoreceptors from *Vibrio cholerae*, *J. Biol. Chem.*, 2003, **278**, 39143–39154.
67. G.B. Sancar, F.W. Smith and P.F. Heelis, Purification of the yeast *PHR1* photolyase from an *Escherichia coli* overproducing strain and characterization of the intrinsic chromophores of the enzyme, *J. Biol. Chem.*, 1987, **262**, 15457–15465.
68. H. Yajima, H. Inoue, A. Oikawa and A. Yasui, Cloning and functional characterization of a eucaryotic DNA photolyase gene from *Neurospora crassa*, *Nucl. Acids Res.*, 1991, **19**, 5359–5362.
69. A. Yasui, M. Takao, A. Oikawa, A. Kiener, C.T. Walsh and A.P.M. Eker, Cloning and characterization of a photolyase gene from the cyanobacterium *Anacystis nidulans*, *Nucl. Acids Res.*, 1988, **16**, 4447–4463.

70. A.P.M. Eker, R.H. Dekker and W. Berends, Photoreactivating enzyme from *Streptomyces griseus*. IV. On the nature of the chromophoric cofactor in *Streptomyces griseus* photoreactivating enzyme, *Photochem. Photobiol.*, 1981, **33**, 65–72.
71. S. Braatsch and G. Klug, ORF90, a gene required for photoreactivation in *Rhodobacter capsulatus* SB1003 encodes a cyclobutane pyrimidine dimer photolyase, *Photosynth. Res.*, 2004, **79**, 167–177.
72. A.P.M. Eker, J.K.C. Hessels and J. van de Velde, Photoreactivating enzyme from the green alga. *Scenedesmus acutus*, Evidence for the presence of two different flavin chromophores, *Biochemistry*, 1988, **27**, 1758–1765.
73. T. Iwasa, S. Tokutomi and F. Tokunaga, Photoreactivation of *Halobacterium halobium*: action spectrum and role of pigmentation, *Photochem. Photobiol.*, 1988, **47**, 267–270.
74. K. Hitomi, S.-T. Kim, S. Iwai, N. Harima, E. Otoshi, M. Ikenaga and T. Todo, Binding and catalytic properties of *Xenopus* (6-4) photolyase, *J. Biol. Chem.*, 1997, **272**, 32591–32598.
75. T. Todo, H. Takemori, H. Ryo, M. Ihara, T. Matsunaga, O. Nikaido, K. Sato and T. Nomura, A new photoreactivating enzyme that specifically repairs ultraviolet light-induced (6-4) photoproducts, *Nature*, 1993, **361**, 371–374.
76. S.-T. Kim, K. Malhotra, C.A. Smith, J.-S. Taylor and A. Sancar, Characterization of (6-4) photoproduct DNA photolyase, *J. Biol. Chem.*, 1994, **269**, 8535–8540.
77. S.-T. Kim, K. Malhotra, J.-S. Taylor and A. Sancar, Purification and partial characterization of (6-4) photoproduct DNA photolyase from *Xenopus laevis*, *Photochem. Photobiol.*, 1996, **63**, 292–295.
78. Y. Kobayashi, T. Ishikawa, J. Hirayama, H. Daiyasu, S. Kanai, H. Toh, I. Fukuda, T. Tsujimura, N. Terada, Y. Kamei, S. Yuba, S. Iwai and T. Todo, Molecular analysis of zebrafish photolyase/cryptochrome family: two types of cryptochromes present in zebrafish, *Genes Cells*, 2000, **5**, 725–738.
79. J.-J. Chen, D.L. Mitchell and A.B. Britt, A light-dependent pathway for the elimination of UV-induced pyrimidine (6-4) pyrimidinone photoproducts in *Arabidopsis*, *Plant Cell*, 1994, **6**, 1311–1317.
80. S. Nakajima, M. Sugiyama, S. Iwai, K. Hitomi, E. Otoshi, S.-T. Kim, C.-Z. Jiang, T. Todo, A.B. Britt and K. Yamamoto, Cloning and characterization of a gene (*UVR3*) required for photorepair of 6-4 photoproducts in *Arabidopsis thaliana*, *Nucl. Acids Res.*, 1998, **26**, 638–644.
81. H.C. Schröder, A. Krasko, D. Gundacker, S.P. Leys, I.M. Müller and W.E.G. Müller, Molecular and functional analysis of the (6-4) photolyase from the hexactinellid *Aphrocallistes vastus*, *Biochim. Biophys. Acta*, 2003, **1651**, 41–49.
82. X. Zhao, J. Liu, D.S. Hsu, S. Zhao, J.-S. Taylor and A. Sancar, Reaction mechanism of (6-4) photolyase, *J. Biol. Chem.*, 1997, **272**, 32580–32590.
83. C. Walsh, Naturally occurring 5-deazaflavin coenzymes: biological redox roles, *Acc. Chem. Res.*, 1986, **19**, 216–221.
84. H.-W. Park, S.-T. Kim, A. Sancar and J. Deisenhofer, Crystal structure of DNA photolyase from *Escherichia coli*, *Science*, 1995, **268**, 1866–1872.
85. T. Tamada, K. Kitadokoro, Y. Higuchi, K. Inaka, A. Yasui, P.E. de Ruiter, A.P.M. Eker and K. Miki, Crystal structure of DNA photolyase from *Anacystis nidulans*, *Nat. Struct. Biol.*, 1997, **4**, 887–891.
86. R. Kort, H. Komori, S.-i. Adachi, K. Miki and A.P.M. Eker, DNA apophotolyase from *Anacystis nidulans*: 1.8 Å structure, 8-HDF reconstitution and X-ray-induced FAD reduction, *Acta Crystallogr., Sect. D: Biol. Crystallogr.*, 2004, **60**, 1205–1213.

87. H. Komori, R. Masui, S. Kuramitsu, S. Yokoyama, T. Shibata, Y. Inoue and K. Miki, Crystal structure of thermostable DNA photolyase: pyrimidine-dimer recognition mechanism, *Proc. Natl. Acad. Sci. USA*, 2001, **98**, 13560–13565.
88. R.J. Roberts and X. Cheng, Base flipping, *Annu. Rev. Biochem.*, 1998, **67**, 181–198.
89. A. Yasui and W. Laskowski, Determination of the number of photoreactivating enzyme molecules per haploid *Saccharomyces* cell, *Int. J. Radiat. Biol.*, 1975, **28**, 511–518.
90. A. Fukui, K. Hieda and Y. Matsudaira, Light-flash analysis of the photoenzymic repair process in yeast cells. I. Determination of the number of photoreactivating enzyme molecules, *Mutat. Res.*, 1978, **51**, 435–439.
91. I.H. Kavakli and A. Sancar, Analysis of the role of intraprotein electron transfer in photoreactivation by DNA photolyase *in vivo*, *Biochemistry*, 2004, **43**, 15103–15110.
92. G. Payne, M. Wills, C. Walsh and A. Sancar, Reconstitution of *Escherichia coli* photolyase with flavins and flavin analogues, *Biochemistry*, 1990, **29**, 5706–5711.
93. G.B. Sancar, M.S. Jorns, G. Payne, D.J. Fluke, C.S. Rupert and A. Sancar, Action mechanism of *Escherichia coli* DNA photolyase. III. Photolysis of the enzyme-substrate complex and the absolute action spectrum, *J. Biol. Chem.*, 1987, **262**, 492–498.
94. I. Husain, G.B. Sancar, S.R. Holbrook and A. Sancar, Mechanism of damage recognition by *Escherichia coli* DNA photolyase, *J. Biol. Chem.*, 1987, **262**, 13188–13197.
95. I. Husain and A. Sancar, Binding of *E. coli* DNA photolyase to a defined substrate containing a single T< >T dimer, *Nucl. Acids Res.*, 1987, **15**, 1109–1120.
96. G. Payne and A. Sancar, Absolute action spectrum of E-FADH$_2$ and E-FADH$_2$-MTHF forms of *Escherichia coli* DNA photolyase, *Biochemistry*, 1990, **29**, 7715–7727.
97. G.B. Sancar, F.W. Smith and A. Sancar, Binding of *Escherichia coli* DNA photolyase to UV-irradiated DNA, *Biochemistry*, 1985, **24**, 1849–1855.
98. G.B. Sancar, F.W. Smith, R. Reid, G. Payne, M. Levy and A. Sancar, Action mechanism of *Escherichia coli* DNA photolyase. I. Formation of the enzyme-substrate complex, *J. Biol. Chem.*, 1987, **262**, 478–485.
99. Y.F. Li and A. Sancar, Active site of *Escherichia coli* DNA photolyase: mutations at Trp277 alter the selectivity of the enzyme without affecting the quantum yield of photorepair, *Biochemistry*, 1990, **29**, 5698–5706.
100. M.E. Baer and G.B. Sancar, The role of conserved amino acids in substrate binding and discrimination by photolyase, *J. Biol. Chem.*, 1993, **268**, 16717–16724.
101. S.-T. Kim, K. Malhotra, C.A. Smith, J.-S. Taylor and A. Sancar, DNA photolyase repairs the *trans-syn* cyclobutane thymine dimer, *Biochemistry*, 1993, **32**, 7065–7068.
102. S.-T. Kim and A. Sancar, Effect of base, pentose, and phosphodiester backbone structures on binding and repair of pyrimidine dimers by *Escherichia coli* DNA photolyase, *Biochemistry*, 1991, **30**, 8623–8630.
103. A. Mees, T. Klar, P. Gnau, U. Hennecke, A.P.M. Eker, T. Carell and L.-O. Essen, Crystal structure of a photolyase bound to a CPD-like DNA lesion after in situ repair, *Science*, 2004, **306**, 1789–1793.
104. J. van Noort, F. Orsini, A.P.M. Eker, C. Wyman, B. de Grooth and J. Greve, DNA bending by photolyase in specific and non-specific complexes studied by atomic force microscopy, *Nucl. Acids Res.*, 1999, **27**, 3875–3880.

105. A. Kiener, I. Husain, A. Sancar and C. Walsh, Purification and properties of *Methanobacterium thermoautotrophicum* DNA photolyase, *J. Biol. Chem.*, 1989, **264**, 13880–13887.
106. M. Baer and G.B. Sancar, Photolyases from *Saccharomyces cerevisiae* and *Escherichia coli* recognize common binding determinants in DNA containing pyrimidine dimers, *Mol. Cell. Biol.*, 1989, **9**, 4777–4788.
107. T. Torizawa, T. Ueda, S. Kuramitsu, K. Hitomi, T. Todo, S. Iwai, K. Morikawa and I. Shimada, Investigation of the cyclobutane pyrimidine dimer (CPD) photolyase DNA recognition mechanism by NMR analyses, *J. Biol. Chem.*, 2004, **279**, 32950–32956.
108. B.J. Van de Berg and G.B. Sancar, Evidence for dinucleotide flipping by DNA photolyase, *J. Biol. Chem.*, 1998, **273**, 20276–20284.
109. K.S. Christine, A.W. MacFarlane IV, K. Yang and R.J. Stanley, Cyclobutylpyrimidine dimer base flipping by DNA photolyase, *J. Biol. Chem.*, 2002, **277**, 38339–38344.
110. D.B. Sanders and O. Wiest, A model for the enzyme–substrate complex of DNA photolyase and photodamaged DNA, *J. Am. Chem. Soc.*, 1999, **121**, 5127–5134.
111. J. Hahn, M.-E. Michel-Beyerle and N. Rösch, Binding of pyrimidine model dimers to the photolyase enzyme: a molecular dynamics study, *J. Phys. Chem. B*, 1999, **103**, 2001–2007.
112. J. Antony, D.M. Medvedev and A.A. Stuchebrukhov, Theoretical study of electron transfer between the photolyase catalytic cofactor $FADH^-$ and DNA thymine dimer, *J. Am. Chem. Soc.*, 2000, **122**, 1057–1065.
113. S. Weber, G. Richter, E. Schleicher, A. Bacher, K. Möbius and C.W.M. Kay, Substrate binding to DNA photolyase studied by electron paramagnetic resonance spectroscopy, *Biophys. J.*, 2001, **81**, 1195–1204.
114. S. Weber, K. Möbius, G. Richter and C.W.M. Kay, The electronic structure of the flavin cofactor in DNA photolyase, *J. Am. Chem. Soc.*, 2001, **123**, 3790–3798.
115. A.W. MacFarlane IV and R.J. Stanley, Evidence of powerful substrate electric fields in DNA photolyase: implications for thymidine dimer repair, *Biochemistry*, 2001, **40**, 15203–15214.
116. M.S. Jorns, B. Wang, S.P. Jordan and L.P. Chanderkar, Chromophore function and interaction in *Escherichia coli* DNA photolyase: reconstitution of the apoenzyme with pterin and/or flavin derivatives, *Biochemistry*, 1990, **29**, 552–561.
117. S.P. Jordan, J.L. Alderfer, L.P. Chanderkar and M.S. Jorns, Reaction of *Escherichia coli* and yeast photolyases with homogeneous short-chain oligonucleotide substrates, *Biochemistry*, 1989, **28**, 8149–8153.
118. Y.M. Gindt, J.P.M. Schelvis, K.L. Thoren and T.H. Huang, Substrate binding modulates the reduction potential of DNA photolyase, *J. Am. Chem. Soc.*, 2005, **127**, 10472–10473.
119. J.P.M. Schelvis, M. Ramsey, O. Sokolova, C. Tavares, C. Cecala, K. Connell, S. Wagner and Y.M. Gindt, Resonance Raman and UV–vis spectroscopic characterization of $FADH^{\cdot}$ in the complex of photolyase with UV-damaged DNA, *J. Phys. Chem. B*, 2003, **107**, 12352–12362.
120. D.H. Murgida, E. Schleicher, A. Bacher, G. Richter and P. Hildebrandt, Resonance Raman spectroscopic study of the neutral flavin radical complex of DNA photolyase from *Escherichia coli*, *J. Raman Spectrosc.*, 2001, **32**, 551–556.
121. A.A. Henry, R. Jimenez, D. Hanway and F.E. Romesberg, Preliminary characterization of light harvesting in *E. coli* DNA photolyase, *ChemBioChem*, 2004, **5**, 1088–1094.

122. T. Ueda, A. Kato, Y. Ogawa, T. Torizawa, S. Kuramitsu, S. Iwai, H. Terasawa and I. Shimada, NMR study of repair mechanism of DNA photolyase by FAD-induced paramagnetic relaxation enhancement, *J. Biol. Chem.*, 2004, **279**, 52574–52579.
123. P.F. Heelis and A. Sancar, Photochemical properties of *Escherichia coli* DNA photolyase: a flash photolysis study, *Biochemistry*, 1986, **25**, 8163–8166.
124. M.S. Jorns, E.T. Baldwin, G.B. Sancar and A. Sancar, Action mechanism of *Escherichia coli* DNA photolyase. II. Role of the chromophores in catalysis, *J. Biol. Chem.*, 1987, **262**, 486–491.
125. G. Payne, P.F. Heelis, B.R. Rohrs and A. Sancar, The active form of *Escherichia coli* DNA photolyase contains a fully reduced flavin and not a flavin radical, both in vivo and in vitro, *Biochemistry*, 1987, **26**, 7121–7127.
126. P.F. Heelis, G. Payne and A. Sancar, Photochemical properties of *Escherichia coli* DNA photolyase: selective photodecomposition of the second chromophore, *Biochemistry*, 1987, **26**, 4634–4640.
127. M.S. Jorns, B. Wang and S.P. Jordan, DNA repair catalyzed by *Escherichia coli* DNA photolyase containing only reduced flavin: elimination of the enzyme's second chromophore by reduction with sodium borohydride, *Biochemistry*, 1987, **26**, 6810–6816.
128. R.S.A. Lipman and M.S. Jorns, Direct evidence for singlet–singlet energy transfer in *Escherichia coli* DNA photolyase, *Biochemistry*, 1992, **31**, 786–791.
129. C. Saxena, A. Sancar and D. Zhong, Femtosecond dynamics of DNA photolyase: energy transfer of antenna initiation and electron transfer of cofactor reduction, *J. Phys. Chem. B*, 2004, **108**, 18026–18033.
130. E. Schleicher, B. Heßling, V. Illarionova, A. Bacher, S. Weber, G. Richter and K. Gerwert, Light-induced reactions of *Escherichia coli* DNA photolyase monitored by Fourier transform infrared spectroscopy, *FEBS J.*, 2005, **272**, 1855–1866.
131. T. Okamura, A. Sancar, P.F. Heelis, T.P. Begley, Y. Hirata and N. Mataga, Picosecond laser photolysis studies on the photorepair of pyrimidine dimers by DNA photolyase. 1. Laser photolysis of photolyase–2-deoxyuridine dinucleotide photodimer complex, *J. Am. Chem. Soc.*, 1991, **113**, 3143–3145.
132. S.P. Jordan and M.S. Jorns, Evidence for a singlet intermediate in catalysis by *Escherichia coli* DNA photolyase and evaluation of substrate binding determinants, *Biochemistry*, 1988, **27**, 8915–8923.
133. S.-T. Kim, P.F. Heelis, T. Okamura, Y. Hirata, N. Mataga and A. Sancar, Determination of rates and yields of interchromophore (folate → flavin) energy transfer and intermolecular (flavin → DNA) electron transfer in *Escherichia coli* photolyase by time-resolved fluorescence and absorption spectroscopy, *Biochemistry*, 1991, **30**, 11262–11270.
134. S.-T. Kim, P.F. Heelis and A. Sancar, Energy transfer (deazaflavin → $FADH_2$) and electron transfer ($FADH_2$ → T< >T) kinetics in *Anacystis nidulans* photolyase, *Biochemistry*, 1992, **31**, 11244–11248.
135. A.W. MacFarlane IV and R.J. Stanley, *Cis-syn* thymidine dimer repair by DNA photolyase in real time, *Biochemistry*, 2003, **42**, 8558–8568.
136. S.-T. Kim, M. Volk, G. Rousseau, P.F. Heelis, A. Sancar and M.E. Michel-Beyerle, Laser flash photolysis on an intermediate in the repair of pyrimidine dimers by DNA photolyase, *J. Am. Chem. Soc.*, 1994, **116**, 3115–3116.
137. T. Langenbacher, X. Zhao, G. Bieser, P.F. Heelis, A. Sancar and M.E. Michel-Beyerle, Substrate and temperature dependence of DNA photolyase repair activity examined with ultrafast spectroscopy, *J. Am. Chem. Soc.*, 1997, **119**, 10532–10536.

138. D. Medvedev and A.A. Stuchebrukhov, DNA repair mechanism by photolyase: electron transfer path from the photolyase catalytic cofactor $FADH^-$ to DNA thymine dimer, *J. Theor. Biol.*, 2001, **210**, 237–248.
139. C.W.M. Kay, R. Feicht, K. Schulz, P. Sadewater, A. Sancar, A. Bacher, K. Möbius, G. Richter and S. Weber, EPR, ENDOR and TRIPLE resonance spectroscopy on the neutral flavin radical in *Escherichia coli* DNA Photolyase, *Biochemistry*, 1999, **38**, 16740–16748.
140. H. Wang, C. Saxena, D. Quan, A. Sancar and D. Zhong, Femtosecond dynamics of flavin cofactor in DNA photolyase: radical reduction, local solvation, and charge recombination, *J. Phys. Chem. B*, 2005, **109**, 1329–1333.
141. M.P. Scannell, D.J. Fenick, S.-R. Yeh and D.E. Falvey, Model studies of DNA photorepair: reduction potentials of thymine and cytosine cyclobutane dimers measured by fluorescence quenching, *J. Am. Chem. Soc.*, 1997, **119**, 1971–1977.
142. A.A. Voityuk, M.-E. Michel-Beyerle and N. Rösch, A quantum chemical study of photoinduced DNA repair: on the splitting of pyrimidine model dimers initiated by electron transfer, *J. Am. Chem. Soc.*, 1996, **118**, 9750–9758.
143. A.A. Voityuk and N. Rösch, Ab initio study on the structure and splitting of the uracil dimer anion radical, *J. Phys. Chem. A*, 1997, **101**, 8335–8338.
144. B. Durbeej and L.A. Eriksson, Thermodynamics of the photoenzymic repair mechanism studied by density functional theory, *J. Am. Chem. Soc.*, 2000, **122**, 10126–10132.
145. N.J. Saettel and O. Wiest, DFT study of the [2+2] cycloreversion of uracil dimer anion radical: waters matter, *J. Am. Chem. Soc.*, 2001, **123**, 2693–2694.
146. M. Schmitz, P. Tavan and M. Nonella, Vibrational analysis of carbonyl modes in different stages of light-induced cyclopyrimidine dimer repair reactions, *Chem. Phys. Lett.*, 2001, **349**, 342–348.
147. A.J. Ramsey, J.L. Alderfer and M.S. Jorns, Energy transduction during catalysis by *Escherichia coli* DNA photolyase, *Biochemistry*, 1992, **31**, 7134–7142.
148. S.-T. Kim and S.D. Rose, Pyrimidine dimer splitting in covalently linked dimer–arylamine systems, *J. Photochem. Photobiol. B Biol.*, 1992, **12**, 179–191.
149. S.-T. Kim, R.F. Hartman and S.D. Rose, Solvent dependence of pyrimidine dimer splitting in a covalently linked dimer–indole system, *Photochem. Photobiol.*, 1990, **52**, 789–794.
150. E. Otoshi, T. Yagi, T. Mori, T. Matsunaga, O. Nikaido, S.-T. Kim, K. Hitomi, M. Ikenaga and T. Todo, Respective roles of cyclobutane pyrimidine dimmers, (6-4) photoproducts, and minor photoproducts in ultraviolet mutagenesis of repair-deficient *Xeroderma Pigmentosum* A cells, *Cancer Res.*, 2000, **60**, 1729–1735.
151. T. Mizukoshi, K. Hitomi, T. Todo and S. Iwai, Studies on the chemical synthesis of oligonucleotides containing the (6-4) photoproduct of thymine–cytosine and its repair by (6-4) photolyase, *J. Am. Chem. Soc.*, 1998, **120**, 10634–10642.
152. K. Hitomi, H. Nakamura, S.-T. Kim, T. Mizukoshi, T. Ishikawa, S. Iwai and T. Todo, Role of two histidines in the (6-4) photolyase reaction, *J. Biol. Chem.*, 2001, **276**, 10103–10109.
153. P.F. Heelis, T. Okamura and A. Sancar, Excited-state properties of *Escherichia coli* DNA photolyase in the picosecond to millisecond time scale, *Biochemistry*, 1990, **29**, 5694–5698.
154. Y.F. Li, P.F. Heelis and A. Sancar, Active site of DNA photolyase: tryptophan-306 is the intrinsic hydrogen atom donor essential for flavin radical photoreduction and DNA repair in vitro, *Biochemistry*, 1991, **30**, 6322–6329.

155. S.-T. Kim, P.F. Heelis and A. Sancar, Role of tryptophans in substrate binding and catalysis by DNA photolyase, *Methods Enzymol.*, 1995, **258**, 319–343.
156. C.C. Page, C.C. Moser, X. Chen and L. Dutton, Natural engineering principles of electron tunnelling in biological oxidation-reduction, *Nature*, 1999, **402**, 47–52.
157. M. Byrdin, A.P.M. Eker, M.H. Vos and K. Brettel, Dissection of the triple tryptophan electron transfer chain in *Escherichia coli* DNA photolyase: Trp382 is the primary donor in photoactivation, *Proc. Natl. Acad. Sci. USA*, 2003, **100**, 8676–8681.
158. M. Byrdin, V. Sartor, A.P.M. Eker, M.H. Vos, C. Aubert, K. Brettel and P. Mathis, Intraprotein electron transfer and proton dynamics during photoactivation of DNA photolyase from *E. coli*: review and new insights from an "inverse" deuterium isotope effect, *Biochim. Biophys. Acta*, 2004, **1655**, 64–70.
159. C. Aubert, M.H. Vos, P. Mathis, A.P.M. Eker and K. Brettel, Intraprotein radical transfer during photoactivation of DNA photolyase, *Nature*, 2000, **405**, 586–590.
160. C. Saxena, H. Wang, I.H. Kavakli, A. Sancar and D. Zhong, Ultrafast dynamics of resonance energy transfer in cryptochrome, *J. Am. Chem. Soc.*, 2005, **127**, 7984–7985.
161. C. Aubert, P. Mathis, A.P.M. Eker and K. Brettel, Intraprotein electron transfer between tyrosine and tryptophan in DNA photolyase from *Anacystis nidulans*, *Proc. Natl. Acad. Sci. USA*, 1999, **96**, 5423–5427.
162. C. Aubert, K. Brettel, P. Mathis, A.P.M. Eker and A. Boussac, EPR detection of the transient tyrosyl radical in DNA photolyase from *Anacystis nidulans*, *J. Am. Chem. Soc.*, 1999, **121**, 8659–8660.
163. D.M. Popovic, A. Zmiric, S.D. Zaric and E.-W. Knapp, Energetics of radical transfer in DNA photolyase, *J. Am. Chem. Soc.*, 2002, **124**, 3775–3782.
164. C. Essenmacher, S.-T. Kim, M. Atamian, G.T. Babcock and A. Sancar, Tryptophan radical formation in DNA photolyase: electron-spin polarization arising from photoexcitation of a doublet ground state, *J. Am. Chem. Soc.*, 1993, **115**, 1602–1603.
165. S.-T. Kim, A. Sancar, C. Essenmacher and G.T. Babcock, Time-resolved EPR studies with DNA photolyase: excited-state $FADH^0$ abstracts an electron from Trp-306 to generate $FADH^-$, the catalytically active form of the cofactor, *Proc. Natl. Acad. Sci. USA*, 1993, **90**, 8023–8027.
166. Y.M. Gindt, E. Vollenbroek, K. Westphal, H. Sackett, A. Sancar and G.T. Babcock, Origin of the transient electron paramagnetic resonance signals in DNA photolyase, *Biochemistry*, 1999, **38**, 3857–3866.
167. S. Weber, C.W.M. Kay, H. Mögling, K. Möbius, K. Hitomi and T. Todo, Photoactivation of the flavin cofactor in *Xenopus laevis* (6-4) photolyase: observation of a transient tyrosyl radical by time-resolved electron paramagnetic resonance, *Proc. Natl. Acad. Sci. USA*, 2002, **99**, 1319–1322.
168. C.W.M. Kay, R. Bittl, A. Bacher, G. Richter and S. Weber, Unambiguous determination of the g-matrix orientation in a neutral flavin radical by pulsed electron–nuclear double resonance at 94 GHz, *J. Am. Chem. Soc.*, 2005, **127**, 10780–10781.
169. M. Fuchs, E. Schleicher, A. Schnegg, C.W.M. Kay, J.T. Törring, R. Bittl, A. Bacher, G. Richter, K. Möbius and S. Weber, The g-tensor of the neutral flavin radical cofactor of DNA photolyase revealed by 360 GHz electron paramagnetic resonance spectroscopy, *J. Phys. Chem. B*, 2002, **106**, 8885–8890.
170. S. Un, X.-S. Tang and B.A. Diner, 245 GHz high-field EPR study of tyrosine-D^0 and tyrosine-Z^0 in mutants of photosystem II, *Biochemistry*, 1996, **35**, 679–684.

Chapter 9

Flavin-Based Photoreceptors in Plants

WINSLOW R. BRIGGS

Department of Plant Biology, Carnegie Institution of Washington 260 Panama St., Stanford, CA 94305, USA

Abstract

The past decade has seen enormous progress in the identification and characterization of plant blue-light receptor molecules. Three blue-light receptor families are currently recognized: the cryptochromes, the phototropins, and members of the ZTL (for the German Zeitlupe for "slow motion") or ADO (Adagio also for "slow motion") family. In the ZTL/ADO family, only one member has been demonstrated to function as a photoreceptor to date. All three families relay on flavins as their chromophores, although the cryptochromes may additionally use either a deazaflavin or a pterin as an antenna chromophore. This article presents the events leading up to the discovery of each photoreceptor family, some of the known roles for the photoreceptors, and where possible reviews their biochemical and photochemical properties. The LOV domain, the photoreceptor module in the phototropins and the ZTK/ADO family, is given special attention as a result of the extensive recent biophysical investigations elucidating its property. The chapter concludes with a review of LOV-domain-containing proteins in bacteria and fungi, and cryptochrome homologs in prokaryotes. The coming years should see a great expansion of our knowledge of the plant photoreceptors and the signal transduction pathways that are active, but should also elucidate what the photoreceptor homologs and chromophore modules are doing in bacteria and fungi.

9.1. Introduction

The study of the specific effects of blue light on plants has a long history. Likely the earliest report was a paper published in 1817 by Poggioli.[1] Using a prism to obtain the different regions of the visible spectrum, Poggioli found that violet light was more effective than red in supporting overall plant growth. He also reported that violet light was the most effective in inducing changes in the leaf position both in *Mimosa pudica* and in *Raphanus rusticana*, possibly solar tracking responses in both cases. Subsequent studies by Payer in 1842 identified blue light as the most effective color in inducing the phototropic curvature of growing stems of *Lepidium sativum*[2] and a year later studies by Zantedeschi showed the same approximate spectral sensitivity for phototropism of *Impatiens balsamina* and *Oxalis multiflora*.[3] In 1836, Daubeny[4] obtained evidence that blue light was most effective in inducing water loss from bean plants, foreshadowing studies over a century later that would demonstrate the important role of blue light in inducing stomatal opening.[5–8]

Studies by Böhm[9] in the mid-nineteenth century clearly demonstrated a role for blue light in inducing chloroplast movements that, we now know, minimize light absorption in potentially damaging high light levels (the chloroplast avoidance response) and subsequent studies by Famintzen[10] showed that dim blue light induced the opposite response, chloroplast movement to maximize light absorption (the chloroplast accumulation response). (The value of the avoidance response in preventing photodamage was only documented a century and a half later.[11]) Discovery of specific roles for blue light in inducing inhibition of stem growth, both slow[12] and rapid[13,14] and promotion of leaf growth,[15,16] all independent of phytochrome, required the technological advances of the late 20th century, and hence were only recently demonstrated. The detailed history of research on the effects of blue light on plants is covered elsewhere.[17] However, it is appropriate to point out the incredibly long period of time during which the physiology of blue-light responses was studied, often in truly elegant detail, in the absence of any knowledge on the nature of the photoreceptor(s) involved or the identity of their chromophores.

The development of molecular genetic techniques toward the end of the 20th century changed things dramatically. The first evidence for an authentic blue-light receptor, a cryptochrome, was published in 1993,[18] followed in 1998 by the first evidence for a second family of blue-light receptors, the phototropins.[19] Most recently there has appeared evidence that at least one member of the three-member ZTL (for the German Zeitlupe for "slow motion") or ADO (Adagio also for "slow motion" family of proteins may also serve as a blue-light receptor.[20] As will be discussed below, there are now three known cryptochromes (cry1, cry2, and cryDASH or cry3) in *Arabidopsis thaliana*, five in the fern *Adiantum capillus-veneris*, two in the moss *Physcomitrella patens*, and one in the unicellular biflagellate alga *Chlamydomonas reinhardtii*.[21] There is also a cryptochrome in the cyanobacterium *Synecchocystis* PCC6803[22] as well as a putative one in the bacterial pathogen *Vibrio cholerae*.[23] *A. thaliana*[24] and *A. capillus-veneris*[25,26] each contain two phototropins, phot1 and phot2; *P. patens* contains four phototropins,[27] and *C. reinhardtii* contains one.[28] Finally, *A. capillus-veneris*[29] and other higher ferns[30] contain a curious chimeric photoreceptor designated phytochrome 3 (phy3): its N-terminal third is a classic red/far-red-photoreversible phytochrome-binding domain whereas its C-terminal domain is a complete phototropin.

In the decades prior to 1993, paper after paper and speaker after speaker referred to the blue-light receptor implying that there was just one. Gressel[31] designated blue-light receptors from different organisms collectively as cryptochromes because (a) they were elusive (cryptic) and (b) they were prevalent in lower plants and fungi (cryptogams). Hence it was logical to refer to the first blue-light receptor characterized as cryptochrome (later cryptochrome 1 or cry1). What was originally an all-encompassing term is now restricted to the cryptochrome family only.

In the following sections we first summarize what is known about the roles of these three classes of photoreceptors and their subcellular distribution. We next discuss their initial characterization and molecular structure and then review what is known of the biophysical and biochemical consequences of their photoexcitation. The emphasis will be on the properties of the photoreceptors themselves and not the various signal-transduction pathways that activate. Finally we treat briefly the broad distribution of the chromophore module from the phototropins and the ZTL/ADO family now found in a wide range of microbial proteins.

Three recently published books[32–34] dealing partially or wholly with higher plant photoreceptors are in press as this chapter is being written. These contain in depth reviews of the cryptochromes, both plant[21,35–38] and animal,[39] phototropins,[40–48] and the ZTL/ADO putative family of photoreceptors.[49,50] Also pertinent are chapters on photoreceptors in the ascomycete *Neurospora crassa*[51,52] and the several classes of photoreceptors in lower plants.[53] These chapters and their extensive bibliographies, will provide a major resource for the interested reader. Two reviews published in 2005[54,55] are relevant to this chapter. See also chapters by A. Losi and J.T.M. Kennis in this volume.

9.2. Photoreceptor Roles

The three families blue-light receptors mediate a wide range of responses both in higher and lower plants. In some cases one member of a family may appear to be the unique photoreceptor for a particular response, but in others, as is the case with some phytochrome responses, they may function redundantly. Here we review briefly the processes that they mediate. Although phytochromes can also mediate responses to blue light through their minor absorption band in the blue and UV-A region of the spectrum, they will not be discussed in this chapter.

9.2.1. Cryptochrome roles

The cryptochromes (or at least sequences of putative cryptochromes) have been found in both photosynthetic[22] and non-photosynthetic[23] eubacteria, the motile green alga *C. reinhardtii*, mosses, ferns, and higher plants.[21] In the latter they mediate inhibition of stem growth, entrainment of circadian rhythms, daylength detection in photoperiodism, and, likely underlying these responses, regulation of gene expression (see Batschauer[21]). In ferns, they play a role in modulating cell division in gametophytes and spore germination, but the absence of mutants and lack of a transformation system has hindered efforts to elucidate their roles more precisely.[53] In the moss *P. patens*, where homologous recombination is possible, studies with knockout transformants indicate a cryptochrome role in controlling branching patterns and gametophore induction and growth.[53] Brudler *et al.*[56] present evidence suggesting that the cryptochrome in *Synecchocystis* PCC6803 may function as a represser of transcription. Eight genes showed over two-fold enhancement in a knockout mutant compared to the wild type. The role of the cryptochrome in *C. reinhardtii* is unknown (see[53]) as is its role in the pathogen *V. cholerae*.[23]

9.2.2. Phototropin roles

In *A. thaliana*, the two phototropins mediate phototropism, blue light-induced chloroplast movement, blue light-induced stomatal opening, blue light-induced leaf expansion, and likely solar tracking, and one of them (phot1) mediates rapid inhibition of stem growth. In the case of phototropism, phot1 is at least an order of magnitude more sensitive than phot2, but both are equally effective at the higher fluence rates. Whereas both phot1 and phot2 can mediate the chloroplast accumulation response, only phot2 mediates the avoidance response (see Christie and Briggs[24,40] for details).

Most recently, Takemiya *et al.*[57] have shown that the phototropins are essential for growth of *A. thaliana* seedlings in very dim light. Extremely low fluence rates of blue light against a strong red light background gave significant increases in growth over control seedlings with red light alone. This blue-light effect was completely lost in a *phot1* mutant. When the blue-light intensity was increased about 100-fold, the difference disappeared. However, even the higher fluence rate failed to affect growth in a *phot1 phot2* double mutant. As with phototropism, the two phototropins showed significantly different sensitivities in their support of growth. The relative contributions of the effects of blue light through phototropins in maximizing chloroplast light exposure (accumulation response), regulating stomatal opening, and inducing leaf expansion are unknown, but must be substantial in each case. In contrast to the cryptochromes, the role of the phototropins in regulation of gene expression is at most relatively trivial,[58] and in most cases indirect – for example through alterations in auxin distribution in phototropism and therefore alterations in auxin regulation of gene expression. (Blakeslee *et al.*[59] recently reported that phot1 is required for the delocalization of PIN1 protein, a protein involved in auxin transport, by blue light.) Folta and Kaufman report one specific case in which phot1 is required for mRNA destabilization induced by high fluence-rate blue light.[60]

In the fern *A. capillus-veneris*, there are also two phototropins, and as in higher plants, they mediate both phototropism and the chloroplast movements. As in higher plants, only phot2 mediates the chloroplast avoidance response.[26] In the moss *P. patens*, the situation is more complicated, with four phototropins. To date, the roles of these phototropins have only been investigated with respect to chloroplast movement. The situation is complicated in that although a certain phototropin may mediate the avoidance response, it may mediate it only in basal cells but not tip cells of the filamentous protonemal stage.[27]

There is only a single phototropin known in *C. reinhardtii*[28,61] and its role is startlingly different in this biflagellate unicellular green alga.[62] RNAi-transformants are impaired in blue light-induced gamete maturation, blue light-induced functional recovery of dark-adapted gametes, and zygote germination. In this organism, the primary consequences of phototropin activation clearly involve transcriptional changes, in striking contrast to the consequences of phototropin activation in higher plants. However, when *C. reinhardtii* phototropin is transformed into a *phot1 phot2* double mutant of *A. thaliana*, it complements phototropism, blue light-induced leaf expansion, both chloroplast accumulation under dim blue light and avoidance under high blue light, and blue light-induced stomatal opening.[63] This is a striking example of how two very similar proteins (chromophore-binding domains 66% identical to those of rice phot1, similar overall structure, see below) can have dramatically different functions in different biological contexts.

9.2.3. Phy3 role

The chimeric phy3 found only in higher ferns is a particularly intriguing photoreceptor. As mentioned above, when produced heterologously in yeast it shows typical phytochrome red/far-red photoreversibility.[29] Mutant studies show clearly that it mediates red light-activated phototropism[30] and red light-activated chloroplast movement[27] in *A. capillus-veneris*. At present, in the absence of suitable phototropin mutants in this fern, it is not possible to determine whether it mediates blue light-activated phototropism redundantly with the two known phototropins, although this seems likely possible.

9.2.4. ZTL/ADO family roles

There are three members of the ZTL/ADO family of chromoproteins in *A. thaliana*, ZTL [64] or ADO[65], LKP2 (LOV, kelch protein 2),[66] and FKF1 (flavin-binding, kelch repeat, F-box[67]). (see below for definition of LOV domains, F-boxes, and kelch domains). Plants with a mutation at the ZTL/ADO locus show a very much lengthened period for their circadian rhythm[64,65] and an altered response of their periodicity to changes in light intensity, both red and blue light.[64] When LKP2 is overexpressed, day-length control of flowering is lost and the plants may be arhythmic.[66] Deletion mutants at the *FKF1* locus are late flowering, and have an abnormal waveform for expression of those genes whose transcription is regulated by the circadian clock.[67] As we shall see below, only one of these chromoproteins has been elevated to the status of photoreceptor.[20] The jury is still out on the exact functions of ZTL/ADO and LKP2.

9.3. Initial Discovery and Characterization

9.3.1. Cryptochromes

In a classic paper published in 1980, Koornneef *et al.*[68] described a series of mutants of *A. thaliana* that failed to show normal light-mediated inhibition of hypocotyl elongation. Some of these (*hy1–3, hy5*) were insensitive to red light and were subsequently found to involve the phytochromes or elements in their signal-transduction pathways or their chromophore biosynthesis. However, the *hy4* mutant showed normal hypocotyl sensitivity to red light, but failed to show inhibition by blue light. In 1993, Ahmad *et al.*[18] characterized the gene mutated at the *HY4* locus as encoding a protein with high homology to prokaryotic DNA photolyases. These photoactivated DNA repair enzymes containing two chromophores: an antenna chromophore that is either 8-hydroxy-5-deazariboflavin (8-HDF) or the pterin methenyltetrahydrofolate (MTHF), which passes its excitation energy to flavin adenine dinucleotide (FAD). The FAD mediates the repair of cyclopyrimidine dimers (CPDs) through an electron-exchange mechanism characterized in detail by Sancar.[69] Absorption of light by the FAD itself can mediate the DNA repair reaction, but far less efficiently than if it receives its excitation energy from the antenna chromophore. On the basis of the sequence homology of the putative protein with the photolyases, Ahmad *et al.*[18] proposed that the HY4 protein was a blue-light receptor.

Lin *et al.*[70] then expressed the *HY4* gene in an insect-cell system and obtained a protein that bound FAD. As mentioned above, they named the HY4 protein cryptochrome.[70] They showed that unlike photolyase, photoreduction of cryptochrome produced a spectral form that was likely a semiquinone. They speculated that a stable semiquinone form of the chromophore could account for the low level of green light-induced inhibition of hypocotyl elongation that they found. They failed to detect a second chromophore and speculated on the basis of amino acid sequence that a second chromophore, if present, might be a deazaflavin. However, at about the same time, Malhotra *et al.*[71] succeeded in producing both HY4 protein from *A. thaliana* and a highly homologous protein from *Sinapis alba* that bound both FAD and MTHF. A plasmid expressing *cry1* failed to complement an *Escherichia coli* mutant strain that lacked its endogenous DNA photolyase[71] and in neither case did heterologously produced cry1 show DNA photolyase activity.[70,71] Overexpression of cry1 in tobacco caused hypersensitivity of the hypocotyls to blue, UV-A, and green light[72] and the authors again suggested that the FAD might exist in the semiquinone form to account for the induced green sensitivity.

The protein encoded by the HY4 gene differs from all known photolyases in that it has a lengthy C-terminal extension. The importance of the C-terminal extension was highlighted by a subsequent study by Ahmad *et al.*[73] showing that many of the *hy4* mutant alleles had lesions in the C-terminal extension. We will return to this C-terminal extension below.

In 1996, Lin *et al.* reported the gene sequence and putative amino-acid sequence for a second cryptochrome, designated cryptochrome 2 or cry2.[74] It also had a C-terminal extension although with only low homology to the C-terminal extension of cry1. In a subsequent study, Lin *et al.*[75] showed definitively that over-expression of cry2 in *A. thaliana* enhanced the blue-light sensitivity of hypocotyl inhibition. They also showed that the cry2 protein itself was down-regulated by blue light, although its mRNA synthesis remained unchanged. Blue light somehow destabilized the photoreceptor protein.

Two studies[76,77] showed that *A. thaliana* cry2 is constitutively localized to the nucleus. The nuclear-localization signal is located in the C-terminus, which by itself is sufficient for nuclear localization. Indeed Cutler *et al.*[78] found that cry2 was closely associated with the chromosomes in dividing cells of etiolated *A. thaliana*. Blue-light irradiation led to the formation of nuclear speckles that co-localized with phytochrome B (phyB) in *A. thaliana*,[79] suggesting a direct physical interaction between these two photoreceptors. By contrast, cry1 was localized to the nucleus in darkness but moved out into the cytoplasm in blue light. Since the non-chromophoric C-terminus itself shows this behavior,[80] cry1 itself is clearly not the photoreceptor for the response and the responsible photoreceptor remains unidentified. Most recently, Kleine *et al.*[81] reported that a third cryptochrome, cry3, previously identified on the basis of sequence by Brudler *et al.*,[54] was targeted both to chloroplasts and mitochondria (cry-DASH, see below).

The fern *A. capillus-veneris* was shown to contain five cryptochrome genes,[82,83] by far the largest number reported for any species so far. Imaizumi *et al.*[83] determined their intracellular distribution using GUS constructs in darkness or under red or blue light. The results indicated some nuclear AcCry1 and 2 in the nucleus in darkness, and its exclusion from the nucleus under both red and blue light. Blue light somewhat increased the nuclear fraction for AcCry3, but the distribution AcCry4, with both cytoplasmic and nuclear fractions, was light independent. AcCry5 was essentially cytoplasmic and like AcCry4, was not affected by light. AcCry1 and 2, with limited nuclear distribution have somewhat truncated C-termini, and AcCry5 completely lacks a C-terminus. Given the importance of the C-terminus in targeting the *A. thaliana* Cry1 and Cry2 to the nucleus[80] this distribution pattern is perhaps not surprising.

Hitomi *et al.*[22] have described two genes encoding proteins in the cryptochrome/photolyase superfamily in *Synecchocystis* PCC6803. One of these, sll0854, clearly serves as a true photolyase as a knockout mutant shows severe UV-light photosensitivity. However, the other, sll1629, has higher homology with the cryptochromes and a knockout mutant shows wild-type photosensitivity to UV light. The authors tentatively designate it a cryptochrome of unknown function. Subsequently Ng and Pakrasi[84] investigated what were likely the same two genes from *Synecchocystis* PCC6803 and again showed experimentally that one provided strong UV protection while the contribution of the other was trivial. They concluded that the latter could be a 6-4 photolyase or a cryptochrome. Like one of the cryptochromes from the fern *A. capillus-veneris,* it lacked the C-terminal extension found in the higher plant cry1 and cry2.

Most recently Brudler *et al.*[56] crystallized the cryptochrome from *Synechocystis* PCC6803 to 1.9 Å resolution and found its structure similar to that of the CPD photolyases in DNA binding. However, consistent with earlier results indicating a lack of photolyase activity,[22,84] structural analysis indicated that important features essential for photolyase function are missing from the cyanobacterial cryptochrome. Phylogenetic analysis led Brudler *et al.*[56] to establish a new subfamily of the cryptochrome/photolyase group in which the cyanobacterial photolyase was most closely related to a putative cryptochrome from *A. thaliana* (cry3, see below) and animal cryptochromes and (6–4) photolyases. On the basis of a photolyase from **D***rosophila*, the putative cryptochrome from **A***rabidopsis*, the cryptochrome from **S***ynecchocystis* PCC6803, and a cryptochrome from **H***omo sapiens*, they named the subfamily cryptochrome **DASH** (cryDASH).

Worthington and colleagues[23] have characterized one photolyase, VcPhr, and two putative cryptochromes, VcCry1 and VcCry2, from *V. cholorae*. The purified cryptochromes heterologously produced in *E. coli* had no photolyase activity and insertional mutation of either of them did not affect UV sensitivity. However, insertional mutation of VcPhr completely abolished photoprotection against UV light, a result confirmed by *in vitro* assay of photolyase activity in extracts from cells with either wild-type or mutant VcPhr. All three proteins bound both FAD and MTHF and, in the case of VcCry1, in stoichiometric amounts. The authors were unable to assign any specific photoreceptor role to either of the putative cryptochromes.

Kleine *et al.*[81] expressed the cryDASH from *A. thaliana* in *E. coli* and tested it for photolyase activity. It neither complemented the *E. coli* photolyase mutant nor showed photolyase activity *in vitro*. In agreement with the analysis of Brudler *et al.*,[56] they concluded that it was more closely related to the *Synecchocystis* cryptochrome (50% identity over a stretch of 400 amino acids) than to cry1 and cry2 (less than 20%). The cryDASH in *A. thaliana* differs substantially from the other two cryptochromes in that the initial translated protein has a targeting sequence for import into chloroplasts and mitochondria.[81] Organelle import studies indicated that it was indeed targeted to these two organelles. In addition, *A. thaliana* cryDASH lacks the C-terminal extension present in cry1 and cry2. The protein heterologously produced in *E. coli* bound FAD, but not a second chromophore. In common with the other cryptochromes, it also failed to complement an *E. coli* mutant lacking photolyase activity. Finally, it bound DNA, although the binding was neither sequence-specific nor double-strand specific.[81]

The identification of the cryptochrome chromophores requires a few caveats. First, FAD and MTHF have only been identified from cry1 produced in a heterologous system, *E. coli*.[71] Though it is likely that these same chromophores are associated with cry1 *in planta*, chromophore-bearing cryptochromes have not yet been isolated from higher plants so the chromophore issue cannot be considered entirely resolved. Second, although Lin *et al.*[74] reported that cry2, heterologously produced in insect cells, was also a flavoprotein, they did not identify the flavin. Third, to date there is still no evidence regarding the nature of the second chromophore of cry2. Fourth, photochemical studies on heterologously produced cryptochromes that lack the second chromophore (see below) may not yield results that truly reflect the *in vivo* photochemistry. The cyanobacterial cryptochrome heterologously produced bound FAD in all cases[22,56,84] but failed to bind a second chromophore. Curiously only the two putative cryptochromes from *V. cholerae* bound two chromophores, both FAD and MTHF, when produced in *E. coli*.

9.3.2. Phototropins

In 1988, Gallagher *et al.*[85] first reported a plasma-membrane protein from growing etiolated pea (*Pisum sativum*) epicotyls that became phosphorylated on irradiation with blue light. Light-activated phosphorylation could be demonstrated both *in vivo* and *in vitro*. In terms of tissue distribution, spectral sensitivity, and reciprocity relationships, the protein showed a good correlation with phototropism.[89] An *A. thaliana* mutant,[87] *JK224*, with impaired phototropism, was also deficient in the protein.[88] The correlation was not perfect, however, as the light fluences necessary to go from threshold to saturation for phototropism and phosphorylation differed in some cases by almost an order of magnitude.[89] Thus the protein was regarded as a candidate to be the photoreceptor for phototropism but without conclusive proof, and with one major discrepancy. We will return to this discrepancy in a later section.

A great deal was learned about the biochemical and photochemical properties of this protein both from *P. sativum* and *Zea mays*[86] but, despite this information (see next Section), efforts to purify and characterize it by standard biochemical techniques and determine its function unambiguously were unsuccessful. In 1995, Liscum and Briggs[90] reported the characterization of four complementation groups of mutants from *A. thaliana* that were impaired in their phototropic responses. These they designated *nph1* through *nph4* (*non-phototropic hypocotyl*). The mutant *nph1–5,* generated by fast neutron bombardment, lacked any trace of blue light-activated phosphorylation. *JK224* was allelic to *nph1* and therefore was named *nph1–2*. By using the technique of amplified fragment-length polymorphism (AFLP), Huala *et al.*[91] identified the lesion in *nph1–5* and used adjacent sequences to clone the wild-type gene. It encoded a putative protein 996 amino acids in length, with a classic serine/threonine kinase domain in its C-terminal half. It also contained two very similar domains of about 110 amino acids in length that showed homology to domains in a wide range of proteins involved in signaling in response to changes in **L**ight, **O**xygen, or **V**oltage, and Huala *et al.*[91] designated them **LOV** domains (a subfamily of the PAS-domain superfamily, see Taylor and Zhulin[92]). This gene restored phototropism in response to very low fluence-rate blue light when transformed into the *nph1–5* mutant. Clearly *NPH1* was the gene that encoded the light-sensitive plasma membrane protein and the phosphorylation was an autophosphorylation.

Christie *et al.*[93] then expressed the protein in insect cells and demonstrated, first, that it underwent blue light-activated phosphorylation in the absence of any other plant proteins, and, second, that it bound flavin mononucleotide (FMN). They concluded that this single polypeptide chain served as substrate for phosphorylation, the kinase for autophosphorylation, and the photoreceptor driving the process, and that it was therefore the photoreceptor for phototropism. They then expressed the LOV domains either singly or together with the intervening hinge region in *E. coli.*[94] The purified, bright yellow, highly fluorescent product bound FMN stoichiometrically (one flavin per LOV domain). The authors concluded that the NPH1 protein was a dual-chromophoric photoreceptor, and that the LOV domains were the chromophore-binding domains. They designated the protein phototropin (subsequently phot1). Jarillo *et al.*[95] then reported the cloning of a second phototropin from *A. thaliana*, later designated phototropin 2 (phot2). Subsequent studies on both LOV domains of the two phototropins in *A. thaliana*, the two phototropins in rice (*O. sativa*), and both LOV domains from *C. reinhardtii* phototropin, indicated that all of them bound FMN, all had almost identical absorption spectra, and all had very similar extinction coefficients.[27]

As mentioned earlier, phot1 is plasma-membrane associated although it has no hydrophobic membrane-spanning domains and is clearly peripheral on the inside of the plasma membrane.[96] The nature of this association is not currently understood. In etiolated tissues, a small amount is released to the cytoplasm on irradiation with blue light.[97] The significance of this phenomenon is unclear at present. In etiolated hypocotyls it is found uniformly distributed around epidermal cells, but localized predominantly at the apical and basal ends of cortical cells. In green leaves, it is found uniformly distributed on the plasma membrane of both epidermal cells and guard cells, consistent with its roles in leaf expansion and stomatal opening.[97] In the flowering stem, it is associated both with xylem and phloem parenchyma where its distribution is bipolar. The same bipolar distribution is found in the cortical cells of roots. Phot2 is also plasma membrane-localized,[98] but to date no tissue-localization studies have been reported. Given the redundancy of function, however, it is likely that its distribution is not dissimilar to that of phot1.

Huang *et al.*[99] investigated the localization of phototropin in the green *alga C. reinhardtii* and found that in addition to its association with the plasma membrane of the cell body, it was tightly associated with the axonemes of the flagella. Since the localization was not affected in mutants lacking several structural components of the axoneme, the authors suggest that the association is likely with the outer doublet microtubules. In the mutant *fla10*, in which transport within the flagella is impeded at elevated temperatures, the phototropin concentration in the flagella dropped to less than half its initial level, suggesting that it was a cargo molecule for the intraflagellar transport system.

9.3.3. The ZTL/ADO family

Somers *et al.*[64] first identified an *A. thaliana* gene that played some role in regulating the period of the *A. thaliana* circadian rhythm. Mutations at that locus had a lengthened circadian period and they named the gene *ZEITLUPE* (*ZTL*). Expression of *ZTL* is not circadian and is not light-regulated, suggesting that it is not part of the central oscillator, but rather plays some modulating role. The gene encodes a putative protein of 609 amino acids with three highly distinctive domains: near the N-terminal end is a single PAS domain with high homology to the phototropin LOV domains. Just downstream is an F-box domain that is thought to target specific proteins to a ubiquitination complex for degradation. At the C-terminus are six kelch repeats that form a propeller-like structure thought to participate in protein–protein interactions.[100] At this writing, the putative target protein remains unidentified.

Jarillo *et al.*[65] subsequently described the same *A. thaliana* gene under the name *ADO* (for Adagio). They also found that mutations in *ADO* lengthened the circadian period. They further demonstrated both by yeast two-hybrid analysis and *in-vitro* binding that the ADO protein interacted directly both with the C-terminus of phyB and with the C-terminus of cry1. Given the above-mentioned observation that ZTL/ADO mutants show an altered response of their periodicity to changes in the intensity of red and blue light, perhaps this interaction is not surprising. Both studies indicate that ZTL/ADO is an important part of the circadian system. Both groups also suggest the possibility that this LOV-domain-containing protein may itself be a photoreceptor, based on the presence of a LOV domain. At present, however, there is no direct evidence for this suggestion.

The second member of this gene family to receive detailed studies was *FKF1*.[67] As mentioned above, the mutations at the *FKF1* locus caused delayed flowering on long days. They also caused hypersensitivity of suppression of hypocotyl growth both to red and blue light. The gene encodes a protein of 619 amino acids and, like ZTL/ADO, has a LOV domain, an F-box, and a kelch domain. Two key proteins are required for flowering in *A. thaliana* on long days: CONSTANS (CO) and FLOWERING LOCUS-T (FT). CO expression is required to induce expression of the putative transcription factor FT, essential for flowering. Imaizumi *et al.*[20] have shown that FKF1 expression is circadian with peak expression occurring late in the day. CO expression is also circadian but complex with a peak occurring late in the day and a second peak early in the night. Light and the FKF1 protein are both required for the first CO peak and *fkf1* mutants lack it and fail to show early flowering on long days. Given that FKF1 is an F-box protein likely involved in protein degradation, and Imaizumi *et al.*[101] recently demonstrated that it mediates the degradation of a repressor of *CONSTANS* in *A. thaliana*. Evidence that FKF1 protein may be a blue-light receptor is presented in a later section.

The third member of the family, LKP2, also plays a role in the circadian clock.[66] The *LKP2* gene encodes a protein of 612 amino acids, with LOV, F-box, and kelch domains like the other members. Like *ZTL/ADO, LKP2* transcription is not circadian and is not light regulated. When LKP2 was overexpressed, sensitivity of hypocotyl suppression to both red and blue light was reduced, suggesting, as in the case of ZTL/ADO, possible interaction with other photoreceptors. In common with the other two family members, no target protein has yet been specifically identified.

9.4. Biochemical and Photochemical Characterization

9.4.1. Cryptochromes

Although the cryptochromes were the first plant blue-light receptors characterized, information on their biochemical and photochemical properties has only appeared relatively recently. Two laboratories demonstrated that the C-terminal domain of cryptochromes associated physically with COP1, a presumed ubiquitin E3 ligase that is involved in the degradation of the transcriptional regulator HY5[80,102] and there was ample evidence that the cryptochromes mediated blue light-regulation of gene expression.[21] Más *et al.*[79] showed that cryptochrome also interacted physically with phyB and Ahmad *et al.*[103] reported that phyA could phosphorylate cry1 *in vitro*. The latter workers also reported that red light activated phosphorylation of cry1 *in vivo*, an activation that was far-red reversible. However, nothing was known about the immediate consequences of activation of the crys themselves by blue light.

Shalitin *et al.*[104] first reported that blue-light irradiation caused phosphorylation of Arabidopsis cry2 *in vivo*. In the dark, the cry2 protein was inactive and stable, but in the light, it became phosphorylated and unstable. Using a weak mutant allele of *cop1, cop1–4*, they demonstrated that the phosphorylated cry2 was significantly more stable than in wild-type seedlings. They were unable to confirm the results of Ahmad *et al.*[103] that red light, through phytochrome, phosphorylated *cry2 in vivo*. Indeed, single, double, and triple phytochrome mutants all showed blue light-activated cry2 phosphorylation.

It was known that the C-terminal domain of *cry2* transformed into wild type *A. thaliana*, produced a constitutive photomorphogenic phenotype in dark-grown seedlings and Shalitin *et al.*[104] demonstrated that in this case it showed constitutive phosphorylation. Thus it sees likely that light-dependent phosphorylation is required both for cry2 action and for targeting it for degradation.

Bouly *et al.*[105] purified his-tagged cry1 to near homogeneity in an insect-cell/*Baculovirus* system. When they bound the his-tagged protein to a Ni^{2+} column, the eluate lacked a second chromophore but retained FAD. When they bound it to a Co^{2+} column, the eluate lacked both chromophores. In the presence of FAD, blue light activated phosphorylation, but not in its absence, suggesting that cry1 itself was likely the photoreceptor for the reaction. They mention that cry2 also became bound to an ATP-agarose column but did not show the data. Thus, cry1 likely serves as its own light-activated kinase. The purified photoreceptor bound firmly to an ATP-agarose column and [γ-^{32}]P-labeled ATP could be tightly bound to the protein by UV-cross linking. The K_d was determined to be near 20 μM. Hence despite the absence of a recognizable kinase domain, cry1 binds ATP and carries out blue light-activated autophosphorylation. This cannot be the entire story, however, for as mentioned above, the cry1 C-terminal domain alone, minus any chromophores, transformed into *A. thaliana* causes a photomorphogenic phenotype and becomes constitutively phosphorylated in both darkness and light.[100]

Bouly *et al.*[105] also demonstrated that cry1 phosphorylation was on serines. Stronger phosphorylation occurred after a five-minute pre-irradiation rather than during simultaneous irradiation in the presence of the labeled ATP, indicating that light caused formation of a fairly stable state that was then subjected to autophosphoryation. Both KI, a flavin antagonist, and H_2O_2, an oxidant, prevented the reaction, indicating the importance of redox state for the reaction.

Shalitin *et al.*[104] found that when cry2 was phosphorylated *in vivo* its mobility on SDS chromatography decreased. However, Bouly *et al.*[105] could detect no change in mobility on phosphorylation *in vitro*. They hypothesized, first, that the two reactions did not occur on the same sites, and, second, that at least a part of the *in vivo* phosphorylation must be mediated by other plant protein kinases, as must have been the case with the constitutive phosphorylation of the C-terminal domain.[100]

Finally, Bouly *et al.*[105] reported that human cryptochome 1 (Hscry1) became bound to the ATP-agarose and became phosphorylated in the presence of [γ-^{32}]P-labeled ATP and $MgCl_2$. Given that there is considerable controversy over whether mammalian cryptochromes serve as genuine photoreceptors in the entrainment of circadian rhythms, it is of great interest to learn whether this phosphorylation is light-activated as is the case with the *A. thaliana* cryptochromes.

Shalatin *et al.*[106] found that light induced *in vivo* phosphorylation of cry1, as was the case with cry2.[104] Dephosphorylation occurred in about 15 min at room temperature. There was no effect of red light on phosphorylation, and once again, light-activated phosphorylation was normal in a series of single, double, and triple phytochrome mutants. Blue light caused phosphorylation of cry1 purified from an insect-cell/*Baculovirus* system as was found by Bouly *et al.*[105] A whole series of mis-sense *cry1* mutations scattered through the coding region produced full-length protein but yielded seedlings that had long hypocotyls under continuous blue light. They all failed to show blue light-activated phosphorylation, suggesting that light-activated cry1 phosphorylation is an essential component for signal transduction. Since light-activated cry1 phosphorylation was normal in a *cry2* mutant and vice versa, neither serves as a photoactivated kinase to phosphorylate the other. As with cry2, there is no recognizable kinase domain, and unlike the phototropins, the cryptochromes are not typical protein kinases. Shalitin *et al.*[106] also found that cry1, purified from an insect cell/*Baculovirus* system underwent *in-vitro* blue light-induced autophosphorylation.

Several authors had previously suggested that cryptochrome photoactivation likely involved redox changes[21] based on the mechanism of action of the CPD photolyases.[69] Photoactivation of *E. coli* photolyase involves intra-protein transfer of electrons via three tryptophans to reduce the FAD semiquinone to produce fully reduced $FADH^-$ required for DNA repair.[107–110] In *Anacystis nidulans* photolyase a tyrosine is involved in the electron transfer in addition.[111] As mentioned above, the cryptochromes do not repair CPDs. However, as noted by Giovani *et al.*,[112] not only do cryptochromes bind FAD, but they contain three tryptophans in positions comparable to those in *E. coli* photolyase, as indicated by the sequence and the crystal structure of *Synechocystis* PCC6803 cryptochrome.[56] Flash photolysis of purified *A. thaliana* cry1 in the presence of a reducing agent yielded transient absorption changes consistent with the transient formation of a neutral tryptophan radical while spectral changes near 410 nm implicated a tyrosine as well.[112] Although in the photolyases electron transfer occurs from tryptophan to the FAD seminquinone to form $FADH^-$ instead of fully oxidized FAD, to form the FAD semiquinone, the principle is the same.

Most recently Zeugner *et al.*[113] used site-directed mutagenesis to substitute phenylalanine for tryptophan either at W324 or W400 of *A. thaliana* cry1. They first investigated the consequences both for light-activated electron transfer and for light-activated phosphorylation in photoreceptor produced in the insect cell/*Baculovirus* system. Photoreduction of the FAD chromophore in the presence of reductant was completely absent in both mutated proteins and electron transfer as monitored by spectral changes at 520 nm following flash excitation was severely reduced. Light-activated phosphorylation was also completely eliminated in both mutant proteins. The authors also investigated the consequences of the mutations on physiological function in *cry1 cry2* double mutants transformed with the mutated constructs. Both light-activated anthocyanin accumulation and inhibition of hypocotyl elongation were severely impaired, although expression of the transgenes was near wild-type levels. Thus more than a decade after the first characterization of a cryptochrome, we are beginning to learn some of the early photochemical and biochemical consequences of its photoexcitation that are essential for its function.

9.4.2. Phototropins

In dramatic contrast to what occurred with the cryptochromes, where we are only beginning to learn about early events following photoexcitation, there appeared a wealth of biochemical and preliminary photochemical information about phototropins a decade before they were convincingly demonstrated to be blue-light receptors.[40,86] Even without the knowledge that they were photoreceptors, one could study the properties of the light-activated phosphorylation reaction itself. Hence we learned the following: the amount of light required for phosphorylation both *in vivo* and *in vitro*; that phosphorylation occurred on multiple serines; that dark recovery both *in vivo* and *in vitro* required many minutes; that ATP was the preferred nucleotide; that the reaction required magnesium; that the pH optimum was near neutrality; that the reaction was inhibited by known flavin antagonists; that the most effective flavin antagonists were the more hydrophobic ones; that the protein was associated with the inner surface of the plasma membrane; that it was not associated with other membrane fractions; and that it was present in highest amount in growing regions of etiolated seedlings.[86,114] Thus on its ultimate identification as a photoreceptor,[91,93] it was already well characterized at the cellular and biochemical levels.

Recent studies by Salomon *et al.*[115] have resolved the apparent discrepancy between the fluence of blue light needed for phototropism and phosphorylation (see above). The phototropins contain multiple phosphorylation sites, all serines, and the phosphorylation reaction is hierarchical, with some serines becoming phosphorylated at lower fluences than others with the most sensitive phosphorylations occuring in the same fluence range as phototropism. Two other studies bear on this question: Kinoshita *et al.*[116] showed that the activation of a proton ATPase in *Vicia faba* guard cells by blue light involves the phosphorylation of the ATPase and that the phosphorylated ATPase must react with a 14-3-3 protein that recognizes a specific phosphoserine to be functional. They also demonstrated that phot1 reacts with a 14-3-3 protein on photoexcitation and that the 14-3-3 protein interacts specifically with a phosphoserine immediately downstream from LOV1.[117] The reaction is essential for ATPase activation by blue light. The phosphoserines in the *V. faba* phot1a and phot1b are homologous to one of the two that are phosphorylated by the lowest fluence in oat phototropin.[115] It is therefore a reasonable hypothesis that phosphorylation at this site is essential for the physiological response and that the other less sensitive phosphorylations play a different role – for example, desensitization, or uncoupling from the signal transduction pathway. If the hypothesis is correct, it accounts for the apparent discrepancy mentioned above where the fluence-response curve for overall phosphorylation indicates a lower sensitivity to light for phosphorylation than for phototropism. What is required to test the hypothesis is a fluence-response curve for the phosphorylation of that specific serine. At this time it is not known whether this system involves a single 14-3-3 protein or two different ones.

It only became possible to probe the basic photochemistry of phototropins in detail when their domain structure was known and their chromophore-binding domains could be produced in amounts sufficient for structural and spectral analysis. It has not to date been possible to purify full-length phototropins expressed in a heterologous system. However, Salomon *et al.*[118] could purify the FMN-binding chromophore domains (the LOV domains, 110± amino acids each) from *Avena sativa* (oat) phot1 expressed in *E. coli* in amounts sufficient for detailed spectral studies. Irradiation with blue light bleached these domains to form a new spectral species with a single absorption band near 390 nm. Though fluorescent in their dark state, the domains completely lost the fluorescence on formation of the 390 nm species. There were three isosbestic points for the reaction (331, 385, 408 nm) that were nearly identical with those obtained from a cysteinyl adduct to the C(4a) carbon of the FMN, previously shown to be formed as an intermediate in the oxidation of mercuric ion mediated by mercuric oxidase.[119] All of the LOV domains characterized to date contain the conserved amino acid sequence GXNCRFLQ (in oat and other phot1 LOV2 domains this is the only cysteine). Mutating the cysteine to a serine or an alanine (mutants designated C39S or C39A), completely eliminated the photobleaching reaction, a result to be expected if adduct formation was blocked.[118] NMR studies with isotope-labeled FMN[120] subsequently confirmed the light-induced formation of the C(4a) cysteinyl adduct. Hence the light reaction of the LOV domains represents a novel type of photochemistry.

The light-driven formation of the 390 nm form is first order for both LOV domains from *A. sativa* phot1 and fully reversible in darkness.[118] However, the relative quantum efficiency for adduct formation for LOV2 is roughly ten-fold higher than that for LOV1 (0.44 versus 0.045) and the half life of LOV1 at room temperature (11.5 s) is just over one-third that for LOV2 (27 s). Thus at steady state in continuous light *A. sativa* phot1 LOV2 will have a far greater portion of the chromopeptide as adduct than LOV1. The extinction coefficients, ε, for *A. sativa* phot1 LOV1 and LOV2 at peak absorption (LOV1, 448.5 nm; LOV2, 446.5 nm) are 12,200 and 13,800 mol^{-1} cm^{-1} respectively. Circular dichroism (CD) measurements of the ground and adduct states between 260 and 500 nm indicate that adduct formation brings about a major conformational change in the flavin moiety, not surprising as the C(4a) carbon becomes an asymmetric center.[118,121] Molecular modeling indicated that the chromopeptides would likely form a typical PAS fold with five antiparallel β-strands interconnected by α-helices ([118], see below).

The photochemical kinetic properties of chromopeptides containing both LOV domains plus the intervening sequence differed unexpectedly from those of the single LOV domains. Adduct decay in darkness is not first order, is overall much slower that for the single LOV domains, and is not simply the algebraic sum of the decay components found for the individual LOV domains. As the decay kinetics of the two-LOV-domain peptides were similar to those found for the full-length *A. thaliana* phot1 produced in insect cells,[122] it is likely that the behavior of the two-LOV-domain chromopeptides more nearly reflects what may be occurring with phototropins *in vivo*.

Subsequent measurements[61] for LOV, LOV2, and LOV1+2 from *A. thaliana* phot1 and phot2, *O. sativa* (rice) phot1 and phot2, and *C. reinhardtii* phot, yielded extinction coefficients ranging from 11,200 (*O. sativa* phot2 LOV1) and 15,200 mol^{-1} cm^{-1} (*O. sativa* phot1 LOV2), half-lives for the adduct in darkness at room temperature from 5 (A. *thaliana* phot2 LOV2) to 168 s (*C. reinhardtii* phot LOV1), and relative quantum efficiencies (only for the individual LOV domains) from 0.035 (*A. thaliana* phot1 LOV1) to 0.35 (*C. reinhardtii* phot LOV2). The large difference in relative quantum efficiencies noted above for *A. sativa* phot1 LOV domains are also found for phot1 from *A. thaliana* and *O. sativa*, whereas they are far closer to each other for the phot2 LOV domains and those from *C. reinhardtii* phot. Phot1 LOV2 has a longer half life than phot1 LOV1 whereas with phot2 the relationship is reversed. We will return to the matter of two chromophore domains having different properties below.

Dürr *et al.*,[123] using a hydrophobic matrix, were able to release the FMN from the LOV2 domain of *A. sativa* phot1 and replace it with FAD, riboflavin, or 5' malonyl-riboflavin, but not 5-deazaflavin. For the incorporated flavins, the absorption spectra were almost identical to those of the native FMN-binding LOV domain. However, both LOV2-riboflavin and LOV2-malonyl-riboflavin showed more rapid regeneration in the dark. In the case of LOV2-FAD, phosphodiesterase hydrolyzed the FAD to FMN, indicating that the adenine moiety protruded into the aqueous medium. Since riboflavin both became bound and mediated light-activated adduct formation, it is clear that the ribityl phosphate side chain is not essential for the reaction.

The relative ease with which LOV-domain constructs (LOV1 alone, LOV2 alone, both LOV domains plus the intervening sequence, or LOV2 with additional amino acids downstream) could be expressed in *E. coli* quickly spawned a whole series of structural and biophysical studies. X-ray crystallographic studies of phy3 LOV2 from the fern *A. capillus-veneris*[124] and LOV1 from *C. reinhardtii*[125] gave nearly identical structures: a PAS fold constructed of five antiparallel β-strands interconnected by α-helices as hypothesized earlier.[118] FMN appears tightly although not covalently bound within the protein cage and Crosson *et al.*[124] identified 11 highly conserved amino acids involved in chromophore binding either by hydrogen bonding or by hydrophobic interactions. The FMN is held rigidly in place by hydrogen bonding on its pyrimidine side and hydrophobic interactions on its dimethyl-benzene side. The constraints imposed by the cage explains the fine structure in the absorption spectrum. The constraints prevent the broadening of the vibronic structure characteristic of FMN and other flavins in solution. As might be expected from the biochemical studies,[118] the sulfur atom of the cysteine in the crystal structure of LOV2 from *A. capillus-veneris* phy3 is only 4.2 Å away from the FMN C(4a) carbon.[124] Crosson *et al.*[126] subsequently obtained an X-ray structure for the cysteinyl adduct and observed that the cysteine sulfur had moved to within about 2 Å of the C(4a) carbon in the adduct, consistent with the formation of a C-S covalent bond. Other than the movement of the cysteine, however, they detected little other protein conformational change in response to adduct formation.

Resonance Raman and Fourier transform infra-red (FTIR) light-minus-dark difference spectroscopy detected many vibrational-band changes consistent with light-induced formation of the adduct.[127] Most of the difference bands could be attributed to FMN vibrational-mode changes associated with adduct formation. However, as the authors suggest, those changes detected between 1520 and 1609 cm^{-1} could well be ascribed to alterations in the conformation of the peptide backbone. Iwata *et al.*[128] reached a similar conclusion from their FTIR studies of the LOV2 domain of *A. capillis-veneris* phy3. Likewise light-minus-dark CD measurements between 190 and 260 nm indicated a major protein conformational change resulting in approximately a 15% loss in α-helicity,[121] in sharp contrast to the results of the crystallographic study discussed above.[124,126]

NMR spectroscopy resolved this apparent discrepancy. Crosson *et al.*[124,126] obtained their structures from a chromopeptide that included just the LOV2 domain with no C-terminal amino acids. By contrast, Harper *et al.*[129] used a chromopeptide that contained 24 amino acids downstream from the LOV2 domain for NMR analysis. The structure they determined for the LOV domain itself was consistent with that obtained by X-ray crystallography. However, the down-stream amino acids formed an amphipathic α-helix in the dark state (designated Jα-helix), with the hydrophobic side tightly associated with the β-sheets of the domain itself. On light-activated adduct formation the Jα-helix became detached and lost all detectable structure. It also became more accessible to proteolysis. Since Corchnoy *et al.*[121] used a similar preparation, what they probably detected as loss of α-helicity was the unfolding of the Jα-helix on adduct formation. As there do not appear to be similar structures downstream from LOV1, it seems likely that the two domains play sharply different roles (see below). Ataka *et al.*,[130] using infra-red and preresonance Raman spectroscopy, also failed to find evidence for any major protein conformational changes of LOV1 from *C. reinhardtii* phot on adduct formation. This result would be expected if there was no Jα helix downstream from LOV1 as seems to be the case for higher plant phototropins sequenced to date (Kevin Gardner, personal communication), or if there were insufficient amino acid residues downstream from the LOV1 domain for a Jα-helix to form. However, Losi *et al.*[131] using laser-induced optoacoustic spectroscopy, were able to detect small volume decreases in the *C. reinhardtii* phot LOV1 domain on adduct formation.

Corchnoy *et al.*[121] addressed the question as to whether dark decay passed through spectrally detectable intermediates. They monitored the decay of the CD changes at 450 and 290 nm to follow FMN structural changes and 208 and 222 to follow protein structural changes. The decay curves were identical, with a half-life near 46 s at room temperature, consistent with the half life as determined by absorption spectroscopy. Subsequent time-resolved NMR studies indicated that the dark-decay process is likely more complicated.[132] Heterogeneity of lifetimes indicate global changes slower than those measured by

CD, suggesting that reformation of the Jα-helix is followed by tertiary folding not detected in the CD measurements.

Harper *et al.*[133] then investigated the consequences of substituting a hydrophilic residue on the hydrophobic face of the Jα-helix both on the NMR structure of the LOV2 domain plus the modified Jα-helix and on the capacity of full-length phot1 for phosphorylation. Three of the substitutions (I532E, A536E and I539E in oat phot1 LOV2) produced structures that lacked the Jα-helix structure as determined by NMR, and the 24-amino acid region downstream from the LOV domain was equally susceptible to proteolysis with or without light treatment. When expressed in insect cells, the mutated phototropins (in this case *A. thaliana* phot1 LOV2 with comparable substitutions) became constitutively phosphorylated, suggesting that the unfolding of the Jα-helix was an essential step in activating autophosphorylation activity.

Flash photolysis of oat phot1 LOV2 produces an intermediate within a few nanoseconds that is spectrally identical to a flavin triplet state and not to a flavin semiquinone.[134] The flavin triplets decay to form both the adduct and the ground state with equal rates. Both triplet decay to the adduct[121] and dark recovery to the ground state[121,134] are significantly slowed in LOV domains that have been lyophilized and resuspended in D_2O compared to those rehydrated with H_2O, indicating that both processes are rate limited by a proton transfer event. Subsequently Kennis *et al.*[135,136] also using the oat phot1 LOV2 domain, showed that the triplet state was formed by intersystem crossing (ISC) from the excited singlet and that the triplets were largely protonated. Kottke *et al.*[137] studying the LOV1 domain from *C. reinhardtii* phot, obtained a somewhat more complex picture with two triplet species decaying to the adduct with different time constants in addition to a back reaction to the unexcited state.

Using ultra-fast (femto-second) spectroscopy, Kennis *et al.*[136] also showed that the S-C bond in the near UV-A-absorbing adduct (the 390 nm-absorbing intermediate) could be broken by near UV light and the oat phot1 LOV2 domain returned to its dark state. Kottke *et al.*[137] came to the same conclusion based on photoequilibrium measurements on the *C. reinhardtii* phot LOV1 domain. Thus adduct formation is photoreversible, although the biological significance of this photoreversibility is not obvious.[136] Only two thermostable states have been resolved in the LOV-domain photocycle, ground state and adduct. The LOV domain photocycle reaction involves: ISC from the excited singlet to the FMN triplet state; this triplet intermediate then decays to the adduct intermediate on a microsecond time scale. The adduct can be driven back to the dark state by light on an ultrafast time scale (picoseconds) or can return spontaneously in the dark on a time scale best measured in seconds or minutes. At present, very little is known about the pathway of adduct decay in the dark, except that, as mentioned above, it is limited by a proton transfer event.[134]

Several studies have measured the physical parameters of this photocycle. Whereas the fluorescence quantum efficiency of free FMN is about 0.375,[138] the values are much lower for FMN bound in a LOV domain. The fluorescence quantum efficiencies are reported to be 0.17[139] or 0.23[132] for the *C. reinhardtii*

LOV1 and 0.13 for *A. capillus-veneris* phy3 LOV2.[131] These values increase significantly when the cysteine is replaced by a serine or an alanine in oat phot1 LOV2,[134] but apparently not in C. *reinhardtii* phot LOV1.[138] There is in either case a decrease in fluorescence that accompanies FMN binding. This decrease is accompanied by an increase in ISC from singlet to triplet by a factor of about 2.4.[135,140] Quantum efficiencies for triplet yield vary from 0.6 for phy3 LOV2,[135] 0.83 for oat phot1 LOV2, 0.68 for oat phot1 LOV2 C39A,[138] to 0.88 for oat phot1 LOV2.[134] Although one would expect this value to increase on FMN binding and be reduced in the C39A mutant, the data are ambiguous. The quantum efficiency of triplet formation for free FMN is reported to be 0.6[135,139] so further studies will be needed to resolve whether FMN binding in the LOV domain always increases the quantum efficiency for ISC. Swartz and colleagues[46,47] present a more detailed discussion of this apparent ambiguity.

The exact mechanism for adduct formation is currently controversial. Independence of the fluorescence yield on pH between values of 4 and 10 with complete quenching at higher pH suggested that the cysteine sulfur was present as a thiolate (S^-) rather than a thiol (SH).[134] Below pH 4 fluorescence increased, consistent with the protonation of the thiolate. However, two other FTIR studies, one of *A. capillus-veneris* phy3 LOV2[128,141] and the other of *C. reinhardtii* phot LOV1[126] detected S-H vibrations near 2,500 cm^{-1} that disappeared on adduct formation. These showed an expected shift to a lower frequency (from 2567 to 1867 cm^{-1}) in D_2O.[126] Kottke *et al.*[137] summarize the evidence favoring a mechanism involving a cysteine thiol rather than a thiolate. Adduct formation can occur at temperature near 80 *K*[128,142,143] and X-ray crystallography and low-temperature FTIR studies suggest that the cysteine can exist in two different conformations.[142,143] It is possible that the cysteine can exist either as a thiolate or a thiol depending on several different extrinsic and intrinsic factors: presence (LOV2 plus additional amino acids downstream) or absence (LOV2 alone or LOV1) of the Jα helix; use of different sequences to tag domains for purification; species differences (an alga versus a fern versus an angiosperm); different preparative methods; or functional differences between LOV1 and LOV2. In the latter case, biochemical studies indicate that native phototropin is a dimer and that dimerization occurs via LOV1 and not LOV2.[144] Cross-checking samples between laboratories might provide a resolution to the apparent discrepancy.

Adduct formation is not the only reaction induced by blue light. Extended irradiation of *A. sativa* phot1 LOV2 with its cysteine mutated to alanine leads to the formation of a flavin semiquinone, detectable both by absorption spectroscopy, electron paramagnetic resonance, and electron-nuclear double resonance.[145] Decay of the semiquinone occurred on a time scale of many minutes, indicating that the species was quite stable. The results indicate that the FMN radical in the mutant peptide is N(5) protonated, suggesting that the pH near the FMN is sufficiently acidic that the cysteine residue in the wild-type protein might also be protonated. The authors suggest a possible radical-pair mechanism for adduct formation in photoexcited LOV domains. Salomon *et al.*[118] did not observe light-induced formation of the semiquinone in the same oat phototropin chromopeptide on irradiations lasting seconds at most, whereas Kay *et al.*[145] used irradiations lasting many minutes (intensity not specified). As it is clear that adduct formation and consequent unfolding of the Jα helix are required for biological activity, the reaction studied by Kay *et al.*[145] may represent photoinactivation. Whether the reaction has biological relevance remains to be determined. However, further probes of this sort could be very helpful in sorting out the exact mechanism(s) of light-activated C(4a) cysteinyl adduct formation in the LOV domains of the different phototropins. Schleicher *et al.*,[146] using UV-visible spectroscopy and time-resolved electron paramagnetic resonance measurements, also obtained evidence favoring a radical-pair mechanism as the primary reaction leading to adduct formation, at least at low temperatures. Given the number of laboratories addressing the issue, it seems likely that within the near future, we will have a better picture of the nature of the light reaction leading to the formation of the C-S bond within the LOV domain.

When the reactive cysteine of *C. reinhardtii* phot LOV1 was replaced by methionine, blue-light irradiation led to the formation of a species absorbing in the red region of the spectrum.[147] The species was stable both under aerobic and denaturing conditions. On the basis of the known crystal structure of the LOV domain[125] and their own spectroscopic evidence, the authors concluded that the species was an adduct attached to the N5 nitrogen of the FMN instead of the C(4a) carbon, and that the flavin existed as a neutral radical (semiquinone).

An important question is why phototropins have two chromophore-binding domains. As just mentioned, there is evidence that LOV1 serves specifically as a dimerization domain.[144] Furthermore, inactivation of LOV1 has no effect on light-activated autophosphorylation of *A. thaliana* phot1 expressed in the insect-cell/*Baculovirus* system,[122] nor is LOV1 required to complement an *A. thaliana* mutant lacking phot1.[122] Thus to date there is no evidence that the photochemical activity of phot1 LOV1 is required for the biological function of phot1. However, LOV1 alone can mediate a low level of autophosphorylation by *A. thaliana* phot2, and can partially complement a *phot2 phot2* double mutant in restoring phot2-mediated phototropism (H.-Y. Cho, E. Kaiserli, J.M. Christie, T.-S. Tseng, W.R. Briggs, unpublished observations), indicating that at least in phot2, LOV1 can play a photobiological role in the absence of LOV2. It remains to be seen whether LOV1 in phot2 somehow mediates the release and unfolding of the Jα helix downstream from an inactivated LOV2 domain, both activating LOV2 autophosphorylation, and mediating phototropism. It also remains to be determined whether the serines phosphorylated by LOV1 alone in phot2 are the same as those phosphorylated by a fully functional phot2.

Kagawa *et al.*[148] also examined the functional roles of various phototropin domains in mediating the chloroplast high-light avoidance response in the fern *A. capillis-veneris*. As was the case with *A. thaliana* phot1,[122] transgenes expressing Acphot2 minus its LOV1 domain were sufficient to restore biological activity in a mutant lacking Acphot2. In addition, sequences downstream from the kinase domain were shown to contribute to the avoidance response. The authors calculated the rate of signal decay *in vivo* and compared it with the rates of dark reversion of the single LOV domains or a construct that included both *in vitro*. The rate of signal decay fit well with the joint rate of decay of the double LOV-domain chromopeptide, indicating that although not essential for the response, LOV1 might play some role in determining the lifetime of the signal.

It was known that light-activated phosphorylation of oat phot1 *in vivo* resulted both in reduced electrophoretic mobility and loss of reactivity with an anti-phot1 antibody.[115] Kneib *et al.*[149] found, however, that the two processes were not linked. For coleoptile tips held at low temperature, the reduced immunoreactivity could be detected within less than a minute whereas the mobility shift was only detectable after 2 min. In addition, UV-C light (280 nm) induced phosphorylation and a mobility shift, but no change in immunoreactivity. It is of considerable interest to determine whether the same serines become phosphorylated under the different conditions.

9.4.3. The ZTL/ADO family

At this time, very little can be said about the biochemical or photochemical properties of the three members of the ZTL/ADO family. As mentioned earlier, all three have LOV domains and at least FKF1 is likely a true photoreceptor.[20] All three LOV domains expressed in *E. coli* bind FMN, and all three undergo a spectroscopic change on irradiation with blue light indicative of the formation of the cysteinyl adduct (formation of a species with a single absorption maximum near 380 nm). Surprisingly however, the adducts in all three cases are stable. There is no thermal return to the ground state, as found in the phototropin LOV domains. However, the photochemistry work was done on isolated LOV domains and it is unknown if the photochemical kinetics of the full-length proteins *in vivo* also show this truncated photocycle. However, in the case of FKF1, since breakdown of the protein occurs at night and synthesis occurs only late in the day, it could be that a repeated conversion of freshly synthesized FKF1 to the adduct species by light late in the long day is all that is required to activate the system to initiate flowering.[20]

As mentioned earlier, phototropin phosphorylation is activated by the blue light-induced unfolding of a Jα helix downstream from LOV2.[129] It is intriguing to speculate that a similar mechanism may function with FKF1: light could induce the unfolding of the F-box, allowing it to target its substrate repressor for destruction via the ubiquitin pathway. Time will tell whether this model is correct.

9.5. LOV Domains in Fungal Photoreceptors and Prokaryote Proteins

With the rapid increase in the number of genomes sequenced, it soon became apparent that LOV domains were not confined to the phototropins and the ZTL/ADO family, but could also be found in many microbial proteins as well.

9.5.1. Fungi

Blue-light-activated responses in the ascomycete *N. crassa* have been known for many years, and an action spectrum obtained many years ago for carotenogenesis[150] and another for shifting the phase of a circadian rhythm of conidiation[151] clearly placed the photoreceptor within Gressel's "cryptochrome" class.[31] Genetic studies repeatedly came up with two loci, *wc-1* and *wc-2*, that were absolutely required for almost all of the blue-light responses in *N. crassa*. The WC-1 protein was an obvious candidate for the photoreceptor as the *wc-1* gene encodes a putative transcription factor of 1167 amino acids with an N-terminal glutamine-rich domain, a LOV domain, two PAS domains, and at the C-terminus a zinc finger. In the center of the LOV domain is the conserved sequence GRNCRFLQ, identical to the sequences found in the phototropin LOV domains.

Froelich *et al.*[152] showed that WC-1, using FAD as a cofactor, became bound to WC-2, a second zinc finger protein, to form a white-collar complex (WCC) when synthesized in an *in vitro* transcription/translation system. Irradiation of the complex caused it to bind to two light-regulatory elements in the promoter of *frq*, a gene encoding an essential element in the *N. crassa* circadian clock, and was clearly functioning as a photoreceptor. Neither FMN nor riboflavin functioned in this system so the WC-1 LOV domain differs from that of the phototropins where the ligand is FMN. So far, efforts to express the WC-1 LOV domain alone in a heterologous system have not been successful, but presumably it forms a cysteinyl adduct to the FAD on photoexcitation. All 11 amino acids identified by Crosson *et al.*[124] as being involved in FMN binding are conserved in the WC-1 LOV domain (nine identical, two similar). It is not known at present whether there is a Jα helix downstream from the WC-1 LOV domain.

In a related study, He *et al.*[153] found that the purified WCC bound stoichiometric amounts of FAD. They also found that if they introduced a gene encoding a WC-1 protein from which the LOV domain had been removed into a *wc-1* mutant, it failed to complement the mutant phenotype. As WC-1-mediated entrainment of *frq* by temperature shifts was normal, the authors concluded that WC-1 must be the photoreceptor for the circadian and other blue-light responses.

Although WC-1 serves as the photoreceptor for most of the blue-light responses in *N. crassa*, adaptation to extended periods of light, detected as decrease in the transcription of some light-regulated genes, and up-regulation of these genes with an abrupt shift to higher light intensities are mediated by a different blue-light receptor, VIVID (VVD). Schwerdtfeger and Linden[154] showed that VVD, a protein of only 186 amino acids, mediated these responses. VVD had C-terminal LOV domain, and like the LOV domain from WC-1, had all of the conserved amino acids involved in flavin binding. When produced in *E. coli*, VVD bound both FMN and FAD, the ratio depending on conditions during isolation. The LOV domain in the intact protein underwent a typical LOV domain photocycle, although dark recovery was far slower than that reported for the phototropin LOV domains. Substitution of alanine for the conserved cysteine blocked the photocycle. Likewise, *vvd* mutants were not complemented by the VVD protein with the Cys/Ala mutation.

Putative WC-1 homologs have now been identified from sequence analysis in several other ascomycetes (*Aspergillis nidulans*, *Magnaporthe grisea*, *Fusarium graminarium*, *Tubor borchii*.[155] Thus *N. crassa* WC-1 may well be representative of blue-light receptors in many other ascomycetes as well, likely interacting similarly with WC-2 homologs to regulate gene expression in response to blue-light signals in these fungi.

The situation is slightly different with basidiomycetes. Idnurm and Heitman[155] identified a WC1 homolog in the human pathogenic fungus *Cryptococcus neoformans* that they designated BWC1 (basidiomycete WC-1). Light inhibits mating and hyphal development after fusion in this fungus. Deletion of the *BWC1* gene produced a strain that mated equally well in dark or light. Another mutant that developed filaments equally well in light or dark was found to have an insertion in *BWC2*, a homolog of the *N. crassa WC-2*. The two proteins interacted strongly in a yeast two-hybrid assay, suggesting that they likely functioned in a manner similar to WC-1 and WC-2 in *N. crassa*. A major difference from WC-1, however, was that BWC1 lacked a zinc finger domain. Thus DNA binding is presumably mediated exclusively by BWC2. Absence of a zinc-finger domain is a property of BCW1 homologs in all basidiomycetes in which the putative protein has been identified to date (*C. neoformans*, *Coprinus cinereus*, *Ustilago maydis*, *Phanerochaete chrysosporium*,[155]) and likely represents a consistent difference between these two major groups of fungi.

9.5.2. Bacteria

Putative LOV domains have now been identified in a number of different eubacterial proteins, including several from cyanobacteria.[42] These proteins are highly diverse, and include histidine kinases or hybrid histidine-kinase/response regulators, LOV STAS proteins (STAS domains are found in eukaryote Sulfate Transporters and prokaryote Anti Sigma factor regulators), phosphodiesterase/cyclases, and others of unknown function. A recent report[153] listed 29 sequences encoding putative LOV domains in proteins from 24 different bacterial species.

Losi *et al.*[157] purified recombinant YtvA, a LOV STAS protein of 261 amino acids from *Bacillus subtilis*, and demonstrated spectroscopically that it underwent blue light-induced formation of the C(4a) cysteinyl adduct just as did the phototropin LOV domains. In common with the VVD LOV domain, its dark reversion was slower than that reported for the phototropin LOV domains. Biophysical measurements indicated a quantum yield for fluorescence of 0.22, for triplet formation of 0.62, and for adduct formation of 0.49, the latter close to the quantum efficiency for adduct formation reported for LOV2 from oat phot1.[118] In a related study, Losi[156] reported a similar photocycle for a LOV kinase protein from *Caulobacter crescentus*. It seems likely that many, if not all, of the LOV domains in bacterial proteins will show the same photochemistry.

This group subsequently did a comparative FT-IR study of the two LOV domains from *C. reinhardtii* and the LOV domain in YtvA from *B. subtilis*.[146] The reactive cysteine in LOV1 from *C. reinhardtii* was shown to exist in two conformers, but that of LOV2 only a single conformer. Significantly, formation of the cysteinyl adduct in the LOV domain from the *B. subtilis* YtvA protein caused a change in the conformation of the downstream STAS domain.

The situation with these bacterial LOV proteins is quite different from that in the fungi. Despite efforts in several laboratories, convincing evidence that these LOV-domain-containing proteins serve as photoreceptors is still lacking. Thus the question raised here is similar to that with the two *A. thaliana* proteins, ZTL/ADO and LKP2: does the presence of a photoactive LOV domain in a protein serve as evidence that the protein is a photoreceptor? Future research will be required to address this question.

9.6. Epilog

Flavin-based photoreceptors play essential roles in the regulation of the growth and development of green plants and fungi. Since discovery of the very first plant blue-light receptor a little over 10 years ago, the increase in knowledge has been exponential. We are beginning to understand the early events in the photoexcitation of the cryptochromes and we already have a great deal of information, both biochemical and photochemical, on the phototropins and other LOV-domain-containing proteins. Important biochemical and photochemical questions remain, however, as do questions regarding the nature of the interactions of the photoactivated photoreceptors with downstream elements in the various signal-transduction pathways. The coming decade will without doubt produce equally exciting advances in our knowledge of flavin-based photoreceptors.

Acknowledgments

The author is extremely grateful to Dr. Trevor E. Swartz for his careful review of the manuscript. Most of the research from the author's laboratory mentioned in this review was supported by the National Science Foundation. The author is extremely grateful for this support.

References

1. S. Poggioli, Della influenza che ha il raggio magnetico sulla vegetatione delle pianta, *Bologna – Coi Tipi de Annesio Nobili Opusc. Scientif. Fasc I.*, 1817, 9–23.
2. J. Payer, Mémoire sur la tendence des tiges vers la lumiere, *Comte Rendu du Seances de L'Académie des Sciences*, 1842, **XV**, 1194–1196.

3. M. Zantedeschi, De l'influence qu'exercent sur la vegetation des plantes et la germination des graines les rayons transmis à travers des verres colorés, *Comte Rendu du Seances de L'Académie des Sciences*, 1843, **XVI**, 747–749.
4. C. Daubeny, On the action of light upon plants and of plants on the atmosphere, *Phil. Trans. Roy. Soc.*, 1836, **126**, 149–175.
5. T. C. Hsiao and W. G. Allaway, Action spectra for guard cell Rb^+ uptake and stomatal opening in *Vicia faba*, *Plant Physiol.*, 1973, **51**, 82–88.
6. T. Ogawa, H. Ishikawa, K. Shimata and K. Shibata, Synergistic action of red and blue light and action spectra for malate formation in guard cells of *Vicia faba* L., *Planta*, 1978, **142**, 61–65.
7. T. Ogawa, Blue light response of stomata with starch-containing (*Vicia faba*) and starch-deficient (*Allium cepa*) guard cells under background illumination with red light, *Plant Sci. Lett.*, 1981, **22**, 103–108.
8. E. Zeiger and C. Field, Photocontrol of the functional coupling between photosynthesis and stomatal conductance in the intact leaf, *Plant Physiol.*, 1982, **70**, 370–375.
9. J. A. Böhm, Beiträge zur näheren Kentniss des Chlorophylls, *Sitzungber. Mathem. – Naturwiss. Classe kais Akademie Wissenschaften*, 1856, **22**, 479–512.
10. A. Famintzen, Die Wirkung des Lichtes und der Dunkelheit auf die Verteilung der Chlorophyllkörner in den Blättern von Mnium sp.?, *Jahrb. Wiss. Botanik*, 1867, **6**, 49–54.
11. M. Kasahara, T. Kagawa, K. Oikawa, N. Suetsugu, M. Miyao and M. Wada, Chloroplast avoidance movement reduces photodamage in plants, *Nature*, 2002, **420**, 829–832.
12. K. M. Folta and E. P. Spalding, Unexpected roles for cryptochrome 2 and phototropin revealed by high-resolution analysis of blue light-mediated hypocotyl growth inhibition, *Plant J.*, 2001, **26**, 471–478.
13. V. Gaba and M. Black, Two separate photoreceptors control hypocotyl growth in green seedlings, *Nature*, 1979, **278**, 51–54.
14. G. Meijer, Rapid growth inhibition of gherkin hypocotyls in blue light, *Acta Botanica Néerl.*, 1968, **17**, 9–14.
15. E. Van Volkenburgh and R. E. Cleland, Light-stimulated cell expansion in bean (*Phaseolus vulgaris L.*) leaves. I. Growth can occur without photosynthesis, *Planta*, 1990, **182**, 72–76.
16. E. Van Volkenburgh and R. E. Cleland, Light-stimulated cell expansion in bean (*Phaseolus vulgaris L.*) leaves. II. Quantity and quality of light required, *Planta*, 1990, **182**, 77–80.
17. W. R. Briggs, Blue/UV-A receptors: historical overview, in *Photomorphogenesis in plants and Bacteria: function and Signal Transduction Mechanisms*, E. SchaferF. Nagy (ed), Springer, Dordrecht, 2005.
18. M. Ahmad and A. R. Cashmore, *HY4* gene of *A. thaliana* encodes a protein with the characteristics of a blue-light receptor, *Nature*, 1993, **366**, 162–166.
19. J. M. Christie, P. Reymond, P. Bernasconi, G. K. Powell, A. A. Raibekas, E. Liscum, E. Larsen and W. R. Briggs, Arabidopsis NPH1: a flavoprotein with the properties of a photoreceptor for phototropism, *Science*, 1998, **282**, 1698–1701.
20. T. Imaizumi, H. G. Tran, T. E. Swartz, W. R. Briggs and S. A. Kay, FKF1 is essential for photoperiodic-specific signaling in Arabidopsis, *Nature*, 2003, **426**, 302–306.
21. A. Batschauer, Plant cryptochromes their genes, biochemistry, and physiological roles, in *Handbook of Photosensory Receptors*, W.R Briggs andJ.L Spudich (ed), Wiley-VCH, Weinheim, Germany, 2005, 211–246.

22. K. Hitomi, K. Okamoto, H. Diayasu, H. Miyashita, S. Iwai, H. Toh, M. Ishiura and T. Todo, Bacterial cryptochrome and photolyase characterization of two photolyase-like genes of *Synechocystis* PCC6803, *Nucleic Acids Res.*, 2000, **28**, 2353–2362.
23. E. N. Worthington, İ. H. Kavakli, G. Berrocal-Tito, B. E. Bondo and A. Sancar, Purification and characterization of three members of the photolyase/cryptochrome family blue-light photoreceptors from *Vibrio cholerae*, *J. Biol. Chem.*, 2003, **278**, 39143–39154.
24. W. R. Briggs and J. Christie, Phototropins 1 and 2 versatile plant blue-light receptors, *Trends Plant Sci.*, 2002, **7**, 204–210.
25. K. Nozue, J. M. Christie, T. Kiyosue, W. R. Briggs and M. Wada, Isolation and characterization of a fern phototropin (Accession number AB037188), a putative blue-light photoreceptor for phototropism (PGR00-0390), *Plant Physiol.*, 2000, **122**, 1457.
26. T. Kagawa, M. Kasahara, T. Abe, S. Yoshida and M. Wada, Function analysis of phototropin2 using fern mutants deficient in blue light-induced chloroplast avoidance movement, *Plant Cell Physiol.*, 2004, **45**, 416–426.
27. M. Kasahara, T. Kagawa, Y. Sato, T. Kiyosue and M. Wada, Phototropins mediate blue and red light-induced chloroplast movements in *Physcomitrella patens*, *Plant Physiol.*, 2004, **135**, 1388–1397.
28. K.-Y. Huang, T. Merkle and C. Beck, Isolation and characterization of a Chlamydomonas gene that encodes a putative blue-light receptor of the phototropin family, *Physiol. Plantarum*, 2002, **115**, 613–622.
29. K. Nozue, T. Kanegae, T. Imaizumi, S. Fukuda, H. Okamoto, K.-C. Yeh, J. C. Lagarias and M. Wada, A phytochrome from the fern *Adiantum* with features of the putative photoreceptor NPH1, *Proc. Natl. Acad. Sci. USA*, 1998, **95**, 15826–15830.
30. H. Kawai, T. Kanegae, S. Christensen, T. Kiyosue, Y. Sato, T. Imaizumi, A. Kadota and M. Wada, Responses of ferns to red light are mediated by an unconventional photoreceptor, *Nature*, 2003, **421**, 287–290.
31. J. Gressel, Blue light photoreception, *Photochem. Photobiol.*, 1979, **30**, 749–754.
32. W. R. Briggs, J. L. Spudich (eds), *Handbook of Photosensory Receptors*, pp. 1–473, Wiley-VCH, Weinheim, Germany, 2005.
33. E. Schäfer and F. Nagy (eds), Photomorphogenesis in Plants and Bacteria:Function and Signal Transduction Mechanisms, pp. 1–662, Springer, Dordrecht, 2005.
34. M. Wada, K. Shimizaki and M. Iino (eds), *Light Sensing in Plants, Proceedings of the 58th Yamada Conference*, pp. 1–370, M. Wada, K. Shinizaki and M. Iino (eds), Springer, Japan, 2005.
35. J.-P. Bouly, B. Giovani, M. Ahmad, Early events triggered by light activation of the CRY1 blue-light photoreceptor, in *Light Sensing in Plants, Proceedings of the 58th Yamada Conference*, M. WadaK. Shinizaki and M. Iino (eds), Springer, Japan, 2005, 131–138.
36. A.-R. Cashmore, Plant cryptochromes, in *Handbook of Photosensory Receptors*, W.R. Briggs and J.L. Spudich(eds), Wiley-VCH, Weinheim, Germany, 2005, 247–258.
37. A.-R. Cashmore, Cryptochrome overview, in *Light Sensing in Plants, Proceedings of the 58th Yamada Conference*, M. Wada, K. Shinizaki and M. Iino(eds), Springer, Japan, 2005, 121–130.
38. A.R. Cashmore, Cryptochromes, in *Photomorphogenesis in Plants and Bacteria: function and Signal Transduction Mechanisms*, E. Schäfer and F. Nagy (eds), Springer, Dordrecht, 2005, 199–221.

39. R. N. Van Gelder and A. Sancar, Animal cryptochromes, in *Handbook of Photosensory Receptors*, W.R. Briggs and J.L. Spudich (eds), Wiley VCH Weinheim, Germany, 2005, 259–276.
40. W.R. Briggs and J.M Christie, Phototropins, in *Photomorphogenesis in Plants and Bacteria: function and Signal Transduction Mechanisms*, E. Schäfer and F. Nagy (eds), Springer, Dordrecht, 2005, 223–252.
41. R. A. Bogomolni, T. E. Swartz and W. R. Briggs, Proton transfer reactions in LOV-domain photochemistry, in *Light Sensing in Plants, Proceedings of the 58th Yamada Conference*, M. Wada, K. Shinizaki, M. Iino (eds), Springer, Japan, 2005, 147–154.
42. S. Crosson, LOV domain structure, dynamics, and diversity, in *Handbook of Photosensory Receptors*, W.R. Briggs and J.L. Spudich (eds), Wiley-VCH, Weinheim, Germany, 2005, 323–336.
43. J. M. Christie and W. R. Briggs, Blue light sensing and signaling by the phototropins, in *Handbook of Photosensory Receptors*, W.R. Briggs and J.L. Spudich (eds), Wiley-VCH, Weinheim, Germany, 2005, 277–304.
44. J. Heberle, Vibrational spectroscopy explores the photoreactions of LOV domains, in *Light Sensing in Plants, Proceedings of the 58th Yamada Conference*, M. Wada, K. Shinizaki and M. Iino (eds), Springer, Japan, 2005, 155–162.
45. K. Shimizaki, Molecular mechanism of blue light response in stomatal guard cells, in *Light Sensing in Plants, Proceedings of the 58th Yamada Conference*, M. Wada, K. Shinizaki and M. Iino (eds), Springer, Japan, 2005, 185–192.
46. T. E. Swartz and R. Bogomolni, LOV-domain photochemistry, in *Handbook of Photosensory Receptors*, W.R. Briggs and J.L. Spudich (eds), Wiley-VCH, Weinheim, Germany, 2005, 305–322.
47. T. E. Swartz, W. R. Briggs and R. Bogomolni, LOV domain-containing proteins in Arabidopsis, in *Light Sensing in Plants Proceedings of the 58th Yamada Conference*, M. Wada, K. Shinizaki and M. Iino (eds), Springer, Japan, 2005, 163–170.
48. W. Rüdiger, Phototropin phosphorylation, in *Light Sensing in Plants, Proceedings of the 58th Yamada Conference*, M. Wada, K. Shinizaki and M. Iino (eds), Springer, Japan, 2005, 171–178.
49. T. F. Schultz, The ZEITLUPE family of putative photoreceptors, in *Handbook of Photosensory Receptors*, W.R. Briggs and J.L. Spudich (eds), Wiley-VCH, Weinheim, Germany, 2005, 337–347.
50. D. E. Somers, ZEITLUPE and the control of circadian timing, in *Light Sensing in Plants Proceedings of the 58th Yamada Conference*, M. Wada, K. Shinizaki and M. Iino (eds), Springer, Japan, 2005, 247–354.
51. J. C. Dunlap and J. J. Loros, Neurospora photoreceptors, in *Handbook of Photosensory Receptors*, W.R. Briggs and J.L. Spudich (eds), Wiley-VCH, Weinheim, Germany, 2005, 371–389.
52. J.C. Dunlap, Blue light receptors – beyond phototropins and cryptochromes, in *Photomorphogenesis in Plants and Bacteria: Function and Signal Transduction Mechanisms*, E. Schäfer and F. Nagy (eds), Springer, Dordercht, 2005, 253–277.
53. N. Suetsugu and M. Wada, Photoreceptor gene families in lower plants, in *Handbook of Photosensory Receptors*, W.R. Briggs and J.L. Spudich (eds), Wiley-VCH, Weinheim, Germany, 2005, 349–369.
54. R. Banerjee and A. Batschauer, Plant blue-light receptors, *Planta*, 2005, **220**, 498–502.
55. R. B. Celaya and E. Liscum, Phototropins and associated signaling: providing the power of movement in higher plants, *Photochem. Photobiol.*, 2005, **81**, 73–80.

56. R. Brudler, K. Hitomi, H. Daiyasu, H. Toh, K. Kucho, M. Ishiura, M. Kanahisa, V. A. Roberts, T. Todo, J. A. Tainer and E. D. Getzoff, Identification of a new cryptochrome class: structure, function, and evolution, *Mol. Cell*, 2003 **11**, 59–67.
57. A. Takemiya, S.-I. Inoue, M. Doi, T. Kinoshita and K.-I. Shimizaki, Phototropins promote plant growth in response to blue light in low light environments, *Plant Cell*, 2005, **17**, 1120–1127.
58. M. Ohgishi, K. Saji, K. Okada and T. Sakai, Functional analysis of each blue light receptor, cry1, cry2, phot1, and phot2 by using combinatorial multiple mutants in *Arabidopsis*, *Proc. Natl. Acad. Sci. USA*, 2004, **101**, 2223–2228.
59. J. J. Blakeslee, A. Bandytopadhyay, W. A. Peer, S. N. Makam and A. S. Murphy, Relocalization of the PIN1 auxin efflux facilitator plays a role in phototropic responses, *Plant Physiol.*, 2004, **134**, 28–31.
60. K. M. Folta and L. S. Kaufman, Phototropin 1 is required for high-fluence blue-light-mediated mRNA destabilization, *Plant Mol. Biol.*, 2003, **51**, 609–618.
61. M. Kasahara, T. E. Swartz, M. A. Olney, A. Onodera, N. Mochizuki, H. Fukuzawaa, E. Asamizu, S. Tabata, H. Kanegae, M. Takano, J. M. Christie, A. Nagatani and W. R. Briggs, Photochemical properties of the flavin mononucleotide-binding domains of the phototropins from Arabidopsis, Rice, and *Chlamydomonas reinhardtii*, *Plant Physiol.*, 2002, **129**, 762–773.
62. K. Huang and C. Beck, Phototropin is the blue-light receptor that controls multiple steps in the sexual life cycle of the green alga *Chlamydomonas reinhardtii*, *Proc. Natl. Acad. Sci. USA*, 2003, **100**, 6269–6274.
63. A. Onodera, S.-G. Kong, M. Doi, K.-I. Shimizaki, J. Christie, N. Mochizuki and A. Nagatani, Phototropin from Chlamydomonas reinhardtii is functional in Arabidopsis thaliana, *Plant Cell Physiol.*, 2005, **46**, 367–374.
64. D. E. Somers, T. F. Schultz, M. Milnamow and S. A. Kay, ZEITLUPE encodes a novel clock-associated PAS protein from Arabidopsis, *Cell*, 2000, **101**, 319–329.
65. J. A. Jarillo, J. Capel, R.-H. Tang, H.-Q. Yang, J. M. Alonso, J. R. Ecker and A. R. Cashmore, An Arabidopsis circadian clock component interacts with both cry1 and phyB, *Nature*, 2001, **410**, 487–490.
66. T. F. Schultz, T. Kiyosue, M. Yanofsky, M. Wada and S. A. Kay, A role for LKP2 in the circadian clock of Arabidopsis, *Plant Cell*, 2001, **13**, 2659–2670.
67. D. C. Nelson, J. L. Lasswell, L. E. Rogg, M. A. Cohen and B. Bartel, *FKF1*, a clock-controlled gene that regulates the transition to flowering in *Arabidopsis*, *Cell*, 2000, **101**, 331–340.
68. M. Koornneef, E. Rolff and C. J. P. Spruitt, Genetic control of hypocotyl elongation in *Arabidopsis thaliana* (L.) Heynh, *Zeit. Pflanzenphysiol.*, 1980, **100**, 147–160.
69. A. Sancar, Structure and function of DNA photolyase, *Biochemistry*, 1994, **33**, 2–9.
70. C. Lin, D. E. Robertson, M. Ahmad, A. A. Raibekas, M. S. Jorns, P. L. Dutton and A. R. Cashmore, Association of flavin adenine dinucleotide with the *Arabidopsis* blue light receptor CRY1, *Science*, 1995, **269**, 968–970.
71. K. Malhotra, S.-T. Kim, A. Batschauer, L. Dawut and A. Sancar, Putative blue-light photoreceptors from *Arabidopsis thaliana* and *Sinapis alba* with a high degree of sequence homology to DNA photolyase contain two photolyase cofactors but lack DNA repair activity, *Biochemistry*, 1995, **34**, 6892–6899.
72. C. Lin, M. Ahmad, D. Gordon and A. R. Cashmore, expression of an *Arabidopsis* gene in transgenic tobacco results in hypersensitivity to blue, UV-A, and green light, *Proc. Natl. Acad. Sci. USA*, 1995, **92**, 8423–8427.

73. M. Ahmad, C. Lin and A. R. Cashmore, Mutations throughout an Arabidopsis blue-light photoreceptor impair blue-light-responsive anthocyanin accumulation and inhibition of hypocotyl elongation, *Plant J.*, 1995, **8**, 653–658.
74. C. Lin, M. Ahmad, J. Chan and A. R. Cashmore, CRY2: a second member of the Arabidopsis cryptochrome gene family (Accession No. U43397) (PGR96-001), *Plant Physiol.*, 1996, **110**, 1047.
75. C. Lin, Y. Yang, H. Guo, T. Mockler, J. Chen and A. R. Cashmore, Enhancement of blue-light sensitivity of Arabidopsis seedlings by a blue light receptor cryptochrome 2, *Proc. Natl. Acad. Sci. USA*, 1998, **95**, 2686–2690.
76. O. Kleiner, S. Kircher, K. Harter and A. Batschauer, Nuclear localization of the Arabidopsis blue light receptor cryptochrome 2, *Plant J.*, 1999, **19**, 289–296.
77. H. Guo, H. Yang, T. C. Mockler and C. Lin, Regulation of flowering time by *Arabidopsis* photoreceptors, *Science*, 1999, **279**, 1360–1363.
78. S. Cutler, D. W. Ehrhardt, J. S. Griffiths and C. R. Somerville, Random GFP: cDNA fusions enable visualization of subcellular structures in cells of *Arabidopsis* at a high frequency, *Proc. Natl. Acad. Sci. USA*, 2000, **97**, 3718–3723.
79. P. Más, W.-Y. Kim, D. E. Somers and S. A. Kay, Targeted degradation of TOC1 by ZTL modulates circadian function in *Arabidopsis thaliana*, *Nature*, 2003, **426**, 567–570.
80. H. Q. Yang, R. H. Tang and A. R. Cashmore, The signaling mechanism of Arabidopsis CRY1 involves direct action with COP1, *Plant Cell*, 2001, **13**, 2573–2587.
81. T. Kleine, P. Lockhart and A. Batschauer, An *Arabidopsis* protein closely related to *Synechocystis* cryptochrome is targeted to organelles, *Plant J.*, 2003, **35**, 93–103.
82. T. Kanagae and M. Wada, Isolation and characterization of homologues of plant blue-light receptor (cryptochrome) genes from the fern *Adiantum capillus-veneris*, *Mol. Gen. Genet.*, 1998, **259**, 345–353.
83. T. Imaizumi, T. Kanegae and M. Wada, Cryptochrome nucleocytoplasmic distribution and gene expression are regulated by light quality in the fern *Adiantum capillus-veneris*, *Plant Cell*, 2000, **12**, 81–95.
84. W.-O. Ng and H. B. Pakrasi, DNA photolyase homologs are the major UV resistance factors in the cyanobacterium *Synechocystis* sp. PCC 6803, *Mol. Gen. Genet.*, 2001, **264**, 924–930.
85. S. Gallagher, T. W. Short, P. M. Ray, L. H. Pratt and W. R. Briggs, Light-mediated changes in two proteins found associated with plasma membrane fractions from pea stem sections, *Proc. Natl. Acad. Sci. USA*, 1988, **85**, 8003–8007.
86. W. R. Briggs, J. M. Christie and M. Salomon, Phototropins: a new family of flavin-binding blue light receptors in plants, *Antioxidants Redox Signaling*, 2001, **3**, 775–788.
87. J. P. Khurana and K. L. Poff, Mutants of *Arabidopsis thaliana* with altered phototropism, *Planta*, 1989, **178**, 400–406.
88. P. Reymond, T. W. Short, W. R. Briggs and K. L. Poff, Light-induced phosphorylation of a membrane protein plays an early role in signal transduction for phototropism in *Arabidopsis thaliana*, *Proc. Natl. Acad. Sci. USA*, 1992, **89**, 4718–4721.
89. J. M. Palmer, T. W. Short and W. R. Briggs, Correlation of blue light-induced phosphorylation to phototropism in *Zea mays*, *L. Plant Physiol.*, 1993, **102**, 1219–1225.
90. E. Liscum and W. R. Briggs, Mutations in the *NPH1* locus of Arabidopsis disrupt the perception of phototropic stimuli, *Plant Cell*, 1995, **7**, 473–485.

91. E. Huala, P. W. Oeller, E. Liscum, I.-S. Han, E. Larsen and W. R. Briggs, *Arabidopsis* NPH1: a protein kinase with a putatuve redox-sensing domain, *Science*, 1997, **278**, 2120–2123.
92. B. I. Taylor and I. Zhulin, PAS domains: internal sensors of oxygen redox potential, and light,, *Microbiol. Mol. Biol. Rev*, 1999, **63**, 479–506.
93. J. M. Christie, P. Reymond, G. K. Powell, P. Bernasconi, A. A. Raibecas, E. Liscum and W. R. Briggs, *Arabidopsis* NPH1: a flavoprotein with the properties of a photoreceptor for phototropism, *Science*, 1998, **282**, 1698–1701.
94. J. M. Christie, M. Salomon, K. Nozue, M. Wada and W. R. Briggs, LOV (light, oxygen, or voltage) domains of the blue-light photoreceptor phototropin (nph1): binding sites for the chromophore flavin mononucleotide, *Proc. Natl. Acad. Sci. USA*, 1999, **96**, 8779–8783.
95. J. J. Jarillo, M. Ahmad and A. R. Cashmore, NPL1 (Accession No. AF053941): a second member of the serine/threonine kinase family of Arabidopsis. (PGR98-100), *Plant Physiol.*, 1998, **117**, 719.
96. T. W. Short, P. Reymond and W. R. Briggs, A pea plasma membrane protein exhibiting blue light-induced phosphorylation retains photosensitivity following triton solubilization, *Plant Physiol.*, 1993, **101**, 647–655.
97. K. Sakamoto and W. R. Briggs, Cellular and subcellular localization of phototropin 1, *Plant Cell*, 2002, **14**, 1723–1735.
98. Harada, T. Sakai and K. Okada, phot1 and phot2 mediate blue light-induced transient increases in cytosolic Ca^{2+} differently in *Arabidopsis* leaves, *Proc. Natl. Acad. Sci. USA*, 2003, **100**, 8583–8588.
99. K. Huang, T. Kunkel and C. F. Beck, Localization of the blue-light receptor phototropin to the flagella of the green alga *Chlamydomonas reinhardtii*, *Mol. Biol. Cell*, 2004, **15**, 3605–3614.
100. J. Adams, R. Kelso and L. Cooley, The kelch repeat superfamily of proteins: propellers of cell function, *Trends Cell Biol.*, 2000, **10**, 17–24.
101. T. Imaizumi, T.F. Schultz, F.G. Harmon, S. Ho and S.A. Kay, FKF1 F-box protein mediates cyclic degradation of a repressor of *CONSTANS* in *Arabidopsis*, *Science*, 2005, **309**, 293–297.
102. H. Wang, L.-G. Ma, J.-M. Li, H.-Y. Zhao and X.-W. Deng, Direct interaction of Arabidopsis cryptochromes with COP1 in light control of development, *Science*, 2001, **294**, 154–158.
103. M. Ahmad, J. A. Jarillo, O. Smirnova and A. R. Cashmore, The CRY1 blue light photoreceptor of *Arabidopsis* interacts with phytochrome A *in vitro*, *Mol. Cell*, 1998, **1**, 939–948.
104. D. Shalitin, H. Yang, T. C. Mockler, M. Maymon, H. Guo, G. C. Whitelam and C. Lin, Regulation of Arabidopsis cryptochrome 2 by blue light-dependent phosphorylation, *Nature*, 2002, **417**, 763–767.
105. J.-P. Bouly, B. Giovani, A. Djamei, M. Mueller, A. Zeugner, E. A. Dudkin, A. Batschauer and M. Ahmad, Novel ATP-binding and autophosphorylation activity associated with *Arabidopsis* and human cryptochrome-1, *Eur. J. Biochem.*, 2003, **270**, 2921–2928.
106. D. Shalitin, X. Yu, M. Maymon, T. Mockler and C. Lin, Blue light-dependent *in vivo* and *in vitro* phosphorylation of Arabidopsis cryptochrome 1, *Plant Cell*, 2003, **15**, 2421–2429.
107. Y. F. Li, P. F. Heelis and A. Sancar, Active site of DNA photolyase: tryptophan-306 is the intrinsic hydrogen atom donor essential for flavin radical photoreduction and DNA repair *in vitro*, *Biochemistry*, 1991, **30**, 6322–6329.

108. H. W. Park, S. T. Kim, A. Sancar and J. Deisenhofer, Crystal structure of DNA photolyase from *Escherichia coli*, *Science*, 1995, **268**, 1866–1872.
109. M. S. Cheung, I. Daizadeh, A. A. Stuchebrukhov and P. F. Heelis, Pathways of electron transfer in *Escherichia coli* DNA photolyase: trp(sup 306) to FADH, *Biophys. J.*, 1999, **76**, 1241–1249.
110. C. Aubert, M. H. Vos, P. Mathis, A.P. M. Eker and K. Brettel, Intraprotein radical transfer during photoactivation of DNA photolyase, *Nature*, 2000, **405**, 586–590.
111. C. Aubert, P. Mathis, A. P. M. Eker and K. Brettel, Intraprotein electron transfer between tyrosine and tryptophan in DNA photolyase from *Anacystis nidulans*, *Proc. Natl. Acad. Sci. USA*, 1999, **96**, 5423–6527.
112. B. Giovani, M. Byrdin, M. Ahmad and K. Brettel, Light-induced electron transfer in a cryptochrome blue-light receptor, *Nature Structural Biol.*, 2003, **10**, 489–490.
113. A. Zeugner, M. Byrdin, J.-P. Bouly, N. Bakrim, B. Giovani, K. Brettel and M. Ahmad, Light-induced electron transfer in *Arabidopsis* cryptochrome-1 correlates with *in vivo* function, *J. Biol. Chem.*, 2005, **280**, 19437–19440.
114. E. Knieb, M. Salomon and W. Rüdiger, Tissue-specific and subcellular localization of phototropin determined by immunoblotting, *Planta*, 2004, **218**, 843–851.
115. M. Salomon, E. Knieb, T. von Zeppelin and W. Rüdiger, Mapping of low- and high-fluence autophosphorylation sites in phototropin 1, *Biochemistry*, 2003, **42**, 4217–4225.
116. T. Kinoshita and K.-I. Shimizaki, Blue light activates the plasma membrane H^+ ATPase by phosphorylation of the C-terminus in stomatal guard cells, *EMBO J.*, 1999, **18**, 5548–5558.
117. T. Kinoshita, T. Emi, M. Tominaga, K. Sakamoto, A. Shigenaga, M. Doi and K. I. Shimizaki, Blue-light- and phosphorylation-dependent binding of a 14-3-3 protein to phototropins in stomatal guard cells of broad bean, *Plant Physiol.*, 2003, **133**, 1453–1463.
118. M. Salomon, J. M. Christie, E. Knieb, U. Lempert and W. R. Briggs, Photochemical and mutational analysis of the FMN-binding domains of the plant blue light receptor, *phototropin, Biochemistry*, 2000, **39**, 9401–9410.
119. S. M. Miller, V. Massey, D. Ballou, C. H. Williams Jr, M. D. Distefano, M. J. Moore and C. T. Walsh, Use of a site-directed triple mutant to trap intermediates: demonstration that a (C)4-athiol adduct and reduced flavin are kinetically competent intermediates in mercuric ion reductase, *Biochemistry*, 1990, **29**, 2831–2841.
120. M. Salomon, W. Eisenreich, H. Dürr, E. Schleicher, E. Knieb, V. Massey, W. Rüdiger, F. Müller, A. Bacher and G. Richter, An optomechanical transducer in the blue light receptor phototropin from *Avena sativa*, *Proc. Natl. Acad. Sci. USA*, 2001, **96**, 12357–12361.
121. S. B. Corchnoy, T. E. Swartz, J. W. Lewis, I. Szundi, W. R. Briggs and R. A. Bogomolni, Intramolecular proton transfers and structural changes during the photocycle of the LOV2 domain of phototropin 1, *J. Biol. Chem.*, 2003, **278**, 724–731.
122. J. M. Christie, T. E. Swartz, R. A. Bogomolni and W. R. Briggs, Phototropin LOV domains exhibit distinct roles in regulating photoreceptor function, *Plant J.*, 2002, **32**, 205–219.
123. H. Dürr, M. Salomon and W. Rüdiger, Chromophore exchange in the LOV2 domain of the plant photoreceptor phototropin 1 from oat, *Biochemistry*, 2005, **44**, 3050–3055.
124. S. Crosson and K. Moffat, Structure of a flavin-binding plant photoreceptor domain: insights into light-mediated signal transduction, *Proc. Natl. Acad. Sci. USA*, 2001, **98**, 2995–3000.

125. R. Federov, I. Schlichting, E. Hartmann, T. Domratcheva, M. Fuhrmann and P. Hegemann, Crystal structure and molecular mechanism of a light-induced signaling switch: the phot LOV1 domain from *Chlamydomonas reinhardtii*, *Biophys. J.*, 2003, **82**, 2474–2482.
126. S. Crosson and K. Moffat, Photoexcited structure of a plant photoreceptor domain reveals a light-driven molecular switch, *Plant Cell*, 2002, **14**, 1067–1075.
127. T. E. Swartz, P. J. Wenzel, S. B. Corchnoy, W. R. Briggs and R. A. Bogomolni, Vibration spectroscopy reveals light-induced chromophore and protein structural changes in the LOV2 domain of the plant blue-light receptor phototropin 1, *Biochemistry*, 2002, **41**, 7183–7189.
128. T. Iwata, D. Nozaki, S. Tokutomi, T. Kagawa, M. Wada and H. Kandori, Light-induced structural changes in the LOV2 domain of *Adiantum* phytochrome3 studied by low-temperature FTIR and UV-visible spectroscopy, *Biochemistry*, 2003, **42**, 8183–8191.
129. S. M. Harper, L. C. Neil and K. Gardner, Structural basis of a phototropin light switch, *Science*, 2003, **301**, 1541–1544.
130. K. Ataka, P. Hegemann and J. Heberle, Vibrational spectroscopy of an algal phot-LOV1 domain probes the molecular changes associated with blue-light reception, *Biophys. J.*, 2003, **84**, 466–474.
131. A. Losi, T. Kottke and P. Hegemann, Recording of blue light-induced energy and volume changes within the wild-type and mutated phot-LOV1 domain from *Chlamydomonas reinhardtii*, *Biophys. J.*, 2004, **86**, 1051–1060.
132. S. M. Harper, L. C. Neil, I. J. Day, P. J. Hore and K. H. Gardner, Cooperative and chromophore-regulated conformational changes in a phototropin LOV domain monitored by time-resolve NMR spectroscopy, *J. Am. Chem. Soc.*, 2004, **126**, 3390–3391.
133. S. M. Harper, J. M. Christie and K. H. Gardner, Disruption of the LOV/Jα helix interaction activates phototropin kinase activity, *Biochemistry*, 2004, **43**, 16184–16192.
134. T. E. Swartz, S. B. Corchnoy, J. M. Christie, J. W. Lewis, I. Szundi, W. R. Briggs and R. A. Bogomolni, The photocycle of a flavin-binding domain of the blue light photoreceptor phototropin, *J. Biol. Chem.*, 2001, **39**, 36493–36500.
135. J. T. M. Kennis, S. Crosson, M. Gauden, I. H. M. van Stokkum, K. Moffat and R. van Grondelle, Primary reactions of the LOV2 domain of phototropin, a plant blue light photoreceptor, *Biochemistry*, 2003, **42**, 3385–3392.
136. J. T. M. Kennis, I. H. M. van Stokkum, S. Crosson, M. Gauden, K. Moffat and R. van Grondelle, The LOV2 domain of phototropin: a reversible photochromic switch, *J. Am. Chem. Soc.*, 2004, **129**, 4512–4513.
137. T. Kottke, J. Heberle, D. Hehn, B. Dick and P. Hegemann, Phot-LOV1 photocycle of a blue-light receptor domain from the green alga *Chlamydomonas reinhardtii*, *Biophys. J.*, 2003, **84**, 1192–1201.
138. S. D. M. Islam, A. Penzkofer and P. Hegemann, Quantum yield of triplet formation in riboflavin in aqueous solution and of flavin mononucleotide bound to the LOV1 domain of phot1 from *Chlamydomonas reinhardtii*, *Chemical Physics*, 2003, **291**, 97–114.
139. T. A. Schüttrigkeit, C. K. Kompa, M. Salomon, W. Rüdiger and M. E. Michel-Beyerle, Primary photophysics of the FMN binding LOV2 domain of the plant blue light receptor phototropin of *Avena sativa*, *Chem. Physics*, 2003, **294**, 501–508.
140. W. Holzer, A. Penkofer, M. Fuhrmann and P. Hegemann, Spectroscopic characterization of flavin mononucleotide bound to the LOV1 domain of phot1 from *Chlamydomonas reinhardtii*, *Photochem. Photobiol.*, 2002, **75**, 479–487.

141. T. Iwata, S. Tokutomi and H. Kandori, Photoreaction of the cysteine S-H group in the LOV2 domain of *Adiantum* phytochrome3, *J. Am. Chem. Soc.*, 2002, **124**, 11840–11841.
142. M. Gauden, S. Crosson, I. H. M. van Stokkum, R. van Grondelle, K. Moffat and J. T. M. Kennis, Low-temperature and time-resolved spectroscopic characterization of the LOV2 domain of Avena sativa phototropin 1, in *Femtosecond Laser Applications in Biology*, S. Avrillier and J.-M. Tualle (eds), SPIE, Bellingham WA, 2004, 97–104.
143. T. Bednarz, A. Losi, W. Gärtner, P. Hegemann and J. Heberle, Functional variations among LOV domains as revealed by FT-IR difference spectroscopy, *Photochem. Photobiol. Sci.*, 2004, **3**, 575–579.
144. M. Salomon, U. Lempert and W. Rüdiger, Dimerization of the plant photoreceptor phototropin is probably mediated by the LOV1 domain, *FEBS Lett.*, 2004, **572**, 8–10.
145. C. W. M. Kay, E. Schleicher, A. Kuppig, H. Hofner, W. Rüdiger, M. Schleicher, M. Fischer, A. Bacher, S. Weber and G. Richter, Blue light perception in plants, detection and characterization of a light-induced neutral flavin radical in a C450A mutant of phototropin, *J. Biol. Chem.*, 2003, **278**, 10973–110978.
146. E. Schleicher, R. M. Kowalczyk, C. W. M. Kay, P. Hegemann, A. Bacher, M. Fischer, R. Bittl, G. Richter and S. Weber, On the reaction mechanism of adduct formation in LOV domains of the plant blue-light receptor phototropin, *J. Am. Chem. Soc.*, 2004, **126**, 11067–11076.
147. T. Kottke, B. Dick, R. Federov, I. Schlichting, R. Deutzmann and P. Hegemann, Irreversible photoreduction of a mutated phot-LOV1 domain, *Biochemistry*, 2003, **42**, 9854–9862.
148. T. Kagawa, M. Kasahara, T. Abe, S. Yoshida and M. Wada, Function analysis of phototropin 2 using fern mutants deficient in blue light-induced chloroplast avoidance movement, *Plant Cell Physiol.*, 2004, **45**, 416–426.
149. E. Knieb, M. Salomon and W. Rüdiger, Autophosphorylation. electrophoretic mobility and immunoreaction of oat phototropin 1 under UV and blue light, *Photochem. Photobiol.*, 2005, **81**, 177–182.
150. E. C. De Fabo, R. W. Harding and W. Shropshire Jr., Action spectrum between 260 and 800 nanometers for the photoinduction of carotenoid biosynthesis in *Neurospora crassa*, *Plant Physiol.*, 1976, **57**, 440–445.
151. M. L. Sargent and W. R. Briggs, The effects of light on a circadian rhythm of conidiation in Neurospora, *Plant Physiol.*, 1967, **42**, 1504–1510.
152. A. C. Froelich, Y. Liu, J. J. Loros and J. C. Dunlop, White collar-1, a circadian photoreceptor, binding to the *frequency* promoter, *Science*, 2002, **297**, 815–819.
153. Q. He, P. Cheng, Y. Yang, L. Wang, K. H. Gardner and Y. Liu, White collar-1, a DNA binding transcription factor and light sensor, *Science*, 2002, **297**, 840–843.
154. C. Schwerdtfeger and H. Linden, VIVID is a flavoprotein and serves as a fungal blue light photoreceptor for photoadaptation, *EMBO J.*, 2003, 4846–4855.
155. A. Idnurm and J. Heitman, Light controls growth and development via a conserved pathway in the fungal kingdom, *PloS, 3*, 2005, **12**, e95.
156. A. Losi, The bacterial counterparts of plant phototropins, *Photochem. Photobiol. Sci.*, 2004, **3**, 566–574.
157. A. Losi, E. Polverini, B. Quest and W. Gärtner, First evidence for phototropin-related blue-light receptors in prokaryotes, *Biophys. J.*, 2002, **82**, 2627–2634.

Chapter 10

Flavin-Based Photoreceptors in Bacteria

ABA LOSI

Department of Physics-University of Parma, Parco Area delle Scienze, 7/A 43100, Parma, ITALY

Abbreviations

FMN	flavin mononucleotide
FAD	flavin-adenin dinucleotide
phot	phototropin
LOV	Light Oxygen Voltage domains
PAS	PerArntSim domain
Cry	Cryptochrome
PL	photolyase
PAC	photoactivated adenylate cyclase
BLUF	Blue Light-sensing Using FAD domain
STAS	Sulfate Transporter/Anti-Sigma factor antagonist domain
NTP	Nucleoside Triphosphate
HAMP	domain found in Histidine kinases, Adenylyl cyclases, Methyl binding proteins, Phosphatases
Hpt	Histidine phosphotransfer domain
GGDEF, EAL	domains named after their conserved aa motifs
GAF	domain present in phytochrome and cGMP-specific phosphodiesterases
HPK	Histidine Protein Kinase
HB	hydrogen bond
RR	response regulators

Abstract

We review the bacterial proteins related to the flavin-binding plant UVA/Blue-light sensors phototropins (phot), photoactivated adenylyl cyclase (PAC) and cryptochromes

(Cry). Phot and PAC sense light via specialized domains, called respectively LOV (Light, Oxygen, Voltage) and BLUF (Blue Light sensing Using FAD). Genome digging reveals that about 16% of bacteria possess genes encoding for LOV, BLUF or/and Cry proteins. Up to now, their physiological role as photoreceptors has been established only for the BLUF proteins AppA in *Rhodobacter sphaeroides* and Slr1694 in *Synechocystis sp.* PCC 6803, but light-driven reactions have been demonstrated also for LOV proteins and detailed structural information is available for a cyanobacterial Cry. Bacterial LOV and BLUF proteins are highly modular and contain diverse catalytic functions (*e.g.* kinases, phosphodiesterase) associated to the photosensing domain, highlighting their involvement in various signal transduction pathways. This modularity can constitute a powerful tool (and a potential model for protein engineering) in understanding the modalities of interdomain communication in sensor proteins, given that signal transduction can be easily triggered by a light pulse. Thanks to the large spreading of flavin-based bacterial blue-light sensing proteins, phylogenetic analysis may provide important clues to understand how similar proteins have evolved in eukaryotic organisms. Furthermore, investigation of their physiological role is likely to provide us a more comprehensive view of how bacteria "see" their world.

10.1. Introduction

Genome sequencing and annotation projects are revealing that bacteria are equipped with a range of putative photosensor proteins, whose physiological significance remains, in most cases, poorly characterized. The analysis of the different light sensing/light responsive protein modules and the elucidation of their light-triggered reactions, is more and more pointing to a scenario where given photosensing paradigms (*e.g.* retinal-binding proteins, phytochromes, flavin-based photoreceptors) are shared between eukaryots and prokaryots.[1–7]

This parallelism appears to be particularly striking between plant and bacterial photosensors. Phytochromes, the plant sensors for red/far red light, are well represented in both non-photosynthetic and photosynthetic bacteria.[1,7–9] The similarity between plant flavoproteins that function as UVA/blue-light receptors and bacterial putative photosensors has been only recently discovered, but represents a growing and exciting research field.[2,6,10,11] Plant blue-light receptors binding a flavin molecule as chromophore (Figure 1) have been identified during the last years and comprise cryptochromes (Cry), phototropins (phot) and the ZTL/FKF1/LKP2 family[12,13]). In addition, a FAD binding photoactivated adenylyl cyclase (PAC) has been found in *Euglena*.[14] The domain architecture of plant photosensors is depicted in Figure 2. In general, there is already a detailed knowledge of the processes regulated by these photoreceptors in plants and the primary mechanisms of photoreceptor activation and signal transduction are being actively studied. A brief description is given in Section 1.1. The photochemical reactions of flavin-based photosensors are still a matter of debate and necessarily differ from those of photoreceptors containing photoisomerizable chromophores, such as retinal proteins, phytochrome and photoactive yellow protein (PYP).[15] The flavin chromophore and its reactivity are described in Section 1.2.

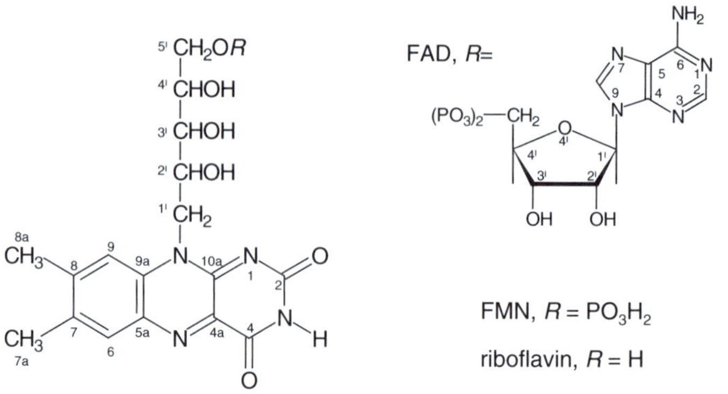

Figure 1. The three flavin chromophores that can be incorporated into known photosensors. Riboflavin consists of a 7,8-dimethyl-isoalloxazine ring linked to D-ribitol. FMN and FAD are riboflavin derivatives

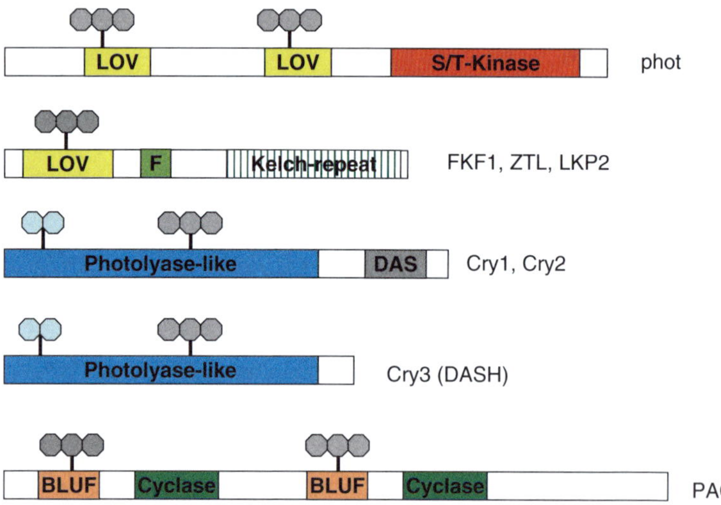

Figure 2. Domain architecture and chromophore binding site of Flavin-based plant blue-light receptors. LOV; BLUF; F; the flavin chromophore is depicted as three annealed rings; the bicyclic symbol represents the antenna chromophore of Cry proteins; see also Section 1.1

In the next sections we will review the bacterial photosensing flavoproteins that are presently detectable through genome digging and sequence comparison, keeping in mind that the scenario is prone to rapidly evolve thanks to ongoing genome projects. The information available up to now on the structure and function of prokaryotic LOV-proteins (phot-related), Cry and BLUF-proteins (PAC related) are reviewed in Section 2. The physiological role of these, in most cases putative, photosensors is discussed in Section 3.1 in view of the known responses to UVA/blue light of bacteria. Finally in Section 3.2 we discuss the evolutionary link between these prokaryotic proteins and their eukaryotic counterparts.

10.1.1. Plant Flavin-Based Blue-Light Receptors: Structure and Function

10.1.1.1. The Phototropin Family

Phototropins (phot1 and phot2) are plant blue-light receptors for phototropism, chloroplast movement, leaf expansion, and stomatal opening.[16,17] In the green alga *Chlamydomonas reinhardtii*, phot controls multiple steps in the sexual life cycle.[18] Phot are membrane associated kinases, typically about 1000 aa in length that undergo UVA/blue light-induced autophosphorylation.[19] Phot proteins possess two *N*-terminal photoactive LOV (Light, Oxygen and Voltage) domains (LOV1 and LOV2), a subset of the PAS (PerArntSim) superfamily,[20,21] and a *C*-terminal serine/threonine kinase domain. Phot-LOV1 and LOV2 bind oxidized flavin mono-nucleotide (FMN) as chromophore and absorb maximally at *ca.* 450 nm (LOV_{447}).[22] Blue-light illumination of phot-LOV domains triggers a photocycle involving the formation of a blue-shifted FMN-cysteine C(4a)-thiol adduct (LOV_{390}) that slowly reverts to LOV_{447} in the dark.[22–26] The photocycle of LOV domains comprises a red-shifted transient species, LOV660 (appearing on the fs time-scale[27]) assigned to the FMN triplet excited state and decaying on the short microsecond time-scale into LOV390.[28,29] Detailed structural information both in the dark (LOV_{447}) and in the photoactivated state (LOV_{390}), are available for the LOV2 domain of phy3[25,30] (a phytochrome-phot hybrid photoreceptor from the fern *Adiantum capillus-veneris*) and for *C. reinhardtii* phot-LOV1.[31] The two structures appear to be very similar, exhibiting a typical α/β PAS fold,[21] and have allowed to identify the residues that interact with FMN, and the structural changes occurring upon light activation. Phot proteins are the first plant blue-light receptors for which identity, function and photochemistry have been elucidated. Their photocycle, being essentially dependent on the formation of the flavin triplet state, has established a new paradigm in the field of photosensory biology, in contrast to chromophore *cis-trans* isomerization.[32] The key questions in the field of phot research remain (a) the mechanism by which the light-induced conformational changes are conveyed from the LOV domains to the effector kinase domain (*vide infra*) and (b) the presence of two LOV domains with the same photochemistry on the same protein, despite the fact that inactivation of LOV1 does not impair the kinase function and photobiological responses.[33]

10.1.1.2. The ZTL/FKF1/LKP2 Family

A LOV domain is also the photosensing unit of the ZTL/FKF1/LKP2 (*Zeitl*upe, *F*lavin-binding *K*elch-repeat *F*-box, *LOV K*elch Protein 2) protein family, a novel group of plant photosensors, involved in the circadian control of periods in plants.[34–36] The LOV domain of ZTL proteins has been shown to bind FMN and to undergo phot-like photochemistry, but the formation of the

FMN-cys adduct was not reversible under the conditions tested.[35] The F-box is known to interact with a ubiquitination complex, thus targeting protein substrates for degradation, and has been found only in eukaryotic protein.[37] The Kelch repeat has typically a β-propeller structure and contains multiple sites for protein–protein contacts.[38] The LOV domains of the ZTL/FKF1/LKP2 family have larger sequence similarity with the LOV domains of the fungal photoreceptor WC-1 (White Collar-1) than with phot-LO*Vs*.[39]

10.1.1.3. Cryptochromes

Cryptochromes (Cry) have large structural and sequence similarity with photolyases (PL).[40] PL catalyze the light-dependent repair of UV-induced cyclobutane pyrimidine dimers or (6–4) pyrimidine-pyrimidone photoproducts in DNA. In PL, the absorption of a photon by the antenna chromophore that can be either a pterin (methenyl tetrahydrofolate, MTHF) or a deazaflavin, is followed by resonance energy transfer to the catalytic FAD chromophore that in turn splits the DNA cyclobutane derivative by non-reductive electron transfer.[41] Albeit the striking homology, Cry lack DNA repair activity and act as blue-light photoreceptors for a variety of growth and development responses, such as inhibition of hypocotyl growth and induction of flowering, circadian rhythms and light-dependent transcriptional regulation.[42] The PL-like core of Cry, of *ca.* 500 amino acids, consists of an *N*-terminal α/β domain and a *C*-terminal α-helical FAD-binding domain, connected by a long loop.[11,43] The "classical" plant Cry1 and Cry2 also possess a 30–250 long *C*-terminal extension, shown to be important for signal transduction,[44] and recently demonstrated to undergo a light-induced conformational change.[45] Overexpression of the *C*-terminal extension of either *Arabidopsis thaliana* Cry1 or Cry2 results in a constitutive photomorphogenic phenotype, in which transgenic plants exhibit the light-driven inhibition of hypocotyl growth even in darkness. However, Cry proteins were identified in plants that have no *C*-terminal extension, but still exhibit photoreceptor function.[46,47] A similar Cry (Sll1629) had been identified in *Synechocystis* PCC 6803[11] and in *A. thaliana* (AtCry3).[48] This new class is more similar to animal Cry than to Type I and Type CPD photolyases and Cry1 or Cry2. To underscore this similarity, the term DASH (*Drosophila*, *Arabidopsis*, *Synechocystis* and *Homo* cryptochrome) was created to designate this novel Cry proteins.[11] Actually, since the recent discovery of true-type Sll1629-like Cry in vertebrates and fungi, they are now thought to form a real new and distinct family,[49] and the term DASH may not be the most appropriate. However, we will use this term in the following as a matter of clearness.

Open key questions for Cry are the mechanism of light activation and the oxidation state of the active chromophore, in other words, the nature of the primary photochemical event. Photo-induced electron transfer from aromatic amino acids to oxidized FAD seems to be important in this process,[50] albeit the same reactions in PL serve merely to regenerate a catalytically active, fully reduced $FADH^-$.[41] Nevertheless, also in PL the photoreduction of FAD by means of intraprotein electron transfer is apparently not part of the photolyase photocycle under physiological conditions.[51]

10.1.1.4. Photoactivated Adenylyl Cyclase in Euglenoids

An additional blue-light receptor flavoprotein, a PAC has been recently identified in the unicellular freshwater protist *Euglena gracilis,* and found to be responsible for the photophobic response[14] and phototaxis.[52] PAC is a dimer formed by two similar subunits, PACα and PACα. Each of the subunits contain two flavin binding domains, originally called F1 and F2, followed by an adenylyl cyclase domain.[14] Besides establishing the long sought identity of the blue-light photosensor responsible for the photophobic response and for phototaxis,[52] this work demonstrated that light-sensing and enzymatic synthesis of a cyclic nucleotide can be coupled on the same protein and that the latter activity is regulated by light-excitation of the flavin-binding domain.[14] Quite surprisingly, F1 and F2 turned to be very similar to the *N*-terminal region of the flavoprotein AppA, involved in the regulation of photosynthesis genes in *Rhodobacter sphaeroides*.[53,54] AppA binds FAD at the *N*-terminal part and has been shown to act as a blue-light receptor and a redox sensor in the regulation of photosynthesis.[55,56] The novel FAD binding domain was named BLUF (Blue Light sensing Using FAD)[2] and, with the exception of PAC proteins found in phototrophic euglenoids,[57] has been detected only in bacteria. Its peculiar photocycle, involving the reversible formation of a red shifted intermediate[55] and its structure[58] represent completely new paradigms in the field of blue-light sensing (*vide infra*).

10.1.2. The Flavin Chromophore: Reactivity and Biosynthesis

The selection of a flavin as a chromophore for light-sensing offers several advantages, at the same time posing some perplexing questions. Flavins are ubiquitous, given their involvement in a variety of enzymatic reactions. They can accomplish one-or two electron transfer reactions and they can function as electrophiles and nucleophiles, with covalent intermediates of flavin and substrate frequently being involved in catalysis.[59] Light excitation obviously results in charge redistribution and altered redox potential, giving rise to a variety of possible photochemical reactions. On the other hand, the same photochemical versatility renders flavins powerful photosensitizers, with a mixed Type I–Type II (electron- and energy transfer, respectively) sensitization mechanism.[60] The excited triplet state of the oxydized form generates singlet oxygen ($^1O_2$) with high quantum yield, $\Phi_\Delta = 0.5$ in aqueous solution, via energy transfer to molecular oxygen.[61] Furthermore, whereas the reduction potential of flavin is about -0.3 *V* in neutral aqueous solution, the redox potential of the triplet flavin is shifted to 1.7 V, and can oxidize several electron rich biological substrates, such as aromatic amino acids and DNA bases[62,63] (and references therein). Flavin photoreceptors must therefore have evolved mechanisms to minimize the photooxidation. In LOV-proteins the triplet state is efficiently quenched by the reactive cysteine, thus avoiding singlet oxygen sensitized formation (Sections 1.1.a and 2.1.a). In BLUF proteins the formation of the signalling state occurs from the excited singlet state of the FAD chromophore with ultrafast kinetics and the triplet state is formed with very low efficiency.[64]

In Cry proteins photo-induced electron transfer reactions involve the flavin chromophore and aromatic amino acids.[50] Aside from their possible role in triggering photosensing and/or in regenerating the active form of the chromophore, these reactions that involve a well defined pool of amino acids within the protein, may prevent energy transfer to O_2 and unspecific electron transfer reactions that would result in photodamage.

Another puzzling question is related to the fact that flavins exists in different oxidation states, having different spectral properties: the absorption maximum of the semiquinone radical, for example, is strongly red-shifted with respect to the fully oxidized state.[65,66] As a consequence, the wavelength sensitivity of a flavin photosensor could vary with the redox properties of the environment. Indeed, it has been suggested that plant Cry may be responsible both for blue- and green-light sensing, depending on the oxidation state of the bound flavin.[67]

Riboflavin is biosynthesized by plants, fungi, bacteria and archaea.[68,69] The biosynthetic pathway of bacteria is summarized in Figure 3 and has been recently reviewed.[68]

Riboflavin is synthetized from GTP and D-ribulose-5-phosphate via a complex enzymatic pathway. Briefly, GTP cyclohydrolase II affords 2,5-diamino-6-ribosylamino-4(3*H*)-pyrimidinone. Reduction of the ribose side-chain, deamination and dephosphorylation afford 5-amino-6-ribitylamino-2,4(1*H*,3*H*)-pyrimidinedione (2) which is converted into 6,7-dimethyl-8-ribityllumazine (5) by condensation with 3,4-dihydroxy-2-butanone 4-phosphate (4), obtained from ribulose 5-phosphate (3) by a mechanistically complex rearrangement reaction. Dismutation of 6,7-dimethyl-8-ribityl-lumazine yields riboflavin and 5-amino-6-ribitylamino-2,4(1*H*,3*H*)-pyrimidinedione (2), which is recycled in the biosynthetic pathway. The conversion of riboflavin to FMN is catalyzed by riboflavin kinase, consuming one molecule of ATP. FMN adenylyltransferase catalyzes the formation of FAD and pyrophosphate from FMN and ATP.[70,71]

Given that flavin-photosensors are generally heterologously expressed, they take up the chromophore from the flavin pool of the host (often *Escherichia coli*). In some cases, if a chromophore preference is not absolute, this can produce heterogenous samples, also because overexpression may lead to depletion of the specific flavin to be incorporated. The BLUF domain of AppA, for example, is capable to bind riboflavin, FMN or FAD, with negligible modifications in the photocycle.[72] Besides the obvious problems that this aspect may cause during biophysical and biochemical studies, it poses the question of which is the physiologically relevant flavin chromophore.

10.2. The Bacterial Blue-Light Sensing Flavoproteins: Structure and Function of LOV Proteins, DASH Cryptochromes, and the BLUF Paradigm

Genes encoding potential flavin-binding photosensors are detectable in a very large and steadily increasing number of bacterial genomes (Figure 4). Presently

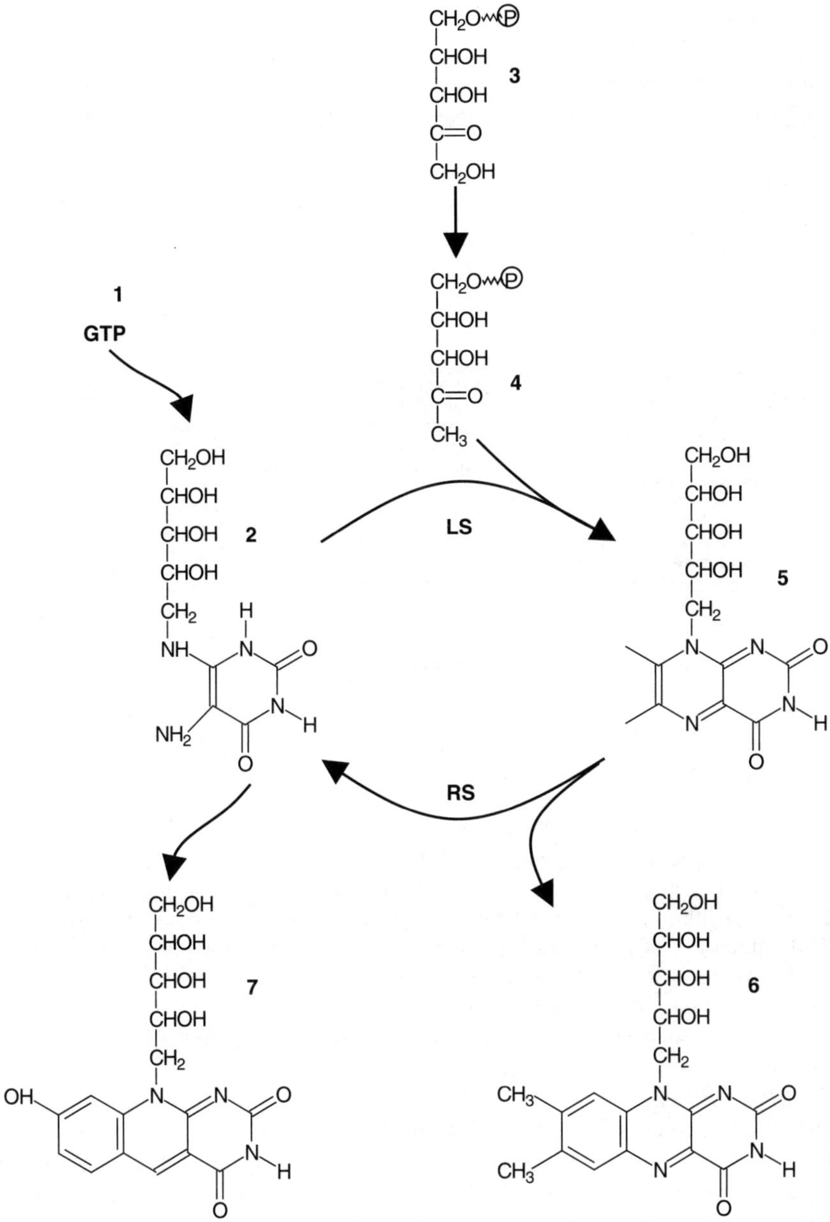

Figure 3. Simplified scheme for biosynthesis of flavins. (2) 5-amino-6-ribitylamino-2,4(1H,3H)-pyrimidinedione; (3) ribulose 5-phosphate; (4) 3,4-dihydroxy-2-butanone 4-phosphate; (5) 6,7-dimethyl-8-ribityllumazine; (6) riboflavin; (7) 7,8-didemethyl-8-hydroxy-5-deazariboflavin. (LS) lumazine synthase; (RS) riboflavin synthase. Modified from ref. [69] The conversion of (1) into (2) is a complex enzymatic pathway (see 1.1.e)

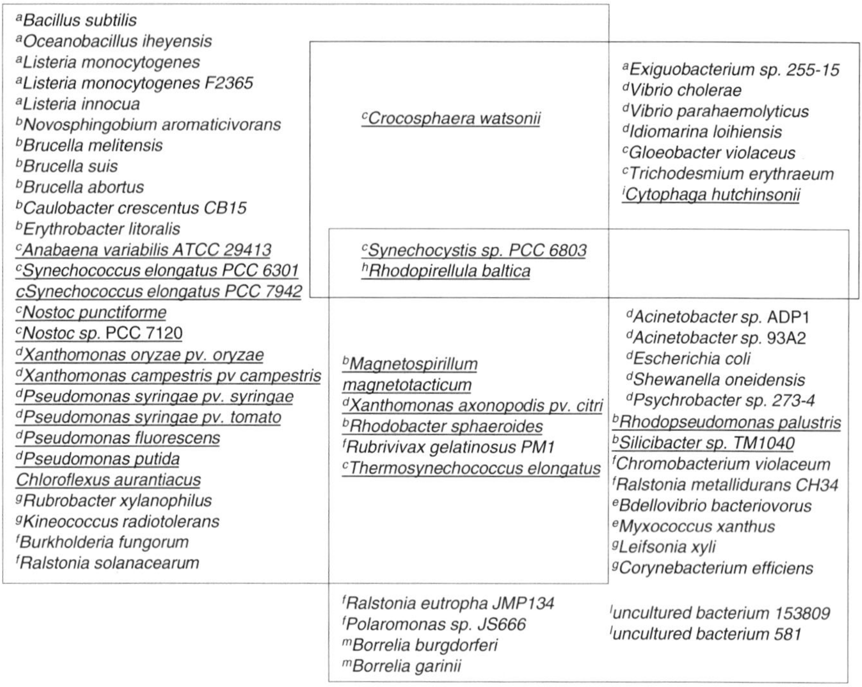

Figure 4. An overview of bacteria with genes encoding for putative flavin-based photosensors. Dark frame, DASH cryptochrome; blue frame, BLUF proteins; purple frame, LOV proteins. Underlined are bacteria with also phytochrome-like genes. Among the listed bacteria, only Nostoc sp. PCC 7120 and Exiguobacterium sp. 255–15 possess genes encoding for opsins. (a) Firmicutes; (b) α-Proteobacteria; (c) Cyanobacteria; (d) γ-Proteobacteria; (e)δ-Proteobacteria; (f) β-Proteobacteria; (g) Actinobacteria; (h) Planctomycetes; (i) Sphingobacteria; (l) unclassified Proteobacteria; (m) Spirochaetes. The BLAST algorithm[73] was run through all gene sequences deposited at EMBL[74] and NCBI databases.[75] The expectation value (E) threshold was set at 0.0001 and sequences were filtered for low-complexity regions. Only sequences with $E \leq 10^{-10}$ were accepted. Visual inspection for key residues was accomplished according to the parameters explained in the text for the three classes of photosensors

a BLAST[73] search through the EMBL[74] and NCBI databases,[75] reveals 46 LOV proteins (Table 1), 36 BLUF proteins (Table 2) and 11 DASH-Cry (Figure 4) (in the genomes of 35, 25 and 11 bacterial species, respectively). Searching through not yet deposited genome sequences, within individual genome projects databanks, may results in an even larger array of potential photosensors.[2]

Albeit their role is largely unexplored, flavin photosensors in bacteria are unequivocally demonstrated to exhibit light-driven reactions.[6,10,55,76,77] The structure of a bacterial DASH-Cry, the first crystallized cryptochrome, shows the bound FAD chromophore[11] and a Cry from *Vibrio cholerae* (VcCry1) has been purified with bound FAD and MTHF.[78] It is conceivable that, in the next

Table 1. LOV proteins in prokaryotes

Organism		Phylum	LOV protein (length)	Domain structure
B. subtilis		F	**O34627-YtvA(261)**	LOV + STAS
Oceanobacillus iheyensis		F	Q8ESN8 (264)	LOV + STAS
Listeria monocytogenes		F	P58724 (253)	LOV + STAS
Listeria monocytogenes F2365		F	Q722B8 (253)	LOV + STAS
Listeria innocua		F	Q92DM1 (253)	LOV + STAS
R. sphaeroides		P(α)	ZP_00007036 (176)	Short LOV
Novosphingobium aromaticivorans	(1)	P(α)	ZP_00302540 (364)	LOV + Kinase
	(2)		ZP_00305604 (196)	LOV + HTH
B. melitensis		P(α)	Q8YC53 (489)	LOV + PAS + Kinase
B. abortus		P(α)	YP_223408 (458)	LOV + PAS + Kinase
B. suis		P(α)	Q8FW73 (463)	LOV + PAS + Kinase
C. crescentus CB15		P(α)	**Q9ABE3 (449)**	LOV + Kinase
M. magnetotacticum	(1)	P(α)	ZP_00051334 (425)	LOV + Kinase
	(2)		ZP_00052303 (358)	LOV + Kinase
E. litoralis HTCC2594	(1)	P(α)	ZP_00376834 (222)	LOV + HTH
	(2)		ZP_00376813 (346)	LOV + Kinase
	(3)		ZP_00376253 (362)	LOV + Kinase
	(4)		ZP_00377191 (368)	LOV + Kinase
Burkholderia fungorum		P(β)	ZP_00034273 (1036)	LOV + multidomain Kinase + RR
Ralstonia solanacearum		P(β)	Q8XT61 (1178)	LOV + multidomain GGDEF + EAL
Rubrivivax gelatinosus PM1		P(β)	ZP_00244844 (1317)	LOV + multidomain Kinase + RR
X. axonopodis pv. citri		P(γ)	Q8PJH6 (540)	LOV + Kinase + RR
X. campestris pv campestris		P(γ)	Q8P827 (540)	LOV + Kinase + RR
X. oryzae pv. oryzae		P(γ)	Q5GY20 (428)	LOV + Kinase + RR
P.s syringae pv. syringae		P(γ)	ZP_00124092	LOV + Kinase + RR
P. syringae pv. tomato		P(γ)	Q881J7	LOV + Kinase + RR
P. fluorescens		P(γ)	ZP_00084650 (158)	Short LOV
P. putida KT2440	(1)	P(γ)	Q88E39 (142)	Short LOV
	(2)		**Q88JB0** (151)	Short LOV
R. xylanophilus		Ac	ZP_00186378 (581)	LOV + Serine Phosphatase
K. radiotolerans	(1)	Ac	ZP_00189651 (743)	GGDEF + LOV + EAL
	(2)		ZP_00226303 (713)	LOV + Serine Phosphatase
C. aurantiacus		Chl	ZP_00020459 (242)	LOV + C-terminus
T. elongatus		Cya	Q8DJE3 (1353)	LOV + multidomain Kinase + RR
Anabaena variabilis ATCC 29413	(1)	Cya	ZP_00161051 (1021)	LOV + multidomain
	(2)		ZP_00351596 (1785)	GGDEF + EAL Multidomain Kinase + RR
Synechocystis sp. PCC 6803		Cya	Q55576	LOV + multidomain GGDEF + EAL

Table 1 (*continued*)

Organism		Phylum	LOV protein (length)	Domain structure
S. elongatus PCC 6301	(1)	Cya	Q5N2F7(578)	LOV + GGDEF + EAL
	(2)		Q5N5M8 (929)	LOV + multidomain GGDEF + EAL
Synechococcus elongatus PCC 7942		Cya	ZP_00202158 (929)	LOV +multidomain GGDEF + EAL
N. punctiforme	(1)	Cya	ZP_00111211 (1043	LOV + multidomain GGDEF + EAL
	(2)		ZP_00105980 (1403)	LOV + multidomain Kinase- RR
Nostoc sp. PCC 7120	(1)	Cya	Q8YSB9 (1021)	LOV + multidomain GGDEF + EAL
	(2)		Q8YT51 (1817)	LOV + multidomain Kinase + RR
C. watsonii		Cya	ZP_00174702 (297)	Short LOV
Rhodopirellula baltica		Pl	Q7USG5	LOV + multidomain Kinase + RR
H. marismortui	(1)	EuA	YP_138072 (2306)	LOV + multidomain Kinase
	(2)		YP_134861 (726)	LOV + multidomain Kinase + RR
	(3)		YP_135535 (748)	LOV + multidomain Kinase + RR

Notes: F: Firmicutes, low G + C Gram positives; P: Proteobacteria (the class is indicated in parenthesis); Ac: Actinobacteria, high G + C Gram-positives; Chl: Chloroflexi; Cya: Cyanobacteria; Pl: Planctomycetes; EuA: Euryarchaeota.

years, we will obtain increasing information about the structure and function of bacterial flavin-based photosensors.

10.2.1. The Prokaryotic LOV Proteins Related to Plant Phototropins

The first evidence for the occurrence of LOV-proteins in bacteria was presented by Huala *et al.* who compared the sequences of phot-LOV domains with those from *Bacillus subtilis* YtvA (O34627, SwissProt-TrEMBL accession number) and *Synechocystis* PCC 6803 Slr0359.[19] Extended lists have been recently reported[3,6] and an update, including 41 prokaryotic proteins, is shown in Table 1. LOV proteins appear to be the most spread among putative blue-light photosensors, in many cases present within the same organism with bacterial phytochromes (Figure 4). They are the sole putative flavin-photosensors up to now represented in an Archaea (*Haloarcula marismortui*, see Table 1). So far phot-like light-induced reactions have been demonstrated for three bacterial proteins: YtvA from *B.subtilis*,[10,79] a LOV-kinase from *Caulobater crescentus*[6] and a short LOV protein from *Pseudomonas putida* (Q88JB0) (Krauss *et al*, submitted). In prokaryots the LOV light sensing module is coupled to diverse

Table 2. BLUF proteins in prokaryotes

Organism	Phylum	BLUF protein (length)	
R. sphaeroides	P(α)	**AppA** (450)	BLUF+C-terminus
		ZP_00007465 (127)	Short BLUF
		ZP_00007954 (140)	Short BLUF
Rhodopseudomonas palustris	P(α)	Q6N4H0 (156)	Short BLUF
		Q6NCF0 (147)	Short BLUF
		Q6N1X0 (157)	Short BLUF
Silicibacter sp. TM1040	P(α)	ZP_00335965 (144)	Short BLUF
		ZP_00335963 (140	Short BLUF
Chromobacterium violaceum	P(β)	Q7NSR1 (141)	Short BLUF
Ralstonia metallidurans CH34	P(β)	ZP_00276275 (147)	Short BLUF
Ralstonia eutropha JMP134	P(β)	ZP_00166643(147)	Short BLUF
Polaromonas sp. JS666	P(β)	ZP_00364277 (136)	Short BLUF
		ZP_00360426 (128)	Short BLUF
Rubrivivax gelatinosus PM1	P(β)	ZP_00242636 (142)	Short BLUF
		ZP_00245230 (97)	Short BLUF
E. coli	P(γ)	**YcgF (403)**	BLUF+EAL
Acinetobacter sp. (strain ADP1)	P(γ)	Q6FAI1 (149)	Short BLUF
		Q6FAJ5 (155)	Short BLUF
		Q6FAH9 (163)	Short BLUF
		Q7BC36 (149	Short BLUF
Acinetobacter sp. 93A2.	P(γ)	Q933L6 (149)	Short BLUF
Shewanella oneidensis	P(γ)	Q8EBA3 (140)	Short BLUF
Psychrobacter sp. 273-4	P(γ)	ZP_00146604 (210)	BLUF + C-terminus
Xanthomonas axonopodis pv. citri	P(γ)	Q8PHH6 (147)	Short BLUF
		Q8PKP9 (158)	Short BLUF
Bdellovibrio bacteriovorus	P(δ)	Q6MPS8 (142)	Short BLUF
M. xanthus	P(δ)	Q93SK7 (253)	BLUF + C-terminus
Leifsonia xyli	Ac	Q6AGS2 (145)	Short BLUF
Corynebacterium efficiens	Ac	Q8FMY2 (158)	Short BLUF
T. elongatus	Cya	**Tll0078** (143)	Short BLUF
Synechocystis sp. PCC 6803	Cya	**Slr1694 (150)**	Short BLUF
Rhodopirellula baltica	Pl	Q7UXU1 (141)	Short BLUF
uncultured bacterium 153809	P	Q8KZ47 (146)	Short BLUF
uncultured bacterium 581	P	Q6SFE3 (159)	Short BLUF
Borrelia burgdorferi	Spi	Q66ZB9 (250)	BLUF + C-terminus
Borrelia garinii	Spi	Q6ASF9 (250)	BLUF + C-terminus

Notes: P: Proteobacteria (the class is indicated in parenthesis); Ac: Actinobacteria; Chl: Chloroflexi; Cya: Cyanobacteria; Pl: Planctomycetes; Spi: Spirochaetes

effectors domains, such as kinases (similar to phot), phosphodiesterases, response regulators, DNA-binding transcription factors and regulators of stress sigma factors (Figure 5, Table 1). These effector modules indicate that bacterial LOV-proteins are part of the cell signaling machinery, although a light-driven regulation of their activity remains to be demonstrated, as well as their light dependent physiological role. Therefore, besides the intrinsic interest regarding their structure, function and physiological role, they also represent a powerful tool to understand fundamental questions in the field of LOV-based photosensors: the light-induced reactions centered on the LOV domain are transmitted to effector partners by means of the same molecular mechanisms?

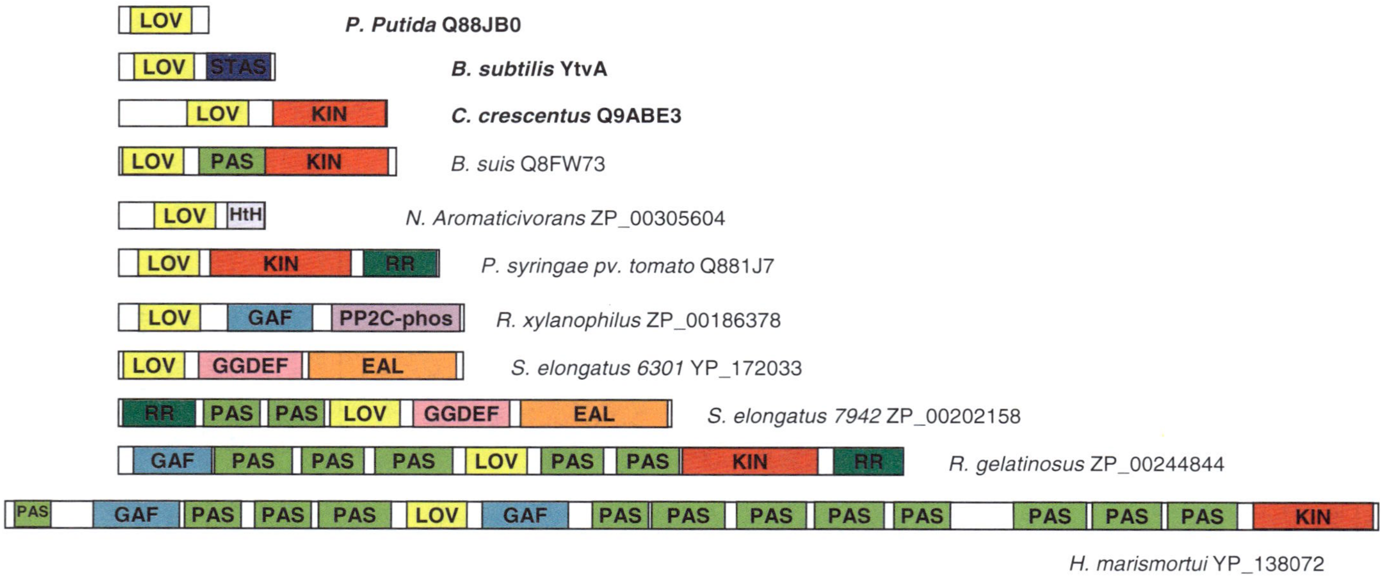

Figure 5. Architecture of selected prokaryotic LOV proteins. In bold the three proteins for which flavin binding at the LOV domain and phot-like photochemical reactions have been demonstrated. Abbreviations are as follows: STAS = domain found in Sulphate Transporters and Anti Sigma factors antagonists; KIN = Kinase; HtH = Helix-turn-Helix DNA-binding motif; RR = response regulator, receiver domain; GGDEF and EAL = domains named after characteristic aa sequences; PAS = PerArntSim domain superfamily; GAF = domain present in phytochrome and cGMP-specific phosphodiesterases. Domain analysis was performed at the European Bioinformatics Institute (EBI) using the InterproScan service.[178]

do LOV domains interact with partner domains by means of the same molecular surface? and last by not least, why only one LOV domain is present in the bacterial proteins, whereas phot possess two of such units organized in tandem?

10.2.1.1. The Bacterial LOV Domains

Sequence comparison with *Avena sativa* and *Arabidopsis thaliana* phot1, *C. reinhardtii* phot and *Adiantum* phy3, shows that bacterial LOV domains present intermediate characteristics between LOV1 and LOV2. In some cases the similarity with LOV2 is markedly higher, with an identity of about 49%, against *ca.* 41% with LOV1 (*e.g.* in *B. subtilis, Pseudomonas syringae, Xanthomonas axonopodis, Xanthomonas campestris* and *Novosphingobium aromaticivorans* ZP_00093141 proteins). Interestingly, the LOV domains from photosynthetic bacteria (*e.g.* cyanobacteria), showe the lowest degree of identity with phot-LOVs (<*ca.* 40%). The alignment of prokaryotic LOV domains with Ac phy3-LOV2 and Cr-phot-LOV1, for which the structure is known,[30,31] is shown in Figure 6.

Secondary structure prediction of bacterial LOV domains does not present any appreciable deviation from LOV1 or LOV2, as expected given the high sequence similarity and the existence of only few gaps in the alignment. Nevertheless, in some cases, aa substitutions that could impair FMN-binding are present. In the following we will refer to the aa numbering and protein structure of *C. reinhardtii* phot-LOV1.[31] The reactive C57 is found in most bacterial LOV domains within the conserved G*X*NCRFLQ motif. In LOV crystal structures, N56, R58 and Q61 are part of the hydrogen bond (HB) network that stabilizes the FMN chromophore (Figure 7).[30,31] The pattern of aa that build the hydrophobic pocket around the dimethyl benzyl ring of FMN is instead conserved in all bacterial LOV domains. In the Hβ strand, L101 in *C. reinhardtii* phot-LOV1 corresponds to F1010 in phy3-LOV2, that is stacked on the *re* face of the FMN isoalloxazine (opposite to the reactive cysteine).[30] An F residue in this position is characteristic of LOV2 and the distance with FMN-C4a becomes considerably larger upon formation of the photoadduct (from 4.1 to 5.2 Å),[25] whereas in LOV1 the movement of L101 is less pronounced (the distance increases from 3.4 to 4.1 Å).[31] As all phot-LOV1, the large majority of the bacterial LOV domains present an L in this position.

A peculiarity of the bacterial LOV-proteins is the presence of N in place of S38, the latter being conserved in all phot-LOVs and shown to be essential for FMN binding (suppressed upon S38F mutation).[23] Interestingly, the S38N substitution does not impair FMN binding in *B. subtilis* YtvA,[10] but could partially account for the lower stability of the isolated YtvA-LOV with respect to full length YtvA and to phot-LO*Vs*.[79] The S38N substitution occurs also in the FMN-binding LOV domain of the *A. thaliana* photoreceptor FKF1.[35] Residue F41, also essential for FMN binding,[23] is conserved in all bacterial LOV domains. Both S38 and F41 are not part of the FMN binding pocket, but could be important for the correct fold.

The few short gaps appearing in the sequence alignment of Figure 6 are confined to the loops between secondary structure elements. The longest

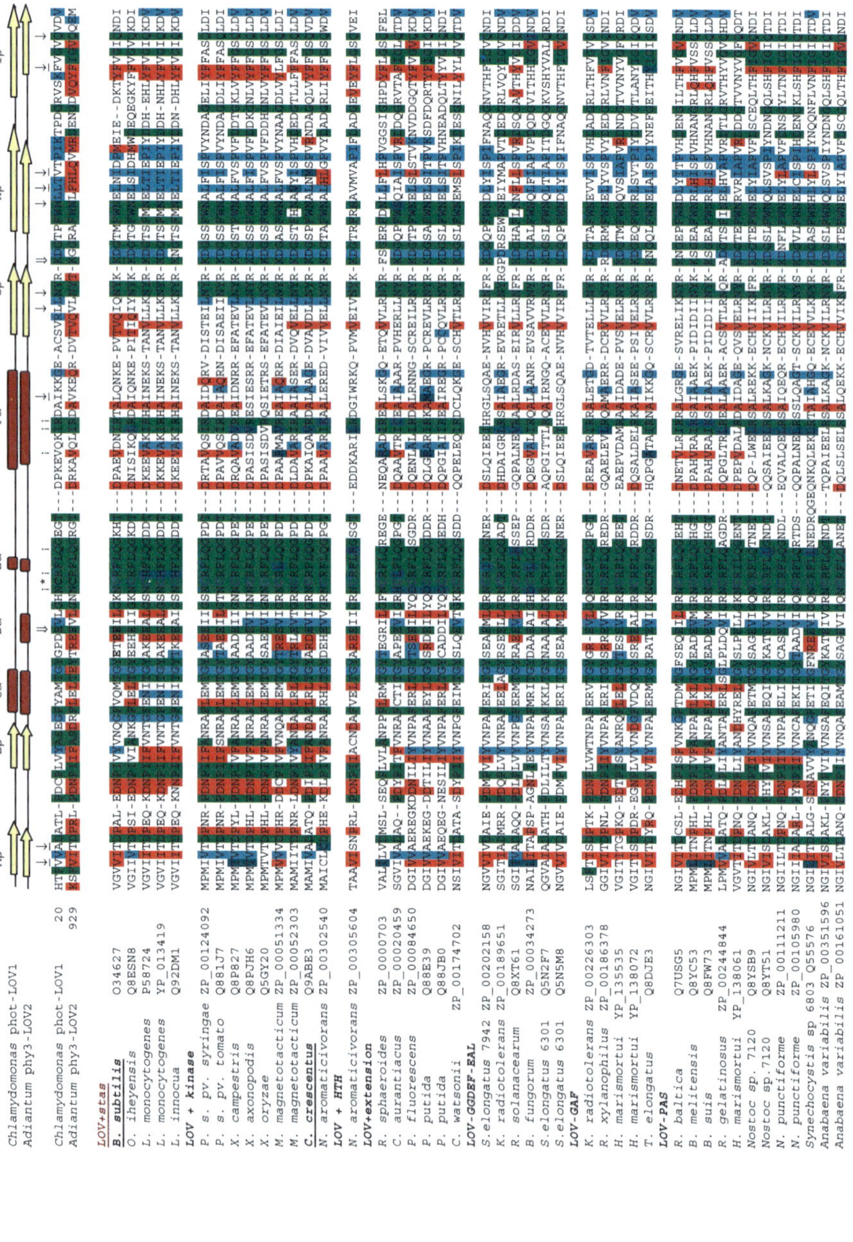

insertion is found in the *Synechocystis sp.* Q55576 protein and localized within the Eα-Fα loop. Insertions in the same region, albeit longer, occur in the LOV domains of the *Neurospora crassa* VIVID and WC-1, *A. thaliana* FKF1 flavin-based photoreceptors and in the WC-1 protein from *Tuber borchii.*[39,80] This extension has been suggested to be important for accommodating a larger chromophore than FMN, given that WC-1 is associated with FAD (Flavin-Adenin Dinucleotide)[81] and purified VIVID can bind both FMN and FAD.[82] The LOV domain of FKF1 has been nevertheless shown to bind FMN and undergo phot-like photochemical reactions, albeit the formation of the photoproduct is irreversible.[35] The extra insertion in the α'A-αC loop could therefore provide a regulatory role on the kinetics. In summary, the majority of the bacterial LOV domains are likely to bind FMN and undergo phot-like photochemistry.

The large number of LOV domains allows to generate a phylogenetic tree, based on the sequence alignment (Figure 8). However, one must keep in mind that the phylogenetic analysis presented here is extremely simple and relies solely on the LOV domains alignment. Alternative alignments are possible and a more sophisticated phylogenetic analysis should take into account the associated domains. An interesting point is that all the LOV domains that are associated to a kinase function (*vide infra*) cluster together and close to phot-LOV (with the exception of multidomain, complex kinases, see Table 1), indicating that an analogy in the *C*-terminally localized enzymatic activity induces a large sequence similarity of the LOV domain.

10.2.1.2. Photochemical Reactions of Bacterial LOV Proteins

YtvA from *B. subtilis* is the first bacterial protein for which phot-like photochemical reactions have been demonstrated.[10] The unphotolyzed (dark) state, YtvA$_{450}$, shows the same absorption spectrum as phot-LOV2 (Figure 9). Modeling studies and chromophore extraction showed that YtvA$_{450}$ binds FMN non covalently via its *N*-terminal LOV domain, YtvA-LOV.[10,79] Blue-light illumination results in bleaching of the FMN feature located around 450

Figure 6. Sequence alignment of bacterial LOV domains (see Table 1 for the primary protein accession numbers) with phot-LOV domains of known structure. Not all the proteins listed in Table 1 are included, for reasons of space. Computation was performed at the ClustalW[179] network service of the EBI. The secondary elements in the crystal structure of C. reinhardtii phot-LOV1[31] and Adiantum phy3-LOV2[30] are shown above the alignment. The residues interacting with FMN are indicated as follows: (¡) interactions with the ribityl chain; (↓) hydrophobic pocket around the dimethyl benzyl ring of FMN; (↓) interactions with the isoalloxazine ring; (*) reactive cysteine. The arrows ⇓ indicate the conserved E and K residues, forming a salt bridge at the surface of LOV domains.[3] In green: residues conserved in all phot-LOV domains; in red: residues conserved in all phot-LOV2 domains; in light blue: residues conserved in all plant phot-LOV1 domains (from Crosson *et al.*[30] note that some residues are changed in C. reinhardtii phot-LOV1). In dark green: residues interacting with FMN and not conserved (neither with any LOV1 nor LOV2) in the bacterial LOV domains

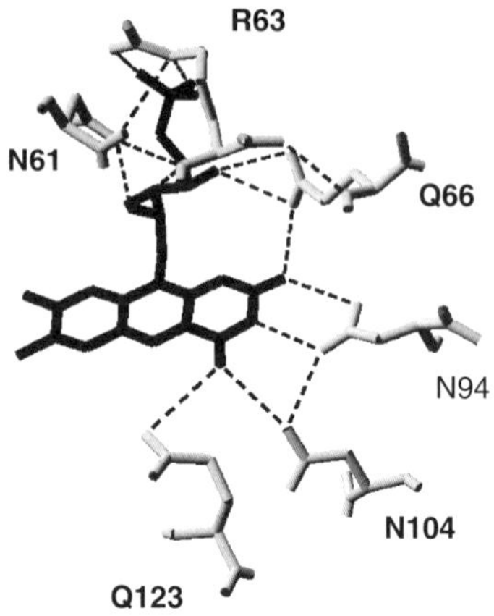

Figure 7. Residues forming HB with FMN in phot-LOV of known structure. The aa shown in bold are changed in some of the bacterial LOV domains. N94 is always conserved. The picture was created from the 3D-model of YtvA-LOV[10] using the beta-version of Deep View/Swiss-PDB viewer 3.7.[180] For the numbering see the accession number of YtvA at the Swiss-Prot/TrEMBL database, O34627.[181] The reactive cycteine is C62 (not shown)

nm, a blue-shift of the absorption spectrum and a loss of fluorescence (Figure 9). These features are typical for phot-LOV domains[22] and are indicative for the formation of the FMN-Cys (Cys62 in YtvA) photoproduct, $YtvA_{390}$. In the dark, $YtvA_{390}$ reverts to $YtvA_{450}$ with a very long recovery lifetime, $\tau_{rec} = 2600$ s at 25°C, strongly dependent on temperature. The isolated YtvA-LOV has an even longer $\tau_{rec} = 3900$ s at 25°C.[79] The photocycle and the absorption spectrum of the dark state and after illumination with blue-light are shown in Figure 9. The τ_{rec} for phot-LOV domains is between 5 and 330 s at room temperature.[22,25] Even if there is no obvious difference in the amino acid sequence that can easily account for this large range for τ_{rec}, one must keep in mind that tiny modifications in the chromophore cavity can affect the high activation energy needed to break the covalent FMN-Cys bond, thus dramatically changing the recovery kinetics at room temperature. We note that there is no systematic characterization of the Arrhenius parameters for the recovery reaction in LOV domains, but data on a subset of LOV proteins show that the value of τ_{rec} is proportional to the activation Gibbs free energy, $\Delta G^{\#}$, with entropic factors (*e.g.* the rigidity of the FMN pocket) playing a large role in determining τ_{rec} (Krauss *et al*, submitted).

Light activation of YtvA results in structural changes readily detected by Laser Induced Optoacoustic Spectroscopy (LIOAS).[10,79] The formation of the photoproduct, on the short microsecond time scale, is accompanied by a volume contraction, $\Delta V_{390} = -12.5$ ml mol^{-1}. The contraction is larger in YtvA-LOV for which $\Delta V_{390} = -17$ ml mol^{-1}. This is probably due to hydrogen bond rearrangements within the FMN binding cavity.[79,83] The relatively large contraction, indicative for an entropy loss, strongly affects the value of τ_{rec} in

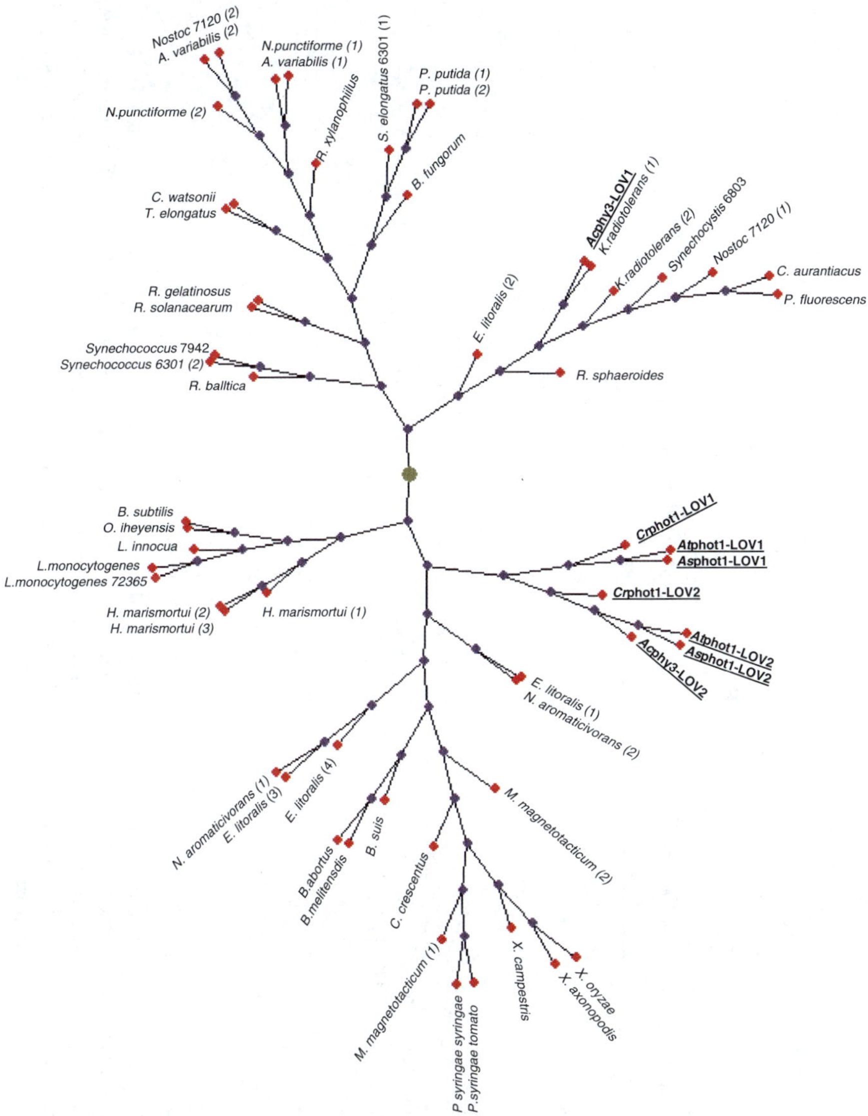

Figure 8. Phylogenetic tree for LOV domains, calculated on the basis of sequence allignment, extended to Avena sativa and Arabidopsis thaliana phot1-LOV1 and LOV2, Adiantum phy3-LOV1, C. reinhardtii phot-LOV2. The phylogenetic tree was drawn using the Phylip output at the ClustalW network service[179]

agreement with the Arrhenius parameters. The high energy level of the photoproduct (136 kJ mol^{-1} in YtvA) is indicative for a strained conformation and drives the completion of the photocycle. Different to phot, the isolated YtvA-LOV is more unstable than the full-length protein, that comprises a *C*-terminal STAS domain (from Sulphate Transporter and AntiSigma factors antagonists)

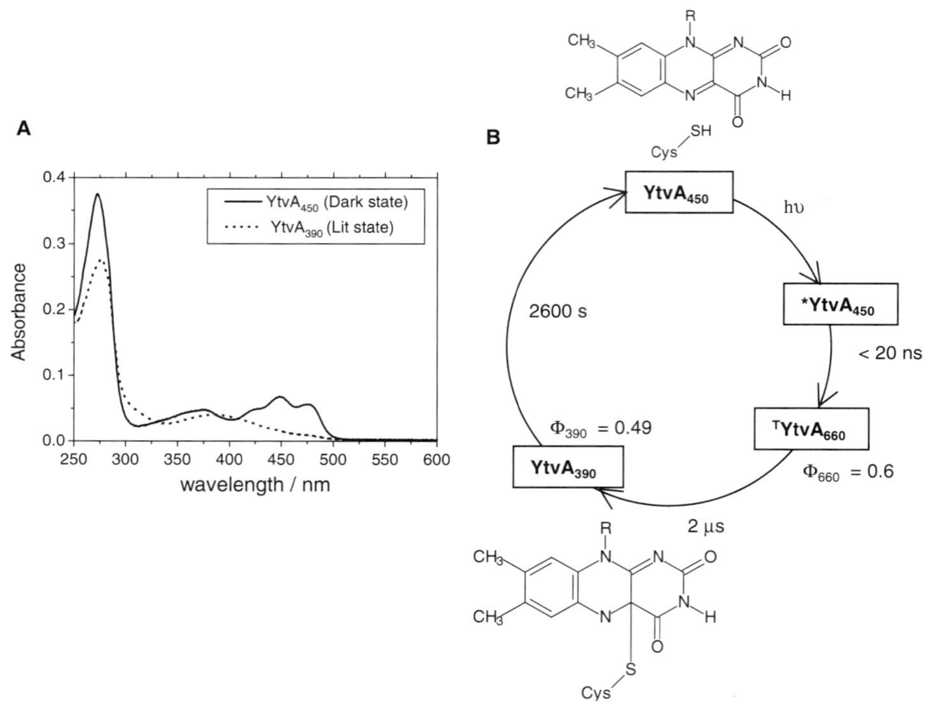

Figure 9. (A) Absorption spectra of YtvA, a LOV-protein from B. subtilis, in the dark ($YtvA_{450}$) and light ($YtvA_{390}$) adapted state. The formation of the photoproduct causes bleaching of the FMN feature in the blue region of the spectrum. (B) Photocycle of YtvA. Light excitation results in a sub-ns formation of a red-shifted transient state, $^{T}YtvA_{660}$, assigned to the triplet state of FMN. $^{T}YtvA_{660}$ decays with $\tau_t = 2$ µs into the photoproduct $YtvA_{390}$ where FMN is covalently bound to C62. The dark recovery kinetics into the unphotolyzed state is strongly temperature dependent with $\tau_{rec} = 2600$ s at 25°C[10,79]

and a short *N*-terminal cap. Given that the two domains have been shown to interact,[79,84,85] this aspect could be important to gain structural information on a full-length LOV protein.

The *C. crescentus* Q9ABE3 protein contains an *N*-terminal LOV domain, associated to a *C*-terminal kinase function (Table 1, Figure 5), therefore it is intrinsically similar to phot. Upon expression and purification of the *C. crescentus* Q9ABE3 kinase in *E. coli* we obtained only a small fraction of soluble protein associated with FMN and undergoing phot-like photochemistry.[6] The protein is also unstable, preventing detailed characterization. A more sophisticated experimental approach is apparently needed in order to fully characterize this system and to test its putative light-regulated kinase activity.

The *P. putida* KT2440 gene PP2739 encodes a 151 aa protein (Q88JB0)[6] that comprises a LOV domain, an 18 amino acids long *N*-terminal sequence and a *C*-terminal extension of *ca.* 30 amino acids, without a fused effector domain. (Table 1). In the LOV core, the canonical G*X*NCRFLQG sequence that contains the active cysteine, is changed into YQDCRFLQG, and is characteristic of a few other bacterial LOV proteins (Figure 6). Despite this difference, the recombinant PpQ88JB0 protein binds a flavin and undergoes phot-like photochemical reactions (Krauss *et al.* submitted), thus defining a novel type of photoactive LOV domain.

10.2.1.3. LOV Proteins Families

The bacterial LOV-proteins can be grouped in families, based on the inferred function of the associated effector domains (Figure 5). In some cases, a high similarity in the primary sequence and domain architecture is observed within the same family, resembling the situation observed among plant phot.

- *STAS-based* σ^B *regulators*

The LOV proteins from the five Firmicutes (low G+C-content, gram-positive bacteria, Table 1), present the YtvA domain structure (*ca.* 260 aa), with a STAS domain located *C*-terminally to LOV and form a highly homologous family. STAS domains have been detected in prokaryotic and eukaryotic proteins, and have been suggested to possess a general NTP (Nucleoside Triphosphate) binding role.[86] The binding of a NTP to YtvA, albeit predictable on the basis of sequence analysis,[10] could not be demonstrated up to now. YtvA was recently included in a family of novel regulators of the general stress transcription factor σ^B (albeit with a relatively modest effect) during salt and ethanol stress conditions.[87] Different from the other members of the regulators family, YtvA enhances the activity of σ^B and cannot be phosphorylated by the RsbT environmental signaling kinase. The mechanism by which YtvA regulates σ^B probably involves its interaction with two other regulators, RsbR and YqhA.[87] It is intriguing to speculate that the interaction might be regulated by light-activation, *i.e.* UVA/blue-light may represent a stress factor sensed by the YtvA photosensing domain, although this remains to be experimentally assessed.

• *Kinases*

A considerably large number (26) of the bacterial LOV proteins are members of the histidine protein kinase (HPK) kinase superfamily (see Table 1).[88] In bacteria, signal transducing HPKs, together with phoshoaspartyl response regulators (RR), are the key elements of two-component signal transduction systems.[89] The HPKs generally contain an *N*-terminal sensing domain (*e.g.* LOV) and a *C*-terminal kinase core, but other domains may be present. The kinase core of HPK features the phospho-accepting histidine box (H-box) within the homodimerization domain and, downstream to it, the highly conserved homology boxes of the nucleotide-binding, catalytic domain (N-, D-, F- and G-boxes).[88] In response to a signal, HPKs autophosphorylate the H-box histidine residue, from which the phosphoryl group is transferred to a conserved aspartic acid residue in the receiver domain of a RR.[89] Generally RRs contain one or more output domains downstream the receiver domain, in some present only the latter, but often also fused to the cognate HPK (hybrid HPK-RR). HPKs can be further divided into subfamilies according to their sequence similarity.[88]

– The LOV proteins from *P. syringae pv. syringae*, *Pseudomonas syringae pv. tomato, X. axonopodis, X. campestris and Xanthomonas oryzae* are highly homologous hybrid HPK-RR (Figure 5). The kinase domain is characteristic of the HPK_4 class: besides a typical H-box, these proteins exhibit the PF-TTK signature in the F-box (Figure 6).[6,88] These proteins are very similar to the rhizobial, PAS-based, heme O_2 sensor-kinase FixL.[90,91] Furthermore their LOV domains are among those that are most homologous to phot-LO*Vs*. Given these features, namely the association of an "ideal" phot-like LOV domain with prototypical kinase and RR motives, they are good candidates to test the molecular properties of this novel, putative blue light-driven two-component signalling system in bacteria.

– The *C. crescentus, Brucella suis, Brucella melitensis, Brucella abortus, N. aromaticivorans* ZP_00095689 LOV-kinases and the three kinases from *Erythrobacter litoralis* show no canonical H-box, rather they are most similar to the HpK_{11} methanobacterial kinases.[88] The two *Magnetospirillum magnetotacticum* LOV-kinases do not fall in any of the described group. In these six proteins, the F-and G2-box appear to be missing or very reduced. This region forms a variable, mobile loop in the structure of diverse HPK, located near the ATP binding site (ATP lid) and its shortening could affect ATP binding or catalytic activity.[89]

– The multidomain/hybrid LOV-kinases (Table 1), belong to class HPK_1, the largest kinase subfamily. The KFT motif in the N-box is typical of the HPK_{1b} subgroup.[92] The *R. baltica* LOV-kinase has a unique domain structure, with a CheB methylesterase and CheR methyltransferase domain in the *N*-terminal. In motile bacteria CheB and CheR control the level of methylation of glutamate residues in methyl-accepting chemotaxis proteins.[93,94] The function of these two catalytic domains fused on the same protein has as yet not been described. The cyanobacterial kinases possess additional GAF (cGMP phosphodiesterase, Adenylate cyclase, FhlA) domains, present also in phytochrome and cGMP-specific phosphodiesterases.[95] The GAF domains do not show the extra-insertion responsible for the binding of a tetrapyrrole chromophore in phytochromes,[95] neither the NNKFDE motif necessary for cAMP or cGMP binding[96] (and references therein). The kinases from *Nostoc sp.*, *Nostoc punctiforme* and *A. variabilis* contain a *C*-terminal Hpt (histidine phosphotransfer) domain, downstream to the two RR units. This feature has been previously characterised in the so-called unorthodox, multistep His-Asp-His-Asp phosphorelay systems where the HPt domain serves as histidine-phosphorylated intermediate during phosphoryl transfer between two RR domains.[97] Accordingly, the genes for these LOV-kinases are coupled with sequences encoding for additional RRs. As a whole, the multidomain LOV kinases from cyanobacteria show striking similarity in their architecture with *E. coli* ArcB (Aerobic respiration control sensor protein), one of the best characterised multi-step phosphorelay sensors.[98]

- *Transcriptional regulators.*

This family is formed by two members, the transcriptional regulators of *N. aromaticivorans* (ZP_00093141) and *E. litoralis* (ZP_00376834) that carry a typical Helix-turn-Helix (HTH) DNA-binding motif. The two species belong to the Sphingomonadaceae family of α-proteobacteria. HLH/PAS proteins are well documented in plants and animals.[99,100] They tend to be ubiquitous, latent transcription factors whose activity is signal regulated, but the molecular mechanism by which they control gene activity is not well understood.

- *Short-LOV proteins*

The small proteins in *P. putida* (Q88E39, Q88JB0), *Pseudomonas fluorescens*, *R. sphaeroides* , *Crocosphaera watsonii* and *Chloroflexus aurantiacus* exhibit a single LOV domain with variable *N*- and *C*-terminal caps. This feature is reminiscent of *N. crassa* VIVID[82] and of the PYP, a small bacterial blue-light photoreceptor that binds 4-hydroxycinnamic acid as chromophore.[101] In the *P. putida* genome, the sequences encoding for the two LOV proteins are associated with putative DNA-binding transcriptional regulators that could represent their molecular partners.

- *Nucleotide cyclases and phosphodiesterases*

This family encompasses LOV bacterial proteins containing the GGDEF domain associated with EAL (GGDEF and EAL are conserved aa sequences), a well documented tandem motif.[102] Although the function of most of the GGDEF proteins has not been characterized, some of them show catalytic activity dependent on a novel bacterial effector molecule, cyclic diguanylate (c-di-GMP, bis(3′,5′)-cyclic diguanylinic acid).[103–105,104] Very recently, six randomly chosen GGDEF proteins have indeed been shown to possess diguanylate cyclase activity involved in the synthesis of c-di-GMP.[106] Together, the GGDEF and EAL domains, regulate the turnover of c-di-GMP, with EAL acting as a phosphodiesterase.[104] The presence of the two domains associated on the same protein, is therefore indicative of a c-di-GMP dependent phosphodiesterase and/or cyclase activity. Should this catalytic role be confirmed for these LOV-proteins, this finding would constitute an unprecedented way of function in bacterial light-signal transduction chains. A similar blue light-driven enzymatic activity has been postulated for a BLUF protein from *E. coli* (*vide infra*).[107] Other PAS-based proteins, binding dioxygen through a heme cofactor, have a similar domain architecture and their catalytic activity as c-di-GMP phosphodiesterases is triggered by hypoxic condition, *i.e.* O_2 dissociation from the heme cofactor.[108,109]

- *LOV S/T-phosphatases 2C*

The Mg^{2+}/Mn^{2+}-dependent protein S/T phosphatase 2C (PP2C) family, is characterized by 11 conserved signature motifs in the catalytic domain and is well represented in all life kingdoms.[110] A light driven 2C phosphatase activity has not yet been described, but a PAS-PP2C is involved in environmental stress signaling in *B. subtilis*.[87,111] LOV PP2C have been solely detected in *Rubrobacter xylanophilus* and *Kineococcus radiotolerans,* two gram-positive bacteria extremely resistant to radiation.[112,113]

10.2.1.4. Signal Transduction Mechanisms and Interdomain Communication in LOV Proteins

As mentioned above, one of the key question in the field of LOV-proteins concerns the molecular mechanisms of light-to-signal conversion, *i.e.* how the formation of the FMN-Cys adduct activates the kinase in phot, and if the mechanisms of activation are similar among the different classes of LOV-proteins illustrated above. Structural data, obtained with different techniques on isolated phot-LOV domains, indicate that the light-driven conformational changes are relatively small and mainly originate in the vicinity of the chromophore.[24–25,31,83,114] For a comprehensive discussion see ref[6]. A major contribution has recently been obtained from NMR studies on an extended LOV2 construct from *A. sativa* phot1 comprising 40 aa downstream the LOV core (AsLOV2-40), showing the existence of a *C*-terminal 20 aa long amphipathic helix (Jα).[115] In AsLOV2-40, the Jα helix interacts with the exposed surface of the LOV core central β-sheet with its apolar face, but the constraint is removed after light activation and the helix becomes disordered.[115] Recently it was demonstrated that mutations removing the LOV-Jα helix interaction results in constitutive activation of the kinase activity in full length phot, supporting the idea that unfolding of the Jα helix is the critical point in the light-dependent self-phosphorylation event.[116]

Another hypothesis is that a conserved E-K salt bridge (E960-K1001 in phy3-LOV2), localized at the surface of the LOV core, might be involved in the function of a LOV domain as light responsive-signaling module[3] This idea is supported by the observation that the E-K salt bridge is the surface end of a set of conserved amino acids, interconnected through van der Waals interactions and extending from the FMN cavity. This conserved connectivity volume undergoes significant conformational changes upon formation of the photo-adduct in a crystallyzed LOV2 domain.[25,30] Albeit it is conceivable that the surface salt bridge plays a regulative role in LOV-mediated signal transduction, no experimental evidence exists yet that its disruption interferes with phot autophosphorylation.

Photoacoustic calorimetry experiments and mutational analysis with *C. reinhardtii* phot-LOV1 have shown that the volume contraction accompanying the formation of LOV_{390} receives large contributions from the rearrangements of the HB network centered on Arg58 (adjacent to the reactive Cys57) and involving the FMN phosphate group.[83] It is possible that this protein region acts as a regulator for the LOV domain itself (*e.g.* by affecting stability of the photoproduct and kinetics of the recovery reaction) rather than constituting the interaction site with partner domains.

Still relatively few functional data are available for bacterial LOV proteins. FTIR measurements on YtvA-LOV have shown that the light-driven conformational changes are similar to those occurring in phot-LOV domains.[84] Nevertheless comparative measurements with full-length YtvA indicate that light-driven conformational changes are transmitted to the remaining part of the protein,[84] in line with spectroscopic and chemo-physical data pointing to an interaction between the LOV and the STAS domain.[79] Additionally, the fluorescence of the W103 conserved within the LOV series and belonging to the central β-sheet of the LOV core has been proven to be sharply different in YtvA-LOV and YtvA.[85] Increasing evidences indicate that W103 is located in a hinge region and interacts with the loop connecting the LOV core to the Jα helix (Krauss *et al.* submitted), similar to the structure adopted in AsLOV2-40.

In summary three possible regulatory or protein-protein interaction sites (not necessarily coinciding) have been identified in LOV domains: the solvent exposed face of the GβC, HD and Iβ strands encompassing the central β-sheet, the conserved E-K salt bridge and the protein region binding the FMN phosphate group. These structural elements are highlighted in Figure 9.

Up to now only the mechanism based on the interaction of the Jα-helix with the central β-sheet and light-driven Jα-helix unfolding has been experimentally confirmed to play a role in signal transduction.[116] But is the mechanism widespread and applicable to bacterial LOV-proteins?

Modeling studies on the *P. putida* Q88JB0 protein, a naturally occurring extended LOV construct, indicate that the *C*-terminal extension is organized as an α-helix that interacts with the central β-sheet in a way similar to the extended AsLOV2-40 construct, but implying a larger number of polar interactions (Krauss *et al.* submitted). Circular dichroism experiments support the assignment of the extension to an α-helix. Secondary structure predictions[117] show that in all LOV proteins the segment *C*-terminal to the LOV core has a helical content of variable length (data not shown) nevertheless, in the bacterial systems up to now investigated, light-activation does not seem to results in dramatic conformational changes[85] (Krauss *et al.* submitted). Therefore, albeit the conformation may be quite similar in the dark state, the activation of the effector domain upon formation of the photoadduct may not require in all cases the unfolding of the Jα-helix.

In the absence of detailed structural information we cannot exclude that the Jα-helix may assume alternative conformations. In the crystal structure of the PAS domain of *B. japonicum* FixL, an O_2 sensor, a rigid α-helix follows, without loops, the Iβ strand and protrudes away from the PAS core.[118] Albeit the latter feature could be an artifact of crystal packing, it shows that the helix can actually assume different orientations. However, a key role for the central β-sheet also in the activation of FixL has been recently proposed on the basis of structural data on the ligand-bound and ligand-free PAS-core.[119]

Recently an interesting helix-swap model has been proposed as a general mechanism for regulation by PAS domains.[109] According to this mechanism, PAS (also LOV) domains alternate between a PAS–PAS and a PAS–transmitter (*e.g.* the Jα-helix) interaction state. This would allow a PAS domain to interact with a helix of a target transmitter in the closed conformation and to swap this interaction for one with a helix of another PAS domain in the open conformation.[109] The open and closed conformation would be mutually exclusive, meaning that they involve (at least partially) the same interaction surface. LOV domains have been indeed shown to form dimers, as reported for the LOV domain of FKF1[120] and for *A. sativa* phot1-LOV1.[121] Interestingly, dimerization does not depend on light. Phot1-LOV2 does not form dimers, albeit LOV1-LOV2 heterodimerization cannot be excluded.[121] This aspect could be characteristic of phot-LOV2 that accounts for its ability to selectively activate the kinase domain.[33] Gel chromatography and mass spectroscopy indicate that YtvA-LOV (Figure 10) and YtvA form non-covalent dimers in solution (Losi *et al.* unpublished data), similar to the *P. putida* Q88JB0 protein. These data suggest that LOV-LOV dimerization must be further explored, in view of its possible role in regulation. We note that dimerization is a

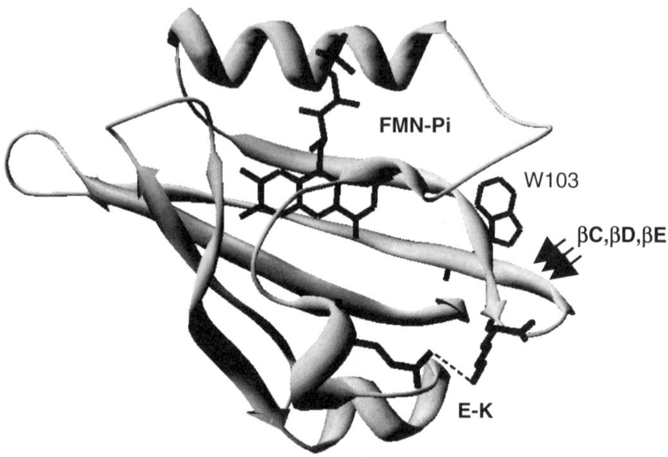

Figure 10. 3-D model of YtvA-LOV[10] with structural elements known or predicted to be involved in interdomain communication or to play regulatory roles in LOV (bold) domains. The three arrows indicate the central β-sheet. The conserved Trp is also shown (see Section 2.1.c)

characteristic feature of PAS domains[21,100] and has recently been highlighted in the crystal structure of diverse PAS systems.[122,123]

An additional regulatory role could be played by loops connecting secondary structure elements. In PYP and Aer (a redox sensor), the cofactor binding and the supposed active site for signalling are centered around the Eα-Fα loop and the PAS core.[124,125] In contrast the Fα-Gβ loop has been suggested to have a regulatory role in the PAS domain of FixL[118] and in the *N*-terminal PAS1 domain of a mammalian kinase.[126] In the latter protein, an exceptionally flexible FG loop serves also as interaction site with the kinase domain. The relevance of the flexible loops in signal transduction or/and regulation has been recently pointed out upon a structure-based redefinition of the PAS-fold.[21]

10.2.2. The Bacterial DASH Cryptochromes

The bacterial DASH-Cry were first identified by Hitomi and co-workers in *Synechocystis sp.* PCC 6803.[127] The chromosome of this strain contains two ORFs, slr0854 and sll1629, encoding for photolyase-like proteins. Disruption of slr0854 results in severe UV sensitivity, in contrast to the sll1629 knock-out mutant. Accordingly, the Slr0854 protein shows DNA repair activity, whereas Sll1629 binds to DNA with low affinity, similar to the binding of *E. coli* PHR to undamaged DNA.[11] The Sll1629 DASH Cry was recently crystallized and the structure was solved at 1.9 Å resolution, showing a typical photolyase folding and a bound FAD chromophore.[11] As mentioned above, different to plant Cry1 and Cry2, the DASH-Cry do not possess a *C*-terminal segment with respect to photolyases. All DASH Cry present a typical photolyase-like architecture, with an *N*-terminal domain (that in PHR binds the folate or deazaflavin antenna pigment) and a FAD-binding domain.

Our BLAST search found that DASH-Cry are potentially present in ten bacteria species: nine are Gram-negative bacteria and the remaining a non-spore forming Gram positive bacterium, *Exiguobacterium* 255–15, isolated from a 2–3 million-year permafrost core[128] (Figure 4).

Sequence alignment (Figure 11) and the crystal structure of *Synechocystis* DASH (Figure 12)[11] allows to identified 17 residues that are unique to the DASH-Crys and distinguish them from photolyases.

These residues are mainly located in the close vicinity of the bound FAD and on the surrounding protein surface that correspond to the active site in PL.[11] One specific hydrogen bond to the adenine moiety of FAD (Asn395 instead of M345 in *E. coli* PL) has been proposed to alter the electron transfer reaction that in photolyases occurs from the isoalloxazine ring to the TT dimer, most probably via the adenine ring.[129] The cavity that hosts the TT dimer in PL, becomes shallower and wider, thus probably impairing binding specificity. The substitution of W394 (*E. coli* PL) with a tyrosine (Y398) may reduce van der Waals interactions with the substrate. Importantly the W394F mutated PL has a very low affinity for the thymidine photodimer.[130]

Little is known about the spectroscopical and photochemical properties of DASH-Cry. The crystallized Sll1629 binds a single FAD molecule, in the same

Figure 11. Sequence alignment of bacterial DASH-Cry, compared to E. coli photolyase. The aa characterizing the DASH family are highlighted. Secondary structures are shown for E. coli PL and Synechocystis DASH-Cry

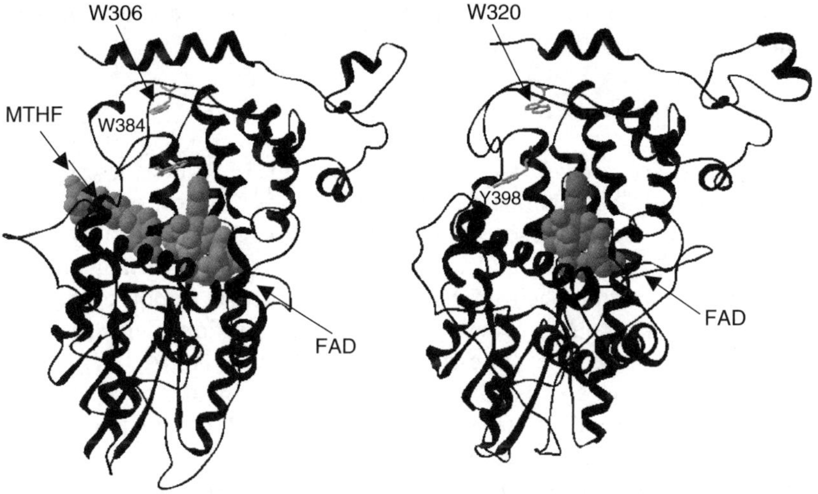

Figure 12. Crystal structure of E. coli photolyase (PDB databank coordinates 1DNP) and of Synechocystis DASH-Cry (1NP7). The figures were drawn with the DeepView/Swiss-PDB viewer software version 3.7 (ref. 180). Key aromatic aa are drawn (see Section 2.2). The associated chromophores are drawn in filled-space

position as PL, without a second deazaflavin or pterin chromophore in the *N*-terminal part. It was suggested that this may be a characteristic of DASH-Cry.[11] However it was noticed that the absence of a second chromophore may be due to the fact that deazaflavin is not synthesized from *E. coli*, and should be supplemented.[78] *V. cholerae* DASH-Cry (Q9KR33) was recently overexpressed in *E. coli* and the purified protein was shown to contain fully oxidized FAD and MTHF.[78] From the fluorescence excitation spectra of FAD, the folate does not seem to perform energy transfer to the flavin. The general presence and role of a second chromophore in DASH-Cry proteins remains therefore to be addressed.

Light-induced electron transfer reactions have been recently described for *A. thaliana* Cry1 (AtCry1).[50] AtCry1 is isolated with a bound FAD in the fully oxydized form, but in the presence of an external electron donor, FAD can be photoconverted to the semireduced radical FADH.[67] The reduction of the flavin proceeds via electron extraction from a tryptophan residue, presumably the conserved W324, corresponding to W320 in *Synechocystis* DASH and W306 in *E. coli* PHR (Figure 4).[50] The external donor (*e.g.* β-mercaptoethanol) seems to reduce a still unidentified tyrosine residue, thus preventing electron back transfer from FADH$^{\cdot}$. A similar mechanism has been described for *Synechococcus elongatus* PCC 6301 and *E. coli* PL.[131,132] The large number of redox-active aromatic residues (tyrosines and tryptophans), located at electron-transfer distances from the FAD chromophore, suggests that similar light-driven reactions may be widespread within PL and Cry proteins. In PL, this kind of reactions are thought to assist enzyme photoreactivation to produce the fully reduced form of the FAD chromophore, without being directly involved in the splitting of the TT dimers (reviewed in[41]). Nevertheless it cannot be excluded that Tyr/Trp mediated electron transfer reactions have a role in Cry signal transduction.

10.2.3. The BLUF Paradigm: Truly Bacterial

AppA from *R. sphaeroides* is a 450 aa protein (Q53119) first described as a redox regulator of photosynthesis gene expression.[53,133] At low oxygen tension, AppA binds to the repressor protein PpsR, whereas under fully aerobic conditions PpsR is released and binds to the promoter of certain photosynthesis genes, repressing their transcription.[53,133] These responses are light-independent, but at intermediate oxygen concentration light determines whether AppA releases the repressor PpsR.[56,134] Thus, AppA integrates and transmits both redox and light signals, the former thanks to a *C*-terminal cysteine-rich sequence, the latter via the *N*-terminal BLUF domain (*ca.* 120 aa) that binds FAD in a 1:1 stoichiometry.[54–55,135]

The BLUF domain seems to be restricted to bacteria and, among eukaryots, to the phylum Euglenida. Increasing information about the activation mechanism of bacterial BLUF proteins is accumulating, also prompted by known physiological functions for AppA and by the description of its novel and unusual photocycle.[55] The majority of BLUF proteins are so called "short-BLUF", where the BLUF domain is not associated to effector domains (Table 2).

10.2.3.1. Photochemical Reactions and Structure of the BLUF Domain

The first photochemical investigation of BLUF-based light-induced reactions has been performed with His-tagged AppA, expressed in *E. coli.*[55] Dark adapted AppA exhibits the two spectral features typical of an oxydized flavin, centered at 365 and 445 nm and vibrationally resolved. Both features undergo a red-shift upon illumination, becoming centered at 371 and 460 nm, respectively. These spectral changes are reversible in the dark, with $t_{1/2} = 15$ min (25°C). Ethanol extraction and thin layer chromatography analysis indicated that a FAD chromophore is bound to AppA in a 1:1 stochiometry.[55] This novel-type photocycle was further investigated with a truncated construct ($AppA_{156}$) corresponding to an extended BLUF domain[136] (Figure 13).

The photoproduct formation upon pulsed-laser irradiation (445 nm) appears to be biphasic with $\tau_1 < 1$ μs and $\tau_2 = 5$ ms and the recovery kinetics is identical to full-length AppA. Formation of the photoproduct also results in loss of flavin fluorescence. Recently, similar light-induced spectral changes have been described for the short-BLUF proteins Slr1694 from *Synechocystis* PCC 6803[76] and Tll0078 from *Thermosynechococcus elongatus*,[77] and also for the *E. coli* YcgF protein.[107]

Further details on BLUF photochemical reactions originate from spectroscopical studies with Slr1694, a short-BLUF protein, that exhibits the same spectroscopical cnahegs as AppA, but with a faster $t_{1/2} = 5$ s.[76] Also in Slr1694 illumination results in strong quenching of the flavin fluorescence. FTIR

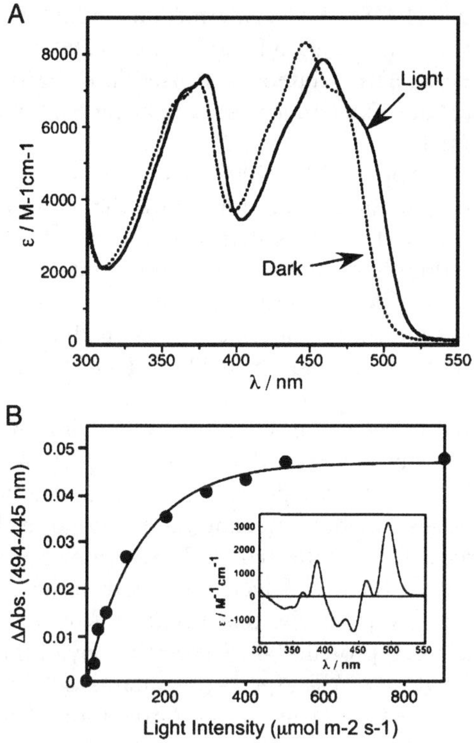

Figure 13. Ultraviolet and visible absorption spectra of the FAD-binding domain of AppA. (A) Absorption spectra of the dark-adapted and light-excited (for 30 s at 900 μmol m^{-2} s^{-1}) FAD-binding domain of AppA. (B) A fluence response curve for the photocycle of the FAD-binding domain of AppA that was generated by plotting the difference between 495 and 445 nm peaks in the light-excited minus dark difference spectrum versus exciting light intensity. Inset: light-excited minus dark (for 30 s at 900 μmol m^{-2} s^{-1}) difference spectrum of the FAD-binding domain of AppA. Reprinted with permission from.[136] Copyright (2003) American Chemical Society

difference spectroscopy indicates that the chromophore undergoes subtle light-induced conformational changes, *i.e.* the C4 = O and C2 = O bonds are weakened, the N1C10a and/or C4aN5 bonds are strengthened and hydrogen bonding at C2 = O and C4 = O increases.[76] Further analysis at low temperature (−35°C) or partial dehydration showed that in these conditions only the changes at C4 = O persist; this has been interpreted as a strengthening of HB to O4 that occurs before the formation of the red-shifted photoproduct.[137] The "C4 = O" precursor represents therefore a potential reaction intermediate in the photocycle of Slr169 (Figure 14). Furthermore it shows a red-shifted absorption spectrum, similar (but not identical) to the photoproduct.

An intermediate state, trapped at low temperature (10 *K*) has recently been detected also in the photocycle of Tll0078 by means of UV/VIS steady state and transient spectroscopy.[77] This intermediate, called $Tll0078_I$ and probably coincident with the "C4 = O" FTIR intermediate in Slr1694, is only 5 nm red-shifted with respect to the dark adapted state ($Tll0078_D$). At temperatures larger than 50 K the I intermediate is converted into the 10 nm red-shifted lit state readily detected at room temperature.[77] The model for Tll0078 photocycle is depicted in Figure 14.

The photocycle of $Appa_{5-125}$ at 20°C has been investigated with ultrafast optical techniques.[64] In these conditions only the 10 nm red shifted state, $AppA_{red}$, formed within few ns, is detected (Figure 14). The triplet state of FAD, FAD^T, is produced as a side reaction and is not a precursor of $AppA_{red}$. The I intermediate is not detectable at this temperature. The quantum yield for the formation of the red-shifted intermediate (signalling state) in BLUF proteins has been measured to be $\Phi_{RED} = 0.29$ for Tll0078[77] and $\Phi_{RED} = 0.24$ for $Appa_{5-125}$.[64]

But which are the molecular mechanisms underlying the formation of the red-shifted signalling state? The conserved Y21 (AppA numbering) is largely involved in the light-driven reactions: the Y21F substitution abolishes the absorption changes, also slightly affecting the spectrum in the dark, albeit the global protein structure is unaffected.[136] Furthermore, the changes in both tertiary and secondary structure, detected with size exclusion chromatography and CD spectroscopy, respectively, that occur in $AppA_{156}$ photoproduct are not observed in the Y21F protein. NMR spectroscopy of the aromatic spectral region, indicated that the WT $AppA_{156}$ contains significant π–π stacking interactions, that are perturbed upon photoexcitation. The stacking interactions are lost upon Y21F mutation.[136] Other mutations indicate that the residue in this position must have both aromatic and polar properties to ensure chromophore binding and to allow the photocycle to proceed.

The recently determined structure of Tll0078 nevertheless showed that neither Y8 (Y21 in AppA) nor any other aromatic amino acid in the vicinity of the flavin form π–π stacking interactions with the isoalloxazine ring.[58] The hydroxyl group of Y8 is hydrogen bonded to Q50, that directly interacts with FAD (Figure 15) by forming a HB with O4.

Mutation of the conserved Q50 into alanine (Q50A) abolishes the photocycle and behaves similar to the Y21F mutation in AppA. The latter phenomenon might therefore arise from the breakage of FAD-Q50-Y8 (Y21 in AppA) hydrogen bonding network rather than from the perturbation of π–π stacking interactions involving FAD. This idea is supported by the FTIR measurements on the *Synechocystis* Slr1694 protein mentioned above.[76,137] Given that O4 is hydrogen bonded to N32 and Q50 in the dark state and that N32A mutation does not suppress the photocycle, whereas Q50A does, it is concluded that the light–induced strengthening of the hydrogen bond to O4 must arise from Q50.[58] Increased HB at the C4 = O oxygen, strengthened by enhanced π–π stacking with Y8 and possibly other aromatic residues has been also suggested for Slr1694.[137] Very recently, FTIR measurements and theoretical calculations have reinforced the hypothesis that HB at C4 = O is responsible for the observed light-induced red shift.[138]

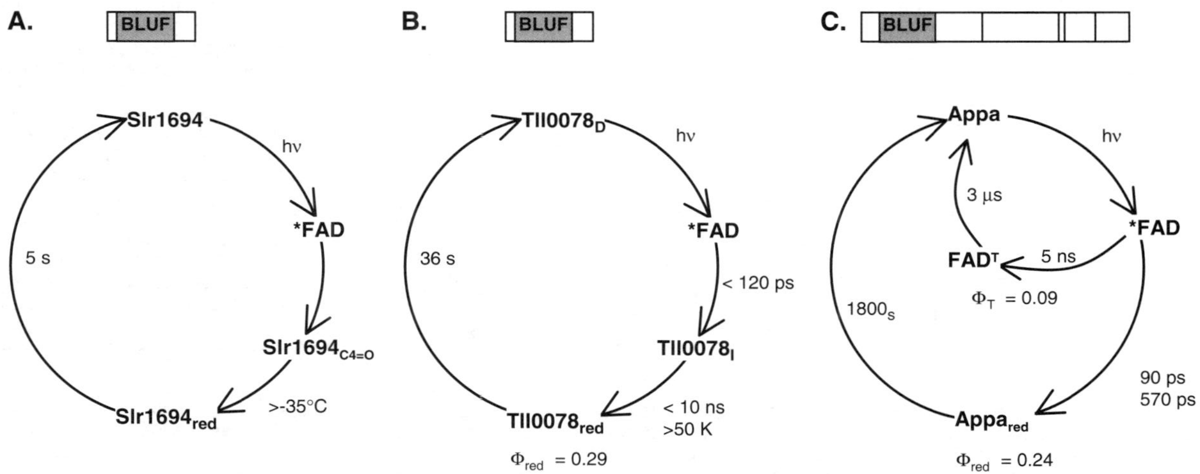

Figure 14. Photocycle of BLUF proteins. (A) Photocycle of Slr1694.[76,137] For the recovery of $Slr1694_{red}$, $t_{1/2} = 5$ s at 20°C (B) Photocycle of Tll0078.[77] The $Tll0078_I$ intermediate is slightly red-shifted with respect to the dark state (5 nm) and can be trapped at low temperature. Is is the direct precursor of the signaling state $Tll0078_{red}$.(C) Photocycle of truncated AppA, $AppA_{5-125}$, determined at 20°C.[64] The triplet state of FAD, FAD^T, represents a side reaction in the photocycle and is not a precursor of the signalling state $AppA_{red}$. The domain architecture of the three BLUF proteins is shown above the corresponding photocycle scheme. See also Table 2

A.

```
Thermosynechococcus elongatus Tll0078  1   MGLHRLIYLS----CATDGLSYPDLRDIMAKSEVNNLRDGITGMLCYGNGMFLQTLEGDRQKVSETYAR
Synechocystis 6803 Slr1694            1   MSLYRLIYSS----QGIPNLQPQDLKDILESSQRNNPANGITGLLCYSKPAFLQVLEGECEQVNETYHR
Rhodobacter sphaeroides Q53119 (AppA) 14  SDLVSCCYRS----LAAPDLTLRDLLDIVETSQAHNARAQLTGALFYSQGVFFQWLEGRPAAVAEVMTH
Escherichia coli P75990 (YCGF)        1    MLTTLIYRS----HIRDDEPVKKIEEMVSIANRRNMQSDVTGILLFNGSHFFQLLEGPEEQVKMIYRA
Corynebacterium efficiens Q8FMY2      14  GKLKYLVYTS----IASPDLTEESVQSMLTTAREHNASAEITGLLLFRTGTFIQFLEGAAEEIDQLISR
Bdellovibrio bacteriovorus Q6MPS8     2   SEIFHLVYFS----KAAEDLSYTDIREILEVSRRNNARLGITGLLIFRDGFFLQLLEGEQASVHKILGA
Rhodopseudomonas palustris Q6N4H0     11  DELFRVLYVSRNRIEGPLDRLSSEIAGILATSRINNARAGITGALIVNSGFFAQVLEGPRSAVEKVFET
uncultured bacterium 581  Q6SFE3      17  DELIHLAYVS----TQTRDMKAADLISLLTEVRGLNESRNISGLLLHKNQSFFQVIEGSRARVQETFNN
Leifsonia xyli Q6AGS2                 1     MMSIAYVS----EATAPMSEQGIVSLLKRARANNDVRQRTGALLYHEGRFVQILEGPDEAVEECFEA
Xanthomonas axonopodis citri Q8PHH6   4   GPLHAIAYVS----EVRRPWTLGEIMDILTEAINFNSLNDITGVLYYGGQRFFQYIEGAPAPLRDIYDR
Acinetobacter sp. ADP1      Q6FAI1    1          MYAS--KTNSEHSQIKQDLMDILTEAINFNSLNDITGVLYYGNGYFLQYLEGEKHQVESLFHH
Chromobacterium violaceum Q7NSR1      1   MALQHLIYVS----TARRELDNADLEALLESCVRHNLRNDISGVLLYGPGNFIQVLEGEADAVAETYQR
uncultured bacterium 153809 Q8KZ47    7   EKLFHLGYVS----IETSDLGSDGMVSLLDEARRTNAERNLTGILLHRDRSFYQVLEGEESEVRRTFEG
```

```
-ILKDPRHHSAEIVEFKAIEERTFINWSMRL
-IVQDERHHSPQIIECMPIRRRNFEVWSMQA
-IQRDRRHSNVEILAEEPIAKRRFAGWHMQL
-ICQDPRHYNIVELLCDYAPARRFGKAGMEL
-IRLDDRHQDVRVIIEEPVDHRRFPDWRMGY
-IREDDRNYSVKVLIEASGQERLFAEWSMAF
-IQQDHRHSDVQVLLASPITVRAFPNWAMAF
-IMRDQRHEGVEVLFDEPLEAREFSNWQMGF
-ICADPRHRSIQKISEEQIEQRRFAEWTMAF
-IKRSSRHDLVAELYNDQITERYFPNWQMAC
MILKDQRHHNCEIIFSKPSENRSFERWSMKF
-ILDDPLHHNLIEILLENIERRSFADWRMGF
-IGKDKRHKAIDVLFEGHVEQREFADWQMGF
```

B.

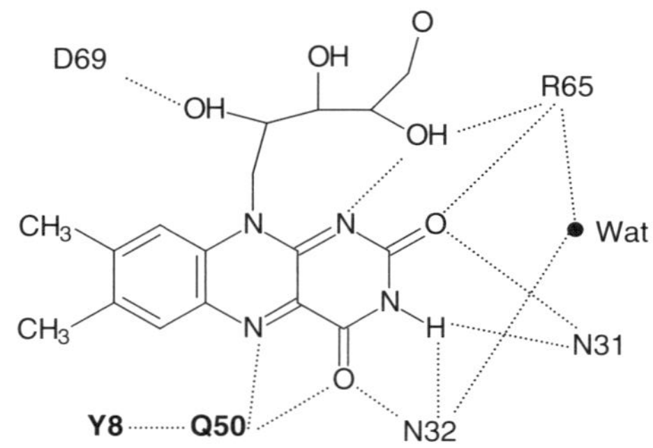

Figure 15. (A) Alignment of selected BLUF domains; the conserved residues are shadowed. The sequence of the BLUF domain from the T. elongatus Tll0078 protein is shown is bold. The secondary structure elements are derived from the structure of Tll0078,[58] for which only the BLUF-core is shown (aa 1–94). Amino acids in the close vicinity of the flavin chromophore are indicated with triangles. Y8 and Q50 are underlined. (B) Hydrogen bond network around the flavin chromophore. Y8 and Q50, the latter interacting directly with N5 and O4, are shown in bold (modified from.[58] Mutation of these two aa prevents the formation of the red-shifted intermediate in the BLUF photocycle (see Section 2.3.a)

A detailed analysis of the dark recovery process has been recently performed on Slr1694 by means of UV/Vis and FTIR spectroscopy, providing a more complex picture of the BLUF photocycle.[138] The largest decay rate of the C4 = O bands indicates that the weakening of HB at this position occurs before the other changes in the flavin and in the protein and coincide with the recovery of the dark state absorption spectrum. This implies that the relaxation of the protein is slower and continues after the UV/Vis absorption changes have been reverted.[138]

The effect of deuteration on the photocycle of BLUF proteins is complex and indicates the involvement of proton transfer reactions during the formation of the red-shifted state.[138] The possibility of hydrogen transfer reactions involving to Y21 has been recently investigated in the truncated $Appa_{5-125}$ from *R. sphaeroides* RK1.[139] Although there are some evidences that a proton transfer occurs form the FAD chromophore to Y21 during the photocycle, this hypothesis has not been definitely confirmed. For Slr1694 there are indications for the fact that FAD is protonated at N3 both in the dark and in the red-shifted states.[76] Therefore it is likely that the putative light-induced proton transfer event occurs between amino acids and does not involve FAD directly. An indirect participation mediated by the increased basicity of N5 in FAD^T has been proposed.[138] Nevertheless, the observation that formation of FAD^T represents a side reaction in the BLUF photocycle[64] is not in agreement with this hypothesis.

It was recently demonstrated that the incubation of cell-extracts from *E. coli* overexpressing $AppA_{5-125}$ with different flavins, results in aspecific incorporation of riboflavin, FAD or FMN.[72] The chromophore identity does not appreciably affect the spectral or molecular changes upon illumination and the photocycle kinetics are only slightly affected. This indicates that neither the adenine moiety nor the phosphate group are required for the formation of the putative signalling state, the photoproduct. It remains to be elucidated if this chromophore heterogeneity is also present in the native protein from *R. sphaeroides*.[72] In the structure of Tll0078 the flavin isoalloxazine ring is located at the center of the BLUF domain, between α_1 and α_2 helices, with the ribityl moiety that elongates towards the molecular surface.[58] The phosphate and adenosine parts are not well defined, but from the unambiguously identified riboflavin part it is conceivable that the chromophore specificity is not absolute: the majority of the conserved residues are in fact in close vicinity of the riboflavin unit. Furthermore, the key residues implied in the photocycle form a HB network again centered on the riboflavin part, and the rest of the FAD chromophore is likely to be poorly involved on the primary photochemical events.

10.2.3.2. BLUF Proteins Families and Signal Transduction

Interestingly the majority of the bacterial BLUF proteins (Table 2) appear to be simply constituted by a BLUF domain flanked by *N*- and *C*-terminal short sequences (short-BLUF), as for example the Tll0078 and Slr1694 proteins discussed above (see also Table 2). In *E. coli* YcgF the BLUF domain, also binding FAD,[54] is coupled to an EAL domain proposed to function as a diguanylate phosphodiesterase site.[2,104,107] In AppA the BLUF domain is linked to a *C*-terminal domain without similarity to any known protein.[2]

In AppA light signaling requires both the BLUF domain and the C-terminus, that in turn is sufficient for redox regulation thanks to a cysteine rich sequence.[56,135] Furthermore, signal transmission does not require covalent binding between the BLUF domain and the C-terminus, in that it occurs even if the two parts of AppA are expressed separately. This result is very important in view of the large number of short-BLUF proteins, that could interact with molecular partners (not yet identified) in a light-regulated way. It has been proposed that the light-induced structural change of the AppA BLUF domain, detected by FTIR, allow the domain to be bound to the *C*-terminal region, producing dissociation of PpsR.[140] Given that similar light-driven protein conformational changes have been measured for short-BLUF proteins,[76] they could affect the direct interaction with their still unidentified partners.

An important result was recently obtained by fusing a BLUF domain from a Euglena PAC protein to the *C*-terminus of AppA. The hybrid protein was fully functional in regulating light-dependent gene expression in *R. sphaeroides*, albeit the sequence similarity between PAC-BLUF and AppA-BLUF is lower than 30%.[135] This implies that a BLUF domain can convey signals to completely different output domains and that an eukaryotic light-sensing module can fully replace its homologue in a prokaryotic cell.

10.3. Bacterial Blue-Light Sensing Flavoproteins in Action: Role and Evolution

For very few of the bacteria depicted in Figure 4 UVA/blue-light regulated responses have been described.[141] Furthermore there is presently no link between these responses and the flavin-binding putative photoreceptors described above, with AppA from *R. sphaeroides* representing an important exception. Nevertheless, given their high level of occurrence, we can suppose that they have some physiological role: for 34 of the 59 bacteria listed in Figure 4 the genome sequencing is completed; they represent 16% of the 208 bacterial whole genomes deposited at NCBI (http://www.ncbi.nlm.nih.gov/genomes).

One working model is related to the ability of blue-light to cause photosensitized damages to living organisms. Blue light is absorbed by porphyrins, ubiquitous photosensitisers for singlet oxygen (see Ghetti *et al.* and references therein).[142] All bacteria synthesize porphyrins as heme or chlorophyll precursors, and it would not be surprising that both photosynthetic and non-photosynthetic microorganisms that experience a changing environment during their life course, have evolved blue-light sensors able to elicit a protective response, *e.g.* negative phototaxis, photophobic responses, activation of metabolic pathways related to stress stimuli. Interconnection between light stimuli and the presence of oxygen, necessarily affecting metabolic pathways, could be a key feature to understand the physiological role of bacterial blue-light sensing.

Last but not least, the discovery of plant-like or plant-related putative photosensors in prokaryots rises questions about their origin and evolution: plants may have received their genes for flavin photosensors from endosymbiotic events, but horizontal gene transfer may have occurred also from plants versus environmental bacteria. A link of specialized metabolic pathways to flavin photosensors in bacteria may explain the limited diffusion of the BLUF paradigm among prokaryots. The increasing number of LOV- and BLUF-proteins and Cry, emerging from genome projects, definitely improves the statistical parameters of phylogenetic analysis and important results are beginning to emerge.

10.3.1. UVA/Blue-Light Triggered Responses in Bacteria and their Link to Flavin-Based Photosensors

10.3.1.1. R. sphaeroides and AppA

The interconnection between blue-light and oxygen stimuli is paradigmatic in the facultatively phototrophic purple bacteria of the genus *Rhodobacter*, for which both parameters regulate the formation of photosynthetic complexes.[141] At high O_2 tension, photosynthesis genes are inhibited, whereas at very low oxygen concentrations formation of the photosynthetic complexes significantly increases, even in the absence of light stimuli. In *R. sphaeroides*, only in semianaerobic conditions, blue-light co-represses photosynthesis. AppA participates in photosynthesis inhibition, when O_2 tension is high in a light-independent way, by releasing the repressor PpsR. This process, redox controlled, is mediated by the cysteine rich *C*-terminus of AppA.[53,133] In semianaerobic conditions, instead, the BLUF domain comes into the play, and blue-light illumination results in inhibition of photosynthesis via the release of PpsR from AppA.[56,135,143] This process has probably a protective role against photosensitized reactive oxygen species, but results in the same metabolism switching from phototrophy to heterotrophy as occurs under high O_2 tension.

As many other motile phototrophic bacteria, *R. sphaeroides* exhibits phototactic responses, that seem to be primarily regulated by the photosynthetic apparatus rather than by specific photosensors.[141,144] Nevertheless, a motility photophobic response to blue-light under anaerobic conditions has been reported to be independent of the photosynthesis apparatus, and ascribed to a separate photosensor.[145] The nature of this photosensor remains elusive, and must be probably searched within the rich pool of putative light-sensing proteins of this bacterium. Besides AppA and a LOV protein (Figure 4), *R. sphaeroides* also contains genes encoding for a bacterio-phytochrome and a PYP.[146] This large number of putative photosensors is up to now a unique feature among the bacteria listed in Figure 4.

10.3.1.2. Myxococcus xanthus and Blue-Light Regulated Carotenoids Production

M. xanthus is a non-photosynthetic δ-proteobacterium that responds to blue light either by lysing or by producing carotenoids, depending on the growth phase[147] (see also ref. 141 and references therein). Carotenoids are thought to have a protective role against singlet oxygen, photosensitized by non-heme porphyrins.[148] In *M. xanthus* carotenoids production is under the control of two sigma-factors, CarR and CarQ. According to the current model for light-induced carotenogenesis, CarR and CarQ form in dark grown cells a transcriptionally inactive complex.[149] Blue light is absorbed by protoporphyrin IX with subsequent production of singlet oxygen that destroys the complex, allowing CarQ to initiate transcription of carotenoids genes. The induced carotenoids quench singlet oxygen allowing recovery of the inhibitory CarR-CarQ complex. Whether any specific photosensor (*e.g.* the BLUF protein, Table 2) is implied in this process, has not been investigated, but the mechanism depicted above does not actually require any. The BLUF protein could alternatively trigger a second protective mechanism, such as the one suggested below for YcgF in *E. coli.*

10.3.1.3. The E. coli Photophobic Paradigm and the BLUF Protein YcgF

E. coli shows a typical photophobic response, called phototumbling, upon exposure to blue-light of high intensity.[141,144,150] The addition of photosensitizers or the occurrence of mutations that lead to protoporphyrin IX accumulation, decrease the light threshold for phototumbling. The chemotactic machinery is needed for this response and no tumbling response was observed under anaerobic conditions.[151,152] As in the case of *M. xanthus*, it is generally believed that no specific blue-light sensor is implied in the phototumbling, a process most probably mediated by singlet oxygen. Accordingly, phototumbling is observed also in *Salmonella typhimurium*[150] that does not possess genes encoding for flavin photosensors.

Again the BLUF protein from *E. coli,* YcgF, could trigger a second protection mechanism. Interestingly YcgF is routinely co-purified with two proteins involved in pathways regulating the covering of the plasma membrane with sugars.[107] This process could have a photoprotective role, controlled by YcgF in a light-dependent manner and deserves definitely further investigation.

10.3.1.4. Light Responses in Bacillaceae and Possible Relation to Blue-Light Sensors

Reports for UVA/Blue-light driven processes in Firmicutes are spurious in the literature and, to our knowledge, restricted to the genus *Bacillus*. Light-induced inhibition of sporulation has been observed with broad-band light and shown to be primarily a UVA/blue-light effect.[153] This inhibition does not imply a block in sporulation, that solely takes a longer time-period in UVA/Blue-light grown cells. No further investigation was performed to ascertain the nature of the photoreceptor and/or photosensitizing molecules implied in this process. A second phenomenon observed in *Bacillus* was the inactivation of spores by

UV/visible light, probably mediated by photooxydation,[154] *i.e.* singlet oxygen production.

Two growth patterns induced by long-wavelength UV (310–400 nm) irradiation in *B. subtilis* have been recently described. In plates with an illuminated boundary region, the bacterial colonies grow only in the dark area of the plate when kept below a "crossover" temperature of *ca.* 22°C.[155] A second UVA response was observed on semi-solid agar plates, where a colony was incoluted in the center of the plate, let first grown in the dark and then subjected to UVA light conditions. The bacteria migrate towards the colony edge, forming a characteristic ring-shaped growth pattern.[156]

Are these four phenomena somehow linked to the LOV-protein Ytva? The inactivation of spores with visible light and the inhibition of sporulation have been observed also for *Bacillus* species that do not possess LOV-proteins or other flavin putative photosensors (*e.g. B. cereus*, *B. licheniformis*),[153,156] so that it is improbable that YtvA is involved. As for the two growth patterns, only a spectroscopic match between the UVA light used and the absorption spectrum of YtvA can be established.

It was recently reported that YtvA is a positive regulator in the activation of the general stress transcription factor σ^B,[157] but the environmental input sensed by its LOV domain has remained undetermined and its precise function unclarified.[87] The environmental signal sensed by YtvA could be UVA/Blue-light that triggers the photochemical cycle. The *ytva* knock-out *B. subtilis* mutant, grown under different conditions combined with blue-light illumination, is being currently investigated and is likely to provide some hints on the photophysiological role of this protein.

10.3.1.5. Cyanobacteria: Blue-Light Effects and Involved Photosensors

As many photosynthetic organisms, cyanobacteria have evolved the ability to respond to light quality and quantity to efficiently capture light energy and at the same time minimizing photodamage. Cyanobacteria show many responses to light quality and intensity, and have evolved several light-sensing systems and signal-transduction pathways.[158] Among putative photosensors the most known are phytochrome-like proteins, detected in all the cyanobacteria species listed in Figure 4 and exhibiting photochemical reactions similar to plant phytochromes.[1,159] Their physiological function remains largely obscure, but may vary from regulation of complementary chromatic adaptation[160] to that of sensors for phototaxis.[161,162]

Synechocystis sp. PCC 6803, one of the best studied cyanobacteria in the field of light sensing, exhibits a positive phototaxis towards light ranging from yellow to red (560–720 nm) and negative phototactic mobility away from UV light (reviewed in[161]). Negative phototaxis is also induced by intense blue-(470 nm) or red-light (600–700 nm). Positive phototaxis is lost in a deletion mutant of the phytochrome-like gene *pixJ1* that shows negative phototaxis away from yellow and red-light.[161,163] These observations suggest that multiple receptors are responsible for phototaxis in *Synechocystis sp.* PCC 6803. A further complication is given to the peculiar spectroscopical properties of the PixJ1 phytochrome that absorbs in the blue-region (435 nm) in the dark-adapted state, suggesting that a blue-light signal inactivates positive phototaxis via PixJ1.[164] Also a second phytochrome-like protein, Cph2, is involved in blue-light signalling in that the mutant in *cph2* gene exhibits positive phototaxis towards blue-light.[165] Both Cph2 and PixJ1 phytochrome-like proteins therefore participate in preventing positive phototaxis towards blue-light in wild type cells, but do not account for negative phototaxis at these wavelengths. Notably, the photomovements of *Synechocystis sp.* PCC 6803 and other cyanobacteria, do not require energy coming from photosynthesis, in that they take place even when photosynthesis in inhibited.[67,166]

Do flavin-based photosensors play any role in cyanobacterial phototaxis? The possible involvement of the BLUF protein Slr1694 was recently investigated. Slr1694 appears necessary for positive phototaxis towards white light, as demonstrated with the *slr1694* mutant that shows negative phototaxis.[167] Nevertheless, it does not seem to be specifically involved in blue-light sensing. In *Synechocystis sp.* PCC 6803, transfer of the cells from the dark to UVA or blue light (380 or 450 nm) conditions results in an increase in cellular cAMP.[168] cAMP stimulates cell mobility, by forming a complex with the SYCRP1 receptor that promotes the biogenesis of pili.[169,170] The adenylate cyclase Cya1 is directly involved in this process, but the purified protein is light insensitive and does not contain any chromophore.[167] The inactivation of the flavin pools with phenylacetic acid, results in inhibition of the light-induced increase in cAMP, suggesting that a flavoprotein is the photoreceptor involved.[168] Nevertheless the BLUF containing protein Slr1694 is apparently not involved in the regulation of Cya1-mediated cAMP signal transduction *in vivo*. Furthermore, *in vitro*, Slr1694 does not influence the enzymatic activity of Cya1.[167] A possible role for the LOV protein Q55576 and the DASH-Cry Sll1629 in phototaxis has not been elucidated.

The phototactic movements of other cyanobacteria, do not always resemble those in *Synechocystis sp.* PCC 6803. For example, negative phototaxis is never observed for *T. elongatus* and, in general, action spectra fo photomovements are quite variable in the relative peak amplitudes.[67] Several results strongly suggest that multiple photoreceptors are involved in phototactic responses, and that these sensors may have different spectral properties depending on the light quality and intensity to which the organism has adapted.[67]

Cyanobacteria can switch from phototrophic to heterotrophic metabolism in darkness.[171] A glucose tolerant strain of Synechocystis sp. 6803 does not grow on glucose under complete darkness unless given a daily pulse of white light, insufficient for photoautotrophy.[172] The action spectrum of this response has a maximum at 450 nm, thus matching the absorption maximum of LOV, BLUF and Cry proteins, but no connection to this type of proteins has been established. Wilde *et al.* showed that the *plpa* gene, encoding for a phytochrome-like protein, is required for light-activated heterotrophic growth.[173] Blue-light also regulates the expression of photosynthesis genes in cyanobacteria, but several results suggest that this regulation depends largely from the redox state of photosynthesis components, although a role of (still unidentified) sensors cannot be excluded.[141]

10.3.2. *Evolutionary Aspects*

10.3.2.1. *The Bacterial Origin of Cry: Two Endosymbiotic Events?*

Given that proteins belonging to the PL/Cry family have the same architecture and are extremely spread, the phylogenetic analysis is in this case particularly straightforward. A recently published phylogenetic tree shows that plant Cry1 and Cry2 cannot be grouped with the DASH-Cry, in contrast to AtCry3.[48] The closest relatives of Cry1 and Cry2 are PL sequences from α-proteobacteria. The demonstration that AtCry3 is targeted to chloroplasts, thought to derive from cyanobacteria, and also the fact that α-proteobacteria are considered to be the progenitors of mitochondria, points to the hypothesis that plants might have inherited their Cry gene through two horizontal gene transfer events: Cry1 and Cry2 evolved from endosymbiotic α-proteobacteria, whereas Cry3 could have been inherited from an endosymbiotic cyanobacterial ancestor.[48]

10.3.2.2. *Why is BLUF so Genuinely Bacterial?*

Among bacterial flavin-photosensors BLUF proteins are certainly unique in their diffusion and they appear as a truly bacterial paradigm. Only Euglenida seem to have aquired this domain within the PAC protein family, despite the fact that BLUF is a versatile light-responsive module, able to transfer signals to different effector partners.[135] A recent phylogenetic analysis indicated that the appearance of PAC in Euglenoids occurred later than chloroplast aquisition through an endosymbiotic event.[57] The reasons why BLUF domains have been lost in the large majority of eukaryots are not clear, but may be related to specific metabolic aspects. Bacterial BLUF domains are often linked to c-di-GMP phosphodiesterases,[2,106] and it is possible that BLUF domains were lost in eukaryotes concomitantly with domains involved in c-di-GMP metabolism and c-di-GMP mediated signaling, whereas other blue-light photoreceptors of prokaryotic origin, such as Cry and phot evolved further.[135] *Rhodobacter capsulatus*, a close relative of *R. sphaeroides*, lacks BLUF proteins including AppA, possibly due to the fact that this species possesses an additional defense mechanism against reactive oxygen species.[135] Also this aspect suggests that evolution of BLUF proteins is strictly associated with the functions that they regulate *in vivo*.

10.3.2.3. *The Ubiquitous LOV*

Bacterial LOV proteins are so heterogenous that a phylogenetic analysis of full proteins in view of their possible role as precursors for phot is not possible. Inspection of the simple phylogeny tree based on the LOV alignment (Figures 6 and 8) shows that there are two major groups, one clustering together with phot-LOV domains (lower part of Figure 8) and the second comprising the LOV domains with lower similarity. Phot-LOVs cluster closer to LOV domains from non-photosynthetic α- and γ-Proteobacteria associated, like in phot, to a kinase function (Figure 8). On the contrary, LOV domains from cyanobacteria (bacteria performing oxygenic photosynthesis) are among the less similar to phot-LOV. We note that cyanobacterial LOV domains are often associated to the GGDEF domain, a typical bacterial feature (see Section 3.2.b), whereas the other LOV proteins contain effector domains common to eukaryots and prokaryots (Table 1). In some cases LOV domains very close in the tree, reflect close vicinity of the organisms they belong to (*e.g. E. litoralis* and *N. aromaticivorans*).

The light-sensing LOV domains are well represented in bacteria, fungi and plants,[3,39] reflecting the large spreading of the PAS superfamily,[21] but different to the Cry/PlL proteins they are not found in animals. Their domain association highlights their putative involvement in various metabolic pathways, a versatility paralleled also by other PAS domains, in particular those sensing O_2 via the bound heme cofactor.[109] All the bacteria listed in Table 1 are characterized by the large presence of genes encoding for proteins involved in signal sensing/ signal transduction mechanisms and indeed, as facultative parasites or free-living bacteria, they experience during their life span a variety of environments. Quite interestingly, the only Archaeon that (up to now) possess LOV proteins is *H. marismortui*, intrinsically able to exploit a significantly higher variety of environments that other Archaea.[174]

10.3.2.4. Bacterial Photosensors: Highlighting a Common Photosynthetic Ancestry?

The occurrence of light-sensors in non-photosynthetic bacteria may be related to a common ancestry in which phototrophy played a larger role than in present times. Years ago the 16S ribosomal RNA analysis of the photosynthetic *Heliobacterium chlorum* revealed that this organism belongs to the Gram-positive bacteria, more precisely to the so-called low G + C division, the Firmicutes[175] (the remaining photosynthetic bacteria are all Gram-negative). This discovery suggested that the Firmicutes (the division of *B. subtilis* and other bacteria listed in Figure 4) may be of ancient photosynthetic origin, strengthening the case for a common photosynthetic ancestry for all eubacteria. Genome analysis has recently revealed that non-photosynthetic bacteria, including *B. subtilis*, possess genes coding for proteins similar to the large subunit of RuBisCO.[176] In *B. subtilis* such a protein catalyses an enolase reaction in the methionine salvage pathway. Surprisingly, a growth-defective mutant in which the gene for the RuBisCO-like protein was disrupted, could be rescued by the gene for RuBisCO from the photosynthetic bacterium *Rhodospirillum rubrum*.[176] Phylogenetic analysis and the inspection of the enzymatic mechanisms, suggests that photosynthesis RuBisCO evolved from these bacterial RuBisCO-like proteins.[177] Even if the ability to exploit light as an energy source have been lost in the majority of present bacteria, light-sensors may have survived as a tool to optimize responses to the environment stimuli, among which light is one of the most ancient and ubiquitous.

Acknowledgments

I am indebted with professor Wolfgang Gärtner for his critical review of this manuscript.

References

1. J. E. Hughes, T. Lamparter, F. Mittmann, E. Hartmann, W. Gärtner, A. Wilde and T. Börner, A prokaryotic phytochrome, *Nature*, 1997, **386**, 663.
2. M. Gomelsky and G. Klug, BLUF: a novel FAD-binding domain involved in sensory transduction in microorganisms, *Trends Biochem. Sci.*, 2002, **27**, 497–500.
3. S. Crosson, S. Rajagopal and K. Moffat, The LOV domain family: photoresponsive signaling modules coupled to diverse output domains, *Biochemistry*, 2003, **42**, 2–10.
4. K. H. Jung, V. D. Trivedi and J. L. Spudich, Demonstration of a sensory rhodopsin in eubacteria, *Mol. Microbiol.*, 2003, **47**, 1513–1522.
5. W. Gärtner and A. Losi, Crossing the borders: archaeal rhodopsins go bacterial, *Trends Microbiol.*, 2003, **11**, 405–407.
6. A. Losi, The bacterial counterparts of plants phototropins, *Photochem. Photobiol. Sci.*, 2004, **3**, 566–574.
7. B. L. Montgomery and J. C. Lagarias, Phytochrome ancestry: sensors of bilins and light, *Trends Plant Sci.*, 2002, **7**, 357–366.
8. B. Quest and W. Gartner, Chromophore selectivity in bacterial phytochromes: dissecting the process of chromophore attachment, *Eur. J. Biochem.*, 2004, **271**, 1117–1126.
9. B. Karniol and R. D. Vierstra, The pair of bacteriophytochromes from Agrobacterium tumefaciens are histidine kinases with opposing photobiological properties, *Proc. Nat. Acad. Sci. USA*, 2003, **100**, 2807–2812.
10. A. Losi, E. Polverini, B. Quest and W. Gärtner, First evidence for phototropin-related blue-light receptors in prokaryotes, *Biophys. J.*, 2002, **82**, 2627–2634.
11. R. Brudler, K. Hitomi, H. Daiyasu, H. Toh, K. I. Kucho, M. Ishiura, M. Kanehisa, V. A. Roberts, T. Todo, J. A. Tainer and E. D. Getzoff, Identification of a New Cryptochrome Class: structure, Function, and Evolution, *Mol. Cell*, 2003, **11**, 59–67.
12. W. R. Briggs and E. Huala, Blue-light photoreceptors in higher plants, *Ann. Rev. Cell. Develop. Biol.*, 1999, **15**, 33–62.
13. R. Banerjee and A. Batschauer, Plant blue-light receptors, *Planta*, 2005, **220**, 498–502.
14. M. Iseki, S. Matsunaga, A. Murakami, K. Ohno, K. Shiga, K. Yoshida, M. Sugai, T. Takahashi, T. Hori and A. Watanabe, A blue-light-activated adenylyl cyclase mediates photoavoidance in *Euglena gracilis*, *Nature*, 2002, **415**, 1047–1051.
15. K. J. Hellingwerf, Key issues in the photochemistry and signalling-state formation of photosensor proteins, *J. Photochem. Photobiol. B: Biol.*, 2000, **54**, 94–102.
16. W. R. Briggs and J. M. Christie, Phototropins 1 and 2: versatile plant blue-light receptors, *Trends Plant Sci.*, 2002, **7**, 204–210.
17. M. Ohgishi, K. Saji, K. Okada and T. Sakai, Functional analysis of each blue light receptor, cry1, cry2, phot1, and phot2, by using combinatorial multiple mutants in Arabidopsis, *Proc. Nat. Acad. Sci. USA*, 2004, **101**, 2223–2228.

18. K. Huang and C. F. Beck, Phototropin is the blue-light receptor that controls multiple steps in the sexual life cycle of the green alga *Chlamydomonas reinhardtii*, *Proc. Nat. Acad. Sci. USA*, 2003, **100**, 6269–6274.
19. E. Huala, P. W. Oeller, E. Liscum, I. S. Han, E. Larsen and W. R. Briggs, *Arabidopsis* NPH1: a Protein Kinase with a Putative Redox-Sensing Domain, *Science*, 1997, **278**, 2120–2123.
20. I. B. Zhulin, B. L. Taylor and R. Dixon, PAS domain S-boxes in Archaea, Bacteria and sensors for oxygen and redox, *Trends Biochem. Sci.*, 1997, **22**, 331–333.
21. M. H. Hefti, K. J. Francoijs, S. C. de Vries, R. Dixon and J. Vervoort, The PAS fold: a redefinition of the PAS domain based upon structural prediction, *FEBS J.*, 2004, **271**, 1198–1208.
22. M. Kasahara, T. E. Swartz, M. A. Olney, A. Onodera, N. Mochizuki, H. Fukuzawa, E. Asamizu, S. Tabata, H. Kanegae, M. Takano, J. M. Christie, A. Nagatani and W. R. Briggs, Photochemical properties of the flavin mononucleotide-binding domains of the phototropins from Arabidopsis, rice, and *Chlamydomonas reinhardtii*, *Plant Physiol.*, 2002, **129**, 762–773.
23. M. Salomon, J. M. Christie, E. Knieb, U. Lempert and W. R. Briggs, Photochemical and mutational analysis of the FMN-binding domains of the plant blue light receptor phototropin, *Biochemistry*, 2000, **39**, 9401–9410.
24. M. Salomon, W. Eisenreich, H. Dürr, E. Scleicher, E. Knieb, V. Massey, W. Rüdiger, F. Müller, A. Bacher and G. Richter, An optomechanical transducer in the blue light receptor phototropin from *Avena sativa*, *Proc. Nat. Acad. Sci. USA*, 2001, **98**, 12357–12361.
25. S. Crosson and K. Moffat, Photoexcited Structure of a Plant Photoreceptor Domain Reveals a Light-Driven Molecular Switch, *Plant Cell*, 2002, **14**, 1067–1075.
26. C. W. M. Kay, E. Schleicher, A. Kuppig, H. Hofner, W. Rüdiger, M. Schleicher, M. Fischer, A. Bacher, S. Weber and G. Richter, Blue light perception in plants. Detection and characterization of a light-induced neutral flavin radical in a C450A mutant of phototropin, *J. Biol. Chem.*, 2003, **278**, 10973–10982.
27. J. T. M. Kennis, S. Crosson, M. Gauden, I. H. M. van Stokkum, K. Moffat and R. van Grondelle, Primary Reactions of the LOV2 Domain of Phototropin, a Plant Blue-Light Photoreceptor, *Biochemistry*, 2003, **42**, 3385–3392.
28. T. E. Swartz, S. B. Corchnoy, J. M. Christie, J. W. Lewis, I. Szundi, W. R. Briggs and R. A. Bogomolni, The photocycle of a flavin-binding domain of the blue light photoreceptor phototropin, *J. Biol. Chem.*, 2001, **276**, 36493–36500.
29. T. Kottke, J. Heberle, D. Hehn and P. Hegemann, Phot-LOV1: photocycle of a Blue-Light Receptor Domain from the Green Alga *Chlamydomonas reinhardtii*, *Biophys. J.*, 2003, **84**, 1192–1201.
30. S. Crosson and K. Moffat, Structure of a flavin-binding plant photoreceptor domain: insights into light-mediated signal transduction, *Proc. Nat. Acad. Sci. USA*, 2001, **98**, 2995–3000.
31. R. Fedorov, I. Schlichting, E. Hartmann, T. Domratcheva, M. Fuhrmann and P. Hegemann, Crystal structures and molecular mechanism of a light-induced signaling switch the Phot-LOV1 domain from *Chlamydomonas reinhardtii*, *Biophys. J.*, 2003, **84**, 2492–2501.
32. R. M. Williams and S. E. Braslavsky, *Triggering of photomovement – molecular basis*, in *Photomovement*, D.P. Haeder and M. Lebert (eds), Elsevier Science, Amsterdam, 2001, 16–48.

33. J. M. Christie, T. E. Swartz, R. A. Bogomolni and W. R. Briggs, Phototropin LOV domains exhibit distinct roles in regulating photoreceptor function, *Plant J.*, 2002, **32**, 205–219.
34. T. Mizoguchi and G. Coupland, ZEITLUPE and FKF1: novel connections between flowering time and circadian clock control, *Trends Plant Sci.*, 2000, **5**, 409–411.
35. T. Imaizumi, H. G. Tran, T. E. Swartz, W. R. Briggs and S. A. Kay, FKF1 is essential for photoperiodic-specific light signalling in Arabidopsis, *Nature*, 2003, **426**, 302–306.
36. L. Han, M. Mason, E. P. Risseeuw, W. L. Crosby and D. E. Somers, Formation of an SCF(ZTL) complex is required for proper regulation of circadian timing, *Plant J.*, 2004, **40**, 291–301.
37. C. M. Pickart and M. J. Eddins, Ubiquitin: structures, functions, mechanisms, *Biochim. Biophys. Acta*, 2004, **1695**, 55–72.
38. J. Adams, R. Kelso and L. Cooley, The kelch repeat superfamily of proteins: propellers of cell function, *Trends Cell Biol.*, 2000, **10**, 17–24.
39. R. Ambra, B. Grimaldi, S. Zamboni, P. Filetici, G. Macino and P. Ballario, Photomorphogenesis in the hypogeous fungus *Tuber borchii*: isolation and characterization of Tbwc-1, the homologue of the blue-light photoreceptor of *Neurospora crassa*, *Fungal Genetics Biol.*, 2004, **41**, 688–697.
40. A. Sancar, Photolyase and cryptochrome blue-light photoreceptors, *Adv. Prot. Chem.*, 2004, **69**, 73–100.
41. S. Weber, Light-driven enzymatic catalysis of DNA repair: a review of recent biophysical studies on photolyase, *Biochim. Biophys. Acta-Bioenerg.*, 2005, **1707**, 1–23.
42. A. R. Cashmore, J. A. Jarillo, Y. J. Wu and D. M. Liu, Cryptochromes: blue light receptors for plants and animals, *Science*, 1999, **284**, 760–765.
43. C. A. Brautigam, B. S. Smith, Z. Ma, M. Palnitkar, D. R. Tomchick, M. Machius and J. Deisenhofer, Structure of the photolyase-like domain of cryptochrome 1 from Arabidopsis thaliana, *Proc. Nat. Acad. Sci. USA*, 2004, **101**, 12142 12147.
44. C. Lin and D. Shalitin, Cryptochrome structure and signal transduction, *Ann. Rev. Plant Biol.*, 2003, **54**, 469–496.
45. C. L. Partch, M. W. Clarkson, S. Ozgur, A. L. Lee and A. Sancar, Role of structural plasticity in signal transduction by the cryptochrome blue-light photoreceptor, *Biochemistry*, 2005, **44**, 3795–3805.
46. K. Malhotra, S. T. Kim, A. Batschauer, L. Dawut and A. Sancar, Putative blue-light photoreceptors from *Arabidopsis thaliana* and *Sinapis alba* with a high degree of sequence homology to DNA photolyase contain the two photolyase cofactors but lack DNA repair activity, *Biochemistry*, 1995, **34**, 6892–6899.
47. T. Imaizumi, T. Kanegae and M. Wada, Cryptochrome nucleocytoplasmic distribution and gene expression are regulated by light quality in the fern *Adiantum capillus-veneris*, *Plant Cell*, 2000, **12**, 81–96.
48. T. Kleine, P. Lockhart and A. Batschauer, An Arabidopsis protein closely related to Synechocystis cryptochrome is targeted to organelles, *Plant J.*, 2003, **35**, 93–103.
49. H. Daiyasu, T. Ishikawa, K. I. Kuma, S. Iwai, T. Todo and H. Toh, Identification of cryptochrome DASH from vertebrates, *Genes To Cells*, 2004, **9**, 479–495.
50. B. Giovani, M. Byrdin, M. Ahmad and K. Brettel, Light-induced electron transfer in a cryptochrome blue-light photoreceptor, *Nature Struct. Biol.*, 2003, **10**, 489–490.

51. I. H. Kavakli and A. Sancar, Analysis of the role of intraprotein electron transfer in photoreactivation by DNA photolyase *in vivo*, *Biochemistry*, 2004, **43**, 15103–15110.
52. M. Ntefidou, M. Iseki, M. Watanabe, M. Lebert and D. P. Hader, Photoactivated adenylyl cyclase controls phototaxis in the flagellate *Euglena gracilis*, *Plant Physiol.*, 2003, **133**, 1517–1521.
53. M. Gomelsky and S. Kaplan, appA, a novel gene encoding a trans-acting factor involved in the regulation of photosynthesis gene expression in Rhodobacter sphaeroides 2.4.1, *J. Bacteriol.*, 1995, **177**, 4609–4618.
54. M. Gomelsky and S. Kaplan, AppA, a redox regulator of photosystem formation in Rhodobacter sphaeroides 2.4.1, is a flavoprotein. Identification of a novel fad binding domain, *J. Biol. Chem.*, 1998, **273**, 35319–35325.
55. S. Masuda and C. E. Bauer, AppA Is a Blue Light Photoreceptor that Antirepresses Photosynthesis Gene Expression in *Rhodobacter sphaeroides*, *Cell*, 2002, **110**, 613–623.
56. S. Braatsch, M. Gomelsky, S. Kuphal and G. Klug, A single flavoprotein, AppA, integrates both redox and light signals in Rhodobacter sphaeroides, *Mol. Microbiol.*, 2002, **45**, 827–836.
57. Y. Koumura, T. Suzuki, S. Yoshikawa, M. Watanabe and M. Iseki, The origin of photoactivated adenylyl cyclase (PAC), the *Euglena* blue-light receptor: phylogenetic analysis of orthologues of PAC subunits from several euglenoids and trypanosome-type adenylyl cyclases from *Euglena gracilis*, *Photochem. Photobiol. Sci.*, 2004, **3**, 580–586.
58. A. Kita, K. Okajima, Y. Morimoto, M. Ikeuchi and K. Miki, Structure of a Cyanobacterial BLUF Protein, Tll0078, Containing a Novel Fad-binding Blue Light Sensor Domain, *J. Mol. Biol.*, 2005, **349**, 1–9.
59. V. Massey, The chemical and biological versatility of riboflavin, *Biochem. Soc. T.*, 2000, **28**, 283–296.
60. M. A. Miranda, Photosensitization by drugs, *Pure Appl. Chem.*, 2001, **73**, 481–486.
61. S. Y. Egorov, A. A. Krasnovsky Jr., M. Y. Bashtanov, E. A. Mironov, T. A. Ludnikova and M. S. Kritsky, Photosensitization of singlet oxygen formation by pterins and flavins. Time-resolved studies of oxygen phosphorescence under laser excitation, *Biochemistry Moscow*, 1999, **64**, 1117–1121.
62. C. Y. Lu and Y. Y. Liu, Electron transfer oxidation of tryptophan and tyrosine by triplet states and oxidized radicals of flavin sensitizers: a laser flash photolysis study, *Biochim. Biophys. Acta*, 2002, **1571**, 71–76.
63. C. Y. Lu, S. D. Yao and N. Y. Lin, Photooxidation of 2′-deoxyguanosine 5′-monophosphate (dGMP) by flavin adenine dinucleotide (FAD) via electron transfer: a laser photolysis study, *Chem. Phys. Lett.*, 2000, **330**, 389–396.
64. M. Gauden, S. Yeremenko, W. Laan, I. H. M. van Stokkum, J. A. Ihalainen, R. van Grondelle, K. J. Hellingwerf and J. T. M. Kennis, Photocycle of the Flavin-Binding Photoreceptor AppA, a Bacterial Transcriptional Antirepressor of Photosynthesis Genes, *Biochemistry*, 2005, **44**, 3653–3662.
65. M. Sakai and H. Takahashi, One-electron photoreduction of flavin mononucleotide: time-resolved resonance Raman and absorption study, *J. Mol. Struct.*, 1996, **379**, 9–18.
66. T. Bernt Melo, M. Adriana Ionescu, G. W. Haggquist and K. Razi Naqvi, Hydrogen abstraction by triplet flavins. I: time-resolved multi-channel absorption spectra of flash-irradiated riboflavin solutions in water, *Spectrochim. Acta A*, 1999, **55**, 2299–2307.

67. C. Lin, D. E. Robertson, M. Ahmad, A. A. Raibekas, M. S. Jorns, P. L. Dutton and A. R. Cashmore, Association of flavin adenine dinucleotide with the Arabidopsis blue light receptor CRY1, *Science*, 1995, **269**, 968–970.
68. A. Bacher, S. Eberhardt, M. Fischer, K. Kis and G. Richter, Biosynthesis of vitamin b2 (riboflavin), *Annu. Rev. Nutr.*, 2000, **20**, 153–167.
69. M. Fischer, A. K. Schott, W. Romisch, A. Ramsperger, M. Augustin, A. Fidler, A. Bacher, G. Richter, R. Huber and A. Eisenreich, Evolution of vitamin B2 biosynthesis. A novel class of riboflavin synthase in Archaea, *J. Mol. Biol.*, 2004, **343**, 267–278.
70. H. Katagiri, H. Yamada and K. Imai, Biosynthesis of flavin coenzymes by microorganisms. II. Enzymic synthesis of flavin-adenine dinucleotide by Escherichia coli, *J. Vitaminol.*, 1959, **5**, 307–311.
71. H. Katagiri, H. Yamada and K. Imai, Biosynthesis of flavin coenzymes by microorganisms. I. Bacterial flavokinase, *J. Vitaminol.*, 1959, **5**, 129–133.
72. W. Laan, T. Bednarz, J. Heberle and K. J. Hellingwerf, Chromophore composition of a heterologously expressed BLUF-domain, *Photochem. Photobiol. Sci.*, 2004, **3**, 1011–1016.
73. S. McGinnis and T. L. Madden, BLAST: at the core of a powerful and diverse set of sequence analysis tools, *Nuc. Ac. Res.*, 2004, **32**, W20–W25.
74. C. Kanz, P. Aldebert, N. Althorpe, W. Baker, A. Baldwin, K. Bates, P. Browne, A. van den Broek, M. Castro, G. Cochrane, K. Duggan, R. Eberhardt, N. Faruque, J. Gamble, F. G. Diez, N. Harte, T. Kulikova, Q. Lin, V. Lombard, R. Lopez, R. Mancuso, M. McHale, F. Nardone, V. Silventoinen, S. Sobhany, P. Stoehr, M. A. Tuli, K. Tzouvara, R. Vaughan, D. Wu, W. Zhu and R. Apweiler, The EMBL Nucleotide Sequence Database, *Nuc. Ac. Res.*, 2005, **33**, D29–D33.
75. D. L. Wheeler, T. Barrett, D. A. Benson, S. H. Bryant, K. Canese, D. M. Church, M. DiCuccio, R. Edgar, S. Federhen, W. Helmberg, D. L. Kenton, O. Khovayko, D. J. Lipman, T. L. Madden, D. R. Maglott, J. Ostell, J. U. Pontius, K. D. Pruitt, G. D. Schuler, L. M. Schriml, E. Sequeira, S. T. Sherry, K. Sirotkin, G. Starchenko, T. O. Suzek, R. Tatusov, T. A. Tatusova, L. Wagner and E. Yaschenko, Database resources of the National Center for Biotechnology Information, *Nuc. Ac. Res.*, 2005, **33**, D39–D45.
76. S. Masuda, K. Hasegawa, A. Ishii and T. A. Ono, Light-induced structural changes in a putative blue-light receptor with a novel FAD binding fold sensor of blue-light using FAD (BLUF); Slr1694 of *Synechocystis sp.* PCC6803, *Biochemistry*, 2004, **43**, 5304–5313.
77. Y. Fukushima, K. Okajima, Y. Shibata, M. Ikeuchi and S. Itoh, Primary Intermediate in the Photocycle of a Blue-Light Sensory BLUF FAD-Protein, Tll0078, of Thermosynechococcus elongatus BP-1, *Biochemistry*, 2005, **44**, 5149–5158.
78. E. N. Worthington, I. H. Kavakli, G. Berrocal-Tito, B. E. Bondo and A. Sancar, Purification and characterization of three members of the photolyase/cryptochrome family glue-light photoreceptors from Vibrio cholerae, *J. Biol. Chem.*, 2003, **278**, 39143–39154.
79. A. Losi, B. Quest and W. Gärtner, Listening to the blue: the time-resolved thermodynamics of the bacterial blue-light receptor YtvA and its isolated LOV domain, *Photochem. Photobiol. Sci.*, 2003, **2**, 759–766.
80. P. Cheng, Q. He, Y. Yang, L. Wang and Y. Liu, Functional conservation of light, oxygen, or voltage domains in light sensing, *Proc. Nat. Acad. Sci. USA*, 2003, **100**, 5938–5943.

81. Q. He, P. Cheng, Y. Yang, L. Wang, K. H. Gardner and Y. Liu, White collar-1, a DNA binding transcription factor and a light sensor, *Science*, 2002, **297**, 840–843.
82. C. Schwerdtfeger and H. Linden, VIVID is a flavoprotein and serves as a fungal blue light photoreceptor for photoadaptation, *EMBO J.*, 2003, **22**, 4846–4855.
83. A. Losi, T. Kottke and P. Hegemann, Recording of Blue Light-Induced Energy and Volume Changes within the Wild-Type and Mutated Phot-LOV1 Domain from *Chlamydomonas reinhardtii*, *Biophys. J.*, 2004, **86**, 1051–1060.
84. T. Bednarz, A. Losi, W. Gärtner, P. Hegemann and J. Heberle, Functional variations among LOV domains as revealed by FT-IR difference spectroscopy, *Photochem. Photobiol. Sci.*, 2004, **3**, 575–579.
85. A. Losi, E. Ternelli and W. Gärtner, Tryptophan Fluorescence in the *Bacillus subtilis* Phototropin-related Protein YtvA as a Marker of Interdomain Interaction, *Photochem. Photobiol.*, 2004, **80**, 150–153.
86. L. Aravind and E. V. Koonin, The STAS domain a link between anion transporters and antisigma-factor antagonists, *Curr. Biol.*, 2000, **10**, R53–R55.
87. S. Akbar, T. A. Gaidenko, K. Min, M. O'Reilly, K. M. Devine and C. W. Price, New family of regulators in the environmental signaling pathway which activates the general stress transcription factor of *Bacillus subtilis*, *J. Bacteriol.*, 2001, **183**, 1329–1338.
88. T. W. Grebe and J. B. Stock, The histidine protein kinase superfamily, *Adv. Microb. Physiol.*, 1999, **41**, 139–227.
89. A. H. West and A. M. Stock, Histidine kinases and response regulator proteins in two-component signaling systems, *Trends Biochem. Sci.*, 2001, **26**, 369–376.
90. Y. Shiro and H. Nakamura, Heme-based oxygen sensor protein FixL: its structure and function, *International Congress Series*, 2002, **1233**, 251–257.
91. K. R. Rodgers and G. S. Lukat-Rodgers, Insights into heme-based O2 sensing from structure-function relationships in the FixL proteins, *J. Inorg. Biochem.*, 2005, **99**, 963–977.
92. S. A. Asher, UV Resonance Raman Spectroscopy, *Anal. Chem.*, 1993, **65**, 201–211.
93. G. S. Lukat and J. B. Stock, Response regulation in bacterial chemotaxis, *J. Cell. Biochem.*, 1993, **51**, 41–46.
94. C. Aubert, K. Brettel, P. Mathis, A. P. Eker and A. Boussac, EPR detection of the transient tyrosyl radical in DNA photolyase from *Anacystis nidulans*, *J. Am. Chem. Soc.*, 1999, **121**, 8659–8660.
95. L. Aravind and C. P. Ponting, The GAF domain: an evolutionary link between diverse phototransducing proteins, *Trends Biochem. Sci.*, 1997, **22**, 458–459.
96. S. E. Martinez, S. Bruder, A. Schultz, N. Zheng, J. E. Schultz, J. A. Beavo and J. U. Linder, Crystal structure of the tandem GAF domains from a cyanobacterial adenylyl cyclase: modes of ligand binding and dimerization, *Proc. Nat. Acad. Sci. USA*, 2005, **102**, 3082–3087.
97. F. Janiak-Spens, D. P. Sparling and A. H. West, Novel role for an HPt domain in stabilizing the phosphorylated state of a response regulator domain, *J. Bacteriol.*, 2000, **182**, 6673–6678.
98. A. Matsushika and T. Mizuno, The structure and function of the histidine-containing phosphotransfer (HPt) signaling domain of the Escherichia coli ArcB sensor, *J. Cell. Biochem.*, 1998, **124**, 440–445.
99. M. A. Heim, M. Jakoby, M. Werber, C. Martin, B. Weisshaar and P. C. Bailey, The basic helix-loop-helix transcription factor family in plants: a genome-wide study of protein structure and functional diversity, *Mol. Biol. Evol.*, 2003, **20**, 735–747.

100. R. J. Kewley, M. L. Whitelaw and A. Chapman-Smith, The mammalian basic helix-loop-helix/PAS family of transcriptional regulators, *Int. J. Biochem. Cell. B.*, 2004, **36**, 189–204.
101. K. J. Hellingwerf, J. Hendriks and T. Gensch, Photoactive Yellow Protein, a new type of photoreceptor protein: will this "yellow lab" bring us where we want to go?, *J. Phys. Chem. A*, 2003, **107**, 1082–1094.
102. M. Y. Galperin, A. N. Nikolskaya and E. V. Koonin, Novel domains of the prokaryotic two-component signal transduction systems, *FEMS Microbiol. Lett.*, 2001, **203**, 11–21.
103. N. Ausmees, R. Mayer, H. Weinhouse, G. Volman, D. Amikam, M. Benziman and M. Lindberg, Genetic data indicate that proteins containing the GGDEF domain possess diguanylate cyclase activity, *FEMS Microbiol. Lett.*, 2001, **204**, 163–167.
104. U. Römling, Molecular biology of cellulose production in bacteria, *Res. Microbiol.*, 2002, **153**, 205–212.
105. D. A. D'Argenio and S. I. Miller, Cyclic di-GMP as a bacterial second messenger, *Microbiology*, 2004, **150**, 2497–2502.
106. D. A. Ryjenkov, M. Tarutina, O. V. Moskvin and M. Gomelsky, Cyclic Diguanylate Is a Ubiquitous Signaling Molecule in Bacteria: insights into Biochemistry of the GGDEF Protein Domain, *J. Bacteriol.*, 2005, **187**, 1792–1798.
107. S. Rajagopal, J. M. Key, E. B. Purcell, D. J. Boerema and K. Moffat, Purification and Initial Characterization of a Putative Blue Light-regulated Phosphodiesterase from *Escherichia coli*, *Photochem. Photobiol.*, 2004, **80**, 542–547.
108. A. L. Chang, J. R. Tuckerman, G. Gonzalez, R. Mayer, H. Weinhouse, G. Volman, D. Amikam, M. Benziman and M. A. Gilles-Gonzalez, Phosphodiesterase A1, a regulator of cellulose synthesis in *Acetobacter xylinum*, is a heme-based sensor, *Biochemistry*, 2001, **40**, 3420–3426.
109. M. A. Gilles-Gonzalez and G. Gonzalez, Heme-based sensors: defining characteristics, recent developments, and regulatory hypotheses, *J. Inorg. Biochem.*, 2005, **99**, 1–22.
110. P. Bork, N. P. Brown, H. Hegyi and J. Schultz, The protein phosphatase 2C (PP2C) superfamily: detection of bacterial homologues, *Protein Sci.*, 1996, **5**, 1421–1425.
111. K. Vijay, M. S. Brody, E. Fredlund and C. W. Price, A PP2C phosphatase containing a PAS domain is required to convey signals of energy stress to the sigmaB transcription factor of *Bacillus subtilis*, *Mol. Microbiol.*, 2000, **35**, 180–188.
112. A. C. Ferreira, M. F. Nobre, E. Moore, F. A. Rainey, J. R. Battista and M. S. da Costa, Characterization and radiation resistance of new isolates of *Rubrobacter radiotolerans* and *Rubrobacter xylanophilus*, *Extremophiles*, 1999, **3**, 235–238.
113. R. W. Phillips, J. Wiegel, C. J. Berry, C. Fliermans, A. D. Peacock, D. C. White and L. J. Shimkets, *Kineococcus radiotolerans sp.* nov., a radiation-resistant, gram-positive bacterium, *Int. J. Syst. Evol. Micr.*, 2002, **52**, 933–938.
114. T. E. Swartz, P. J. Wenzel, S. B. Corchnoy, W. R. Briggs and R. A. Bogomolni, Vibration Spectroscopy Reveals Light-Induced Chromophore and Protein Structural Changes in the LOV2 Domain of the Plant Blue-Light Receptor Phototropin 1, *Biochemistry*, 2002, **41**, 7183–7189.
115. S. M. Harper, L. C. Neil and K. H. Gardner, Structural basis of a phototropin light switch, *Science*, 2003, **301**, 1541–1544.
116. S. M. Harper, J. M. Christie and K. H. Gardner, Disruption of the LOV-Jalpha helix interaction activates phototropin kinase activity, *Biochemistry*, 2004, **43**, 16184–16192.

117. C. Combet, C. Blanchet, C. Geourjon and G. Deleage, NPS@: Network Protein Sequence Analysis, *Trends Biochem. Sci.*, 2000, **25**, 147–150.
118. W. Gong, B. Hao, S. S. Mansy, G. Gonzalez, M. A. Gilles-Gonzalez and M. K. Chan, Structure of a biological oxygen sensor: a new mechanism for heme-driven signal transduction, *Proc. Nat. Acad. Sci. USA*, 1998, **95**, 15177–15182.
119. J. Key and K. Moffat, Crystal Structures of Deoxy and CO-Bound bjFixLH Reveal Details of Ligand Recognition and Signaling, *Biochemistry*, 2005, **44**, 4627–4635.
120. M. Nakasako, D. Matsuoka, K. Zikihara and S. Tokutomi, Quaternary structure of LOV-domain containing polypeptide of Arabidopsis FKF1 protein, *FEBS Lett.*, 2005, **579**, 1067–1071.
121. M. Salomon, U. Lempert and W. Rüdiger, Dimerization of the plant photoreceptor phototropin is probably mediated by the LOV1 domain, *FEBS Lett.*, 2004, **572**, 8–10.
122. O. Yildiz, M. Doi, I. Yujnovsky, L. Cardone, A. Berndt, S. Hennig, S. Schulze, C. Urbanke, P. Sassone-Corsi and A. Wolf, Crystal structure and interactions of the PAS repeat region of the Drosophila clock protein PERIOD, *Mol. Cell*, 2005, **17**, 69–82.
123. H. Park, C. Suquet, J. D. Satterlee and C. Kang, Insights into signal transduction involving PAS domain oxygen-sensing heme proteins from the X-ray crystal structure of Escherichia coli Dos heme domain (Ec DosH), *Biochemistry*, 2004, **43**, 2738–2746.
124. J. L. Pellequer, K. A. Wager-Smith, S. A. Kay and E. D. Getzoff, Photoactive yellow protein: a structural prototype for the three-dimensional fold of the PAS domain superfamily, *Proc. Nat. Acad. Sci. USA*, 1998, **95**, 5884–5890.
125. A. Repik, A. Rebbapragada, M. S. Johnson, J. O. Haznedar, I. B. Zhulin and B. L. Taylor, PAS domain residues involved in signal transduction by the Aer redox sensor of *Escherichia coli*, *Mol. Microbiol.*, 2000, **36**, 806–816.
126. C. A. Amezcua, S. M. Harper, J. Rutter and K. H. Gardner, Structure and Interactions of PAS Kinase N-Terminal PAS Domain: Model for Intramolecular Kinase Regulation, *Structure*, 2002, **10**, 1349–1361.
127. K. Hitomi, K. Okamoto, H. Daiyasu, H. Miyashita, S. Iwai, H. Toh, M. Ishiura and T. Todo, Bacterial cryptochrome and photolyase: characterization of two photolyase-like genes of *Synechocystis sp.* PCC6803, *Nuc. Ac. Res.*, 2000, **28**, 2353–2362.
128. T. Vishnivetskaya, S. Kathariou, J. McGrath, D. Gilichinsky and J. M. Tiedje, Low-temperature recovery strategies for the isolation of bacteria from ancient permafrost sediments, *Extremophiles*, 2000, **4**, 165–173.
129. A. Mees, T. Klar, P. Gnau, U. Hennecke, A. P. M. Eker, T. Carell and L. O. Essen, Crystal structure of a photolyase bound to a CPD-like DNA lesion after *in situ* repair, *Science*, 2004, **306**, 1789–1793.
130. Y. F. Li, P. F. Heelis and A. Sancar, Active site of DNA photolyase: tryptophan-306 is the intrinsic hydrogen atom donor essential for flavin radical photoreduction and DNA repair *in vitro*, *Biochemistry*, 1991, **30**, 6322–6329.
131. C. Aubert, P. Mathis, A. P. Eker and K. Brettel, Intraprotein electron transfer between tyrosine and tryptophan in DNA photolyase from *Anacystis nidulans*, *Proc. Natl. Acad. Sci. USA*, 1999, **96**, 5423–5427.
132. C. Aubert, M. H. Vos, P. Mathis, A. P. Eker and K. Brettel, Intraprotein radical transfer during photoactivation of DNA photolyase, *Nature*, 2000, **405**, 586–590.

133. M. Gomelsky and S. Kaplan, Molecular genetic analysis suggesting interactions between AppA and PpsR in regulation of photosynthesis gene expression in Rhodobacter sphaeroides 2.4.1, *J. Bacteriol.*, 1997, **179**, 128–134.
134. C. Bauer, S. Elsen, L. R. Swem, D. L. Swem and S. Masuda, Redox and light regulation of gene expression in photosynthetic prokaryotes, *Philos. T. Roy. Soc. B*, 2003, **358**, 147–153.
135. Y. Han, S. Braatsch, L. Osterloh and G. Klug, A eukaryotic BLUF domain mediates light-dependent gene expression in the purple bacterium Rhodobacter sphaeroides 2.4.1, *Proc. Nat. Acad. Sci. USA*, 2004, **101**, 12306–12311.
136. B. J. Kraft, S. Masuda, J. Kikuchi, V. Dragnea, G. Tollin, J. M. Zaleski and C. E. Bauer, Spectroscopic and mutational analysis of the blue-light photoreceptor AppA: a novel photocycle involving flavin stacking with an aromatic amino acid, *Biochemistry*, 2003, **42**, 6726–6734.
137. K. Hasegawa, S. Masuda and T. A. Ono, Structural Intermediate in the Photocycle of a BLUF (Sensor of Blue Light Using FAD) Protein Slr1694 in a Cyanobacterium *Synechocystis sp.* PCC6803, *Biochemistry*, 2004, **43**, 14979–14986.
138. K. Hasegawa, S. Masuda and T. A. Ono, Spectroscopic Analysis of the Dark Relaxation Process of a Photocycle in a Sensor of Blue Light using FAD (BLUF) Protein Slr1694 of the Cyanobacterium *Synechocystis sp.* PCC6803, *Plant and Cell Physiol.*, 2005, **46**, 136–146.
139. W. Laan, M. A. van der Horst, I. H. van Stokkum and K. J. Hellingwerf, Initial characterization of the primary photochemistry of AppA, a blue-light-using flavin adenine dinucleotide-domain containing transcriptional antirepressor protein from Rhodobacter sphaeroides: a key role for reversible intramolecular proton transfer from the flavin adenine dinucleotide chromophore to a conserved tyrosine?, *Photochem. Photobiol.*, 2003, **78**, 290–297.
140. S. Masuda, K. Hasegawa and T. A. Ono, Light-induced structural changes of apoprotein and chromophore in the sensor of blue light using FAD (BLUF) domain of AppA for a signaling state, *Biochemistry*, 2005, **44**, 1215–1224.
141. B. Stephan and K. Gabriele, Blue Light Perception in Bacteria, *Photosynth. Res.*, 2004, **79**, 45–57.
142. F. Ghetti, G. Checcucci and F. Lenci, Photosensitized reactions as primary molecular events in photomovements of microorganisms, *J. Photochem. Photobiol. B: Biol.*, 1992, **15**, 185–198.
143. S. Braatsch, O. V. Moskvin, G. Klug and M. Gomelsky, Responses of the *Rhodobacter sphaeroides* Transcriptome to Blue Light under Semiaerobic Conditions, *J. Bacteriol.*, 2004, **186**, 7726–7735.
144. J. P. Armitage and K. Hellingwerf, Light-induced behavioral responses ('phototaxis') in prokaryotes, *Photosynth. Res.*, 2003, **76**, 145–155.
145. R. Kort, W. Crielaard, J. L. Spudich and K. J. Hellingwerf, Color-sensitive motility and methanol release responses in Rhodobacter sphaeroides, *J. Bacteriol.*, 2000, **182**, 3017–3021.
146. J. A. Kyndt, T. E. Meyer and M. A. Cusanovich, Photoactive yellow protein, bacteriophytochrome, and sensory rhodopsin in purple phototrophic bacteria, *Photochem. Photobiol. Sci.*, 2004, **3**, 519–530.
147. R. P. Burchard and M. Dworkin, Light-induced lysis and carotenogenesis in Myxococcus xanthus, *J. Bacteriol.*, 1966, **91**, 535–545.
148. A. Martinez-Laborda, J. M. Balsalobre, M. Fontes and F. J. Murillo, Accumulation of carotenoids in structural and regulatory mutants of the bacterium Myxococcus xanthus, *Mol. Gen. Genet.*, 1990, **223**, 205–210.

149. D. F. Browning, D. E. Whitworth and D. A. Hodgson, Light-induced carotenogenesis in Myxococcus xanthus: functional characterization of the ECF sigma factor CarQ and antisigma factor CarR, *Mol. Microbiol.*, 2003, **48**, 237–251.
150. B. L. Taylor and D. E. J. Koshland, Intrinsic and extrinsic light responses of *Salmonella typhimurium* and *Escherichia coli*, *J. Bacteriol.*, 1975, **123**, 557–569.
151. H. Yang, A. Sasarman, H. Inokuchi and J. Adler, Non-iron porphyrins cause tumbling to blue light by an *Escherichia coli* mutant defective in hemG, *Proc. Nat. Acad. Sci. USA*, 1996, **93**, 2459–2463.
152. H. Yang, H. Inokuchi and J. Adler, Phototaxis away from blue light by an *Escherichia coli* mutant accumulating protoporphyrin IX, *Proc. Nat. Acad. Sci. USA*, 1995, **92**, 7332–7336.
153. C. Propst-Ricciuti and L. B. Lubin, Light-induced inhibition of sporulation in *Bacillus licheniformis*, *J. Bacteriol.*, 1976, **128**, 506–509.
154. B. V. Futter and G. Richardson, Inactivation of bacterial spores by visible radiations, *J. Appl. Bacteriol.*, 1967, **30**, 347–353.
155. T. Neicu, A. Pradhan, D. A. Larochelle and A. Kudrolli, Extinction transition in bacterial colonies under forced convection, *Phys. Rev.*, 2000, **62**, 1059.
156. A. M. Delprato, A. Samadani, A. Kudrolli and L. S. Tsimring, Swarming ring patterns in bacterial colonies exposed to ultraviolet radiation, *Phys. Rev. Lett.*, 2001, **87**, 158102.
157. M. M. S. M. Wosten, Eubacterial sigma-factors, *FEMS Microbiol. Rev.*, 1998, **22**, 127–150.
158. C. W. Mullineaux, How do cyanobacteria sense and respond to light?, *Mol. Microbiol.*, 2001, **41**, 965–971.
159. T. Lamparter, Evolution of cyanobacterial and plant phytochromes, *FEBS Lett.*, 2004, **573**, 1–5.
160. D. M. Kehoe and A. R. Grossman, Similarity of a chromatic adaptation sensor to phytochrome and ethylene receptors, *Science*, 1996, **273**, 1409–1412.
161. S. Yoshihara and M. Ikeuchi, Phototactic motility in the unicellular cyanobacterium *Synechocystis sp.* PCC 6803, *Photochem. Photobiol. Sci.*, 2004, **3**, 512–518.
162. B. Fiedler, D. Broc, H. Schubert, A. Rediger, T. Borner and A. Wilde, Involvement of cyanobacterial phytochromes in growth under different light qualities and quantities, *Photochem. Photobiol.*, 2004, **79**, 551–555.
163. W. O. Ng, A. R. Grossman and D. Bhaya, Multiple Light Inputs Control Phototaxis in Synechocystis sp. Strain PCC6803, *J. Bacteriol.*, 2003, **185**, 1599–1607.
164. S. Yoshihara, M. Katayama, X. Geng and M. Ikeuchi, Cyanobacterial Phytochrome-like PixJ1 Holoprotein Shows Novel Reversible Photoconversion Between Blue- and Green-absorbing Forms, *Plant Cell Physiol.*, 2004, **45**, 1729–1737.
165. A. Wilde, B. Fiedler and T. Borner, The cyanobacterial phytochrome Cph2 inhibits phototaxis towards blue light, *Mol. Microbiol.*, 2002, **44**, 981–988.
166. J. S. Choi, Y. H. Chung, Y. J. Moon, C. Kim, M. Watanabe, P. S. Song, C. O. Joe, L. Bogorad and Y. M. Park, Photomovement of the gliding cyanobacterium *Synechocystis sp.* PCC 6803, *Photochem. Photobiol.*, 1999, **70**, 95–102.
167. S. Masuda and T. A. Ono, Biochemical characterization of the major adenylyl cyclase, Cya1, in the cyanobacterium *Synechocystis sp.* PCC 6803, *FEBS Lett.*, 2004, **577**, 255–258.
168. K. Terauchi and M. Ohmori, Blue light stimulates cyanobacterial motility via a cAMP signal transduction system, *Mol. Microbiol.*, 2004, **52**, 303–309.

169. H. Yoshimura, T. Hisabori, S. Yanagisawa and M. Ohmori, Identification and characterization of a novel cAMP receptor protein in the cyanobacterium Synechocystis sp. PCC 6803, *J. Biol. Chem.*, 2000, **275**, 6241–6245.
170. H. Yoshimura, S. Yoshihara, S. Okamoto, M. Ikeuchi and M. Ohmori, A cAMP receptor protein, SYCRP1, is responsible for the cell motility of *Synechocystis sp.* PCC 6803, *Plant and Cell Physiol.*, 2002, **43**, 460–463.
171. L. J. Stal and R. Moezelaar, Fermentation in cyanobacteria, *FEMS Microbiol. Rev.*, 1997, **21**, 179–211.
172. S. L. Anderson and L. McIntosh, Light-activated heterotrophic growth of the cyanobacterium *Synechocystis sp.* strain PCC 6803: a blue-light-requiring process, *J. Bacteriol.*, 1991, **173**, 2761–2767.
173. A. Wilde, Y. Churin, H. Schubert and T. Borner, Disruption of a *Synechocystis sp.* PCC 6803 gene with partial similarity to phytochrome genes alters growth under changing light qualities, *FEBS Lett.*, 1997, **406**, 89–92.
174. N. S. Baliga, R. Bonneau, M. T. Facciotti, M. Pan, G. Glusman, E. W. Deutsch, P. Shannon, Y. Chiu, R. S. Weng and A. Gan, Genome sequence of Haloarcula marismortui: a halophilic archaeon from the Dead Sea, *Genome Res.*, 2004, **14**, 2221–2234.
175. C. R. Woese, B. A. Debrunner-Vossbrinck, H. Oyaizu, E. Stackebrandt and W. Ludwig, Gram-positive bacteria: possible photosynthetic ancestry, *Science*, 1985, **229**, 762–765.
176. H. Ashida, Y. Saito, C. Kojima, K. Kobayashi, N. Ogasawara and A. Yokota, A functional link between RuBisCO-like protein of *Bacillus* and photosynthetic RuBisCO, *Science*, 2003, **302**, 286–290.
177. H. Ashida, A. Danchin and A. Yokota, Was photosynthetic RuBisCO recruited by acquisitive evolution from RuBisCO-like proteins involved in sulfur metabolism?, *Res. Microbiol.*, 2005, **156**, 611–618.
178. N. J. Mulder, R. Apweiler, T. K. Attwood, A. Bairoch, D. Barrel, A. Bateman, M. Biswas, P. Bradley, P. Bucher, R. R. Copley, E. Courcelle, U. Das, R. Durbin, L. Falquet, W. Fleischmann, S. Griffiths-Jones, D. Haft, N. Harte, N. Hulo, D. Kahn, A. Kanapin, M. Krestyaninova, R. Lopez, I. Letunic, D. Lonsdale, V. Silventoinen, S. E. Orchard, M. Pagni, D. Peyruc, C. P. Ponting, J. Selengut, F. Servant, C. J. Sigrist, R. Vaughan and E. M. Zdobnov, The InterPro Database, 2003 brings increased coverage and new features., *Nucl. Acids Res.*, 2003, **31**, 315–318.
179. J. D. Thompson, D. G. Higgins and T. J. Gibson, CLUSTAL W: improving the sensitivity of progressive multiple sequence alignment through sequence weighting, position-specific gap penalties and weight matrix choice, *Nucl. Acids Res.*, 1994, **22**, 4673–4680.
180. N. Guex and M. C. Peitsch, SWISS-MODEL and the Swiss-PdbViewer: an environment for comparative protein modeling, *Electrophoresis*, 1997, **18**, 2714–2723.
181. B. Boeckmann, A. Bairoch, R. Apweiler, M. C. Blatter, A. Estreicher, E. Gasteiger, M. J. Martin, K. Michoud, C. O'Donovan and A. Phan, The SWISS-PROT protein knowledgebase and its supplement TrEMBL in 2003, *Nuc. Ac. Res.*, 2003, **31**, 365–370.

Chapter 11

Photoactivated adenylyl cyclase (PAC), the photoreceptor flavoprotein with intrinsic effector function mediating euglenoid photomovements

MINEO ISEKI,[a,d] SHIGERU MATSUNAGA,[b] AKIO MURAKAMI[c] AND MASAKATSU WATANABE[b,d]

[a] PRESTO, Japan Science and Technology Agency, Honcho, Kawaguchi, Saitama, 332-0012, Japan

[b] Department of Photoscience, School of Advanced Sciences, Graduate University for Advanced Studies (SOKENDAI), Shonan Village, Hayama, Kanagawa, 240-0193, Japan

[c] Kobe University Research Center for Inland Seas, Iwaya, Awaji, Hyogo, 656-2401, Japan

[d] National Institute for Basic Biology, National Institutes of Natural Sciences, Okazaki, Aichi, 444-8585, Japan

Abstract

Photoactivated adenylyl cyclase (PAC) is the blue-light receptor flavoprotein recently identified as a sensor for photomovement in the unicellular flagellate, *Euglena gracilis*. PAC consists of α- and β-subunits, which are similar to each other and contain two FAD-binding domains (BLUF, sensors of Blue-Light Using FAD) each followed by a class III adenylyl cyclase catalytic domain. Purified PAC adenylyl cyclase activity is elevated by UV/blue-light irradiation in a fluence-dependent manner. Heterologously expressed FAD-binding domain of PACα exhibits spectral red shift upon blue-light irradiation, which indicates that the primary mechanism of photoactivation is similar to that seen with other prokaryotic BLUF proteins. Since PAC is directly activated by light without any other intervening components, it is expected to serve as a biological tool for controlling intracellular cAMP levels using light.

11.1. Introduction

11.1.1. Blue-light responses in plants and microorganisms – a gold rush of -flavoprotein-type photoreceptors

Blue-light-induced biological responses have attracted the attention of scientists for over a century, *e.g.* phototropism, bending toward or away from light, in plants[1] and photoaccumulation and photoavoidance in *Euglena*, a unicellular photosynthetic flagellate.[2] The involvement of flavoprotein-type photoreceptors in such blue-light responses has long been suggested on the basis of action spectra studies.[3,4] However, successive unequivocal molecular identification of such 'putative blue-light receptors' as flavoproteins with a variety of structures and functions have only been realized since the milestone work on cryptochromes resulting in photo-suppression of plant growth,[5,6] followed by those on phototropins involved in plant phototropism,[7,8] photoactivated adenylyl cyclase (PAC) causing *Euglena* photoavoidance and phototaxis,[9–11] AppA allowing photocontrol of *Rhodobacter* gene expression,[12,13] and WC-1 providing photocontrol of the *Neurospora* circadian clock.[14,15]

11.1.2. Photoreceptors involved in photomovement in microalgae

Motile microalgae, such as the eukaryotic *Euglena* (Euglenophyta), *Chlamydomonas* (Chlorophyta) as well as the prokaryotic microalga (cyanobacterium) *Synechocystis* (Cyanophyta), use photomovement responses (photocontrol of swimming or gliding movements) to place themselves in optimal light conditions for their photosynthesis and survival.[16] However, action spectroscopic and other physiological studies have indicated that the responsible photoreceptors are generally distinct from those for photosynthesis but are of a flavoprotein (*Euglena*),[17,18] archaerhodopsin (*Chlamydomonas*),[19–21] and even phytochrome (*Synechosistis*)[22] nature. The molecular identity of the flavoprotein-type photoreceptor, PAC, involved in *Euglena* photomovement as well as the archaerhodopsin-type ones used for *Chlamydomonas* photomovement were unequivocally discovered in 2002.[23–25]

On the basis of action spectroscopic and sequence homology studies on photomovement photoreceptors in other groups of eukaryotic microalgae as summarized in Figure 1, it seems that, at least in eukaryotic flagellate algae, most fall into the two major types, the flavoprotein-type and the archaeo-rhodopsin-type photoreceptors. Understanding the evolutionary origin of the apparent presence of two alternative modes of action in view of the relationship with prokaryotic photoreceptors and the various endosymbiotic processes that resulted in eukaryote evolution is a challenging subject for future studies.

In view of the above, the discovery and characterization of PAC as well as future prospects are described and discussed in the following Sections.

11.2. The discovery of PAC

11.2.1. Isolation of the paraflagellar body (PFB), a photosensory organelle of Euglena

The *Euglena* cell rapidly changes its swimming direction upon a sudden increase or decrease in incident light intensity, thus exhibiting step-up or step-down photophobic responses, which are considered to be elemental processes in photoavoidance and photoaccumulation, respectively. In addition, the cell exhibits positive and negative phototaxis: *i.e.* movement toward or away from the light source.[26] The *Euglena* cell has a PFB, a small ellipsoidal structure near the base of its flagellum, which has been considered a photosensory organelle involved in photomovement (Figure 2). Benedetti and Lenci[27] measured the fluorescence emission spectrum of the PFB using a microspectroscope and showed that the emission peak was centered at about 530 nm, which exhibited good agreement with the fluorescence emission spectrum of flavin mononucleotide (FMN). This strongly indicated the presence of flavins in the PFB and suggested the hypothesis that a flavoprotein located in the PFB acts as the photoreceptor molecule. Several attempts to identify the flavoprotein in the PFB have been made,[28,29] but the results were ambiguous, possibly due to low quality PFB preparations. In 2002, Iseki *et al.*[9] succeeded in obtaining a good preparation of PFBs by cell-disruption and subsequent sucrose density gradient centrifugation. The isolated PFBs exhibited bright green autofluorescence under violet-blue excitation (Figure 3A) while still attached to a small fragment of flagellum (Figure 3B). Negative staining of the isolated PFBs revealed a lattice with a repeating pattern of several nanometers in the PFB (Figure 3C), implying a paracrystalline structure as has been suggested by observation in ultra thin section[30,31] and freeze fracture.[32]

11.2.2. Purification and characterization of PAC

Iseki *et al.*[9] purified a flavoprotein with an apparent molecular mass of 400 kDa from the PFB preparation by liquid chromatography. The 400 kDa

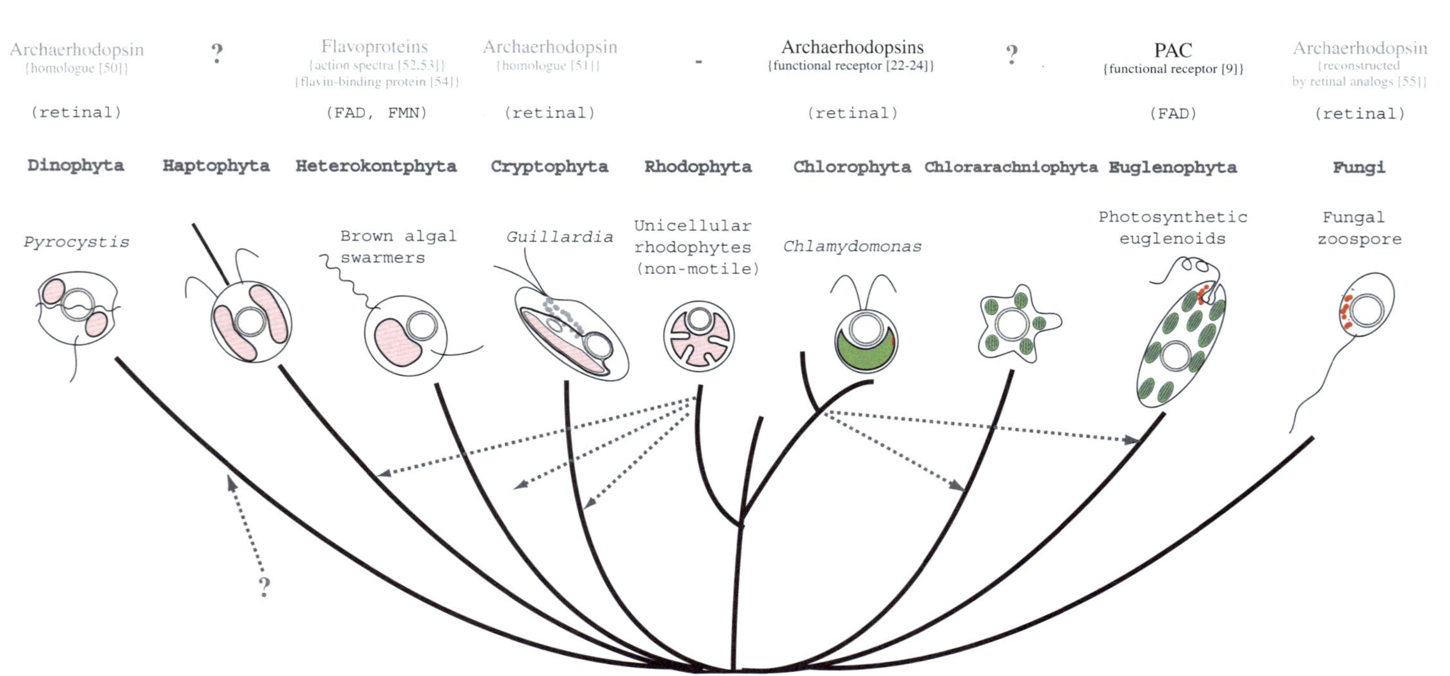

Figure 1. Photoreceptors regulating flagella-mediated photomotile responses. Current understanding of the molecular nature of the photoreceptor in flagellates is shown phylogenetically. The phylogenetic tree is based on Cavalier-Smith.[50] Functionally identified photoreceptors (archaerhodopsins in Chlamydomonas[23–25] and PAC in Euglena[9,10]) are shown in black along the top, while possible candidates proposed by the presence of homologues (archaerhodopsins in Pyrocystis[51] and Guillardia[52]) or by physiological evidence (flavoproteins in brown algal swarmers[53–55] and archaerhodopsins in fungal zoospores[56]) are shown in gray

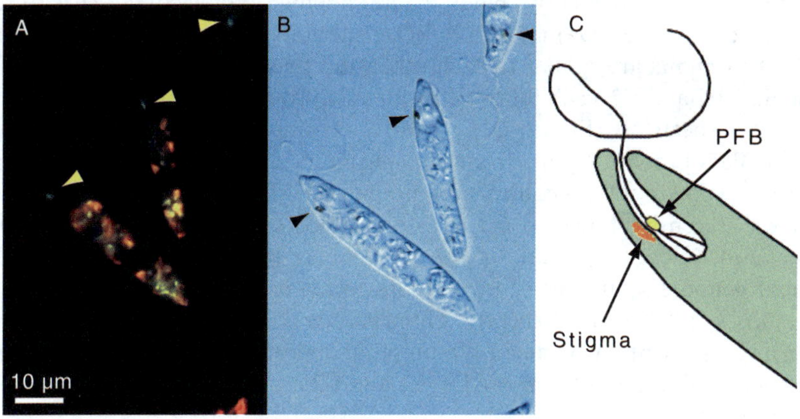

Figure 2. The PFB and the stigma in Euglena gracilis. (A) Epifluorescence photomicrograph of Euglena cells under violet-blue excitation. Yellow arrowheads indicate PFBs. (B) Bright field image of A. Black arrowheads indicate stigma. (C) Diagrammatic representation of the positions of the PFB and the stigma

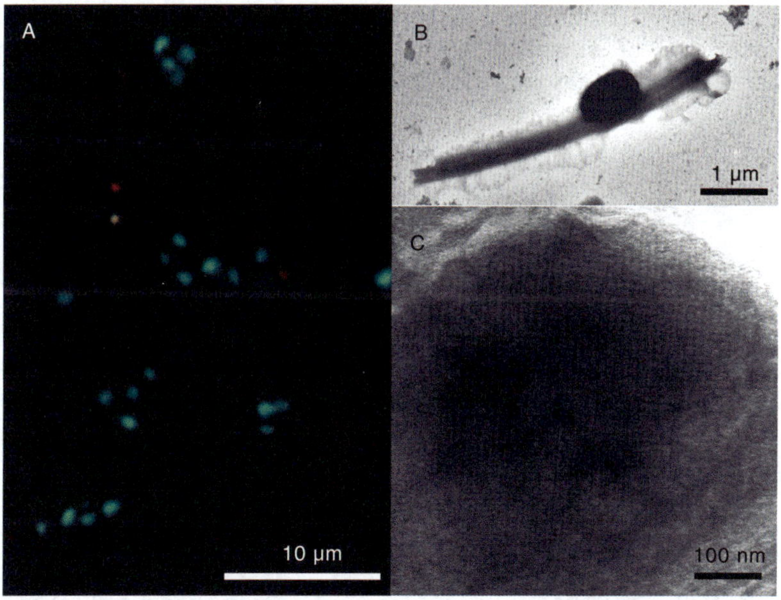

Figure 3. Isolated PFBs. (A) Epifluorescence photomicrograph of the PFB preparation obtained by cell disruption and successive centrifugation. (B) An isolated PFB observed by electron microscopy following negative staining. (C)The negatively stained PFB observed at higher magnification. B and C were taken by Drs. Tomoharu Matsumoto and Kuniaki Nagayama (Center for Integrative Bioscience, Okazaki)

flavoprotein noncovalently binds flavin adenine dinucleotide (FAD) and consists of α– (105 kDa) and β (90 kDa) subunits. The amino acid sequences of both subunits are similar to each other and they contain two flavin binding domains, F1 and F2, each followed by an adenylyl cyclase catalytic domain, C1 and C2, respectively (Figure 4). F1 and F2 are similar to the *N*-terminal region of the flavoprotein AppA from the purple bacterium *Rhodobacter sphaeroides*.[33] AppA binds an equimolar amount of FAD at the *N*-terminal region[34] and was recently shown to act as a photoreceptor for blue-light dependent repression of photosynthetic genes.[12,13] Similar sequences are also found in several genome sequences of prokaryotes, such as *Escherichia coli* and *Synechocystis* sp. PCC6803. Recently, Gomelsky and Klug[35] suggested that these domains represent a new flavin-binding fold and proposed calling them BLUF (sensors of Blue-Light Using FAD). F1 and F2 are the only BLUF domains so far found in eukaryotes.

On the other hand, the adenylyl cyclase catalytic domains, C1 and C2 have similarity with class III adenylyl cyclases which are widely distributed in prokaryotes and eukaryotes.[36] Inspecting the amino acid sequence profiles of the class III adenylyl cyclases, Linder and Schultz[37] divided them into four subclasses, class IIIa-IIId. C1 and C2 belong to class IIIb together with cyanobacterial adenylyl cyclases and mammalian soluble adenylyl cyclases. The 400 kDa flavoprotein purified from the PFB preparation actually showed significant adenylyl cyclase activity that was drastically elevated upon blue light irradiation (Figure 5A). Thus the 400 kDa flavoprotein, probably a heterotetramer of α and β subunits, appears to be a unique adenylyl cyclase whose activity is regulated by blue light and has been named PAC[9,10] (Figure 5B).

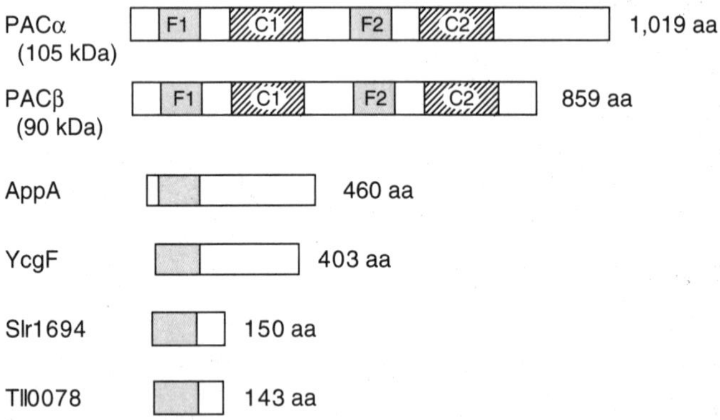

Figure 4. Subunits of PAC. The 105 kDa subunit (PACα) and the 90 kDa subunit (PACβ) contain the flavin binding BLUF domains (F1, F2; gray) and the adenylyl cyclase catalytic domains (C1, C2; shaded). Other members of BLUF family, such as AppA in Rhodobacter sphaeroides,[33,34] YcgF in Escherichia coli,[34,57] Slr1694 in Synechocyctis sp. PCC6803,[58,59] and Tll0078 in Thermosynechococcus elongatus[5,47] are also shown

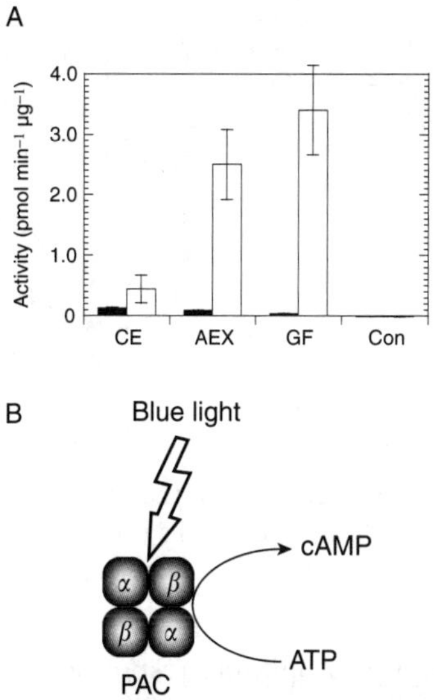

Figure 5. Activation of PAC by blue light. (A) Adenylyl cyclase activity (pmol cAMP/min/μg protein) of the proteins from the PFB preparation at consecutive purification steps: CE, crude extract of PFB; AEX, anion exchange chromatography; GF, gel-filtration chromatography; Control, gel-filtration buffer as a negative control. Average and standard error from 3 measurements in darkness (solid column) or under blue light (open column) at 10 μmol m^{-2} s^{-1} are shown.[9] (B) Illustration of PAC activation. PAC, probably a heterotetramer of α- and β subunits, is activated directly by blue light to produce cAMP

11.3. Cellular functions of PAC

11.3.1. Photophobic responses

RNA interference (RNAi), which was first reported in *Caenorhabditis elegans*[38] and then in *Trypanosoma brucei*,[39] a protozoan parasite that is phylogenetically related with *Euglena*, is now regarded as an excellent method for analyzing cellular functions of genes. The first application of RNAi to *Euglena* was performed by Iseki *et al.*[9,10] to examine cellular functions of PAC. Double stranded RNA containing the sequence of the PAC subunits was synthesized and electroporated into *Euglena* cells, resulting in undetectable levels of endogenous PAC mRNA and the loss of PFB in the cells. The step-up photophobic response disappeared in the RNAi-treated cells even under high intensity light, while the step-down photophobic response remained normal (Figure 6). From these results, they concluded that PAC is the major constituent of the PFB and it acts as the photosensor for the step-up photophobic response (photoavoidance) in *Euglena*, whereas it is not involved in the step-down photophobic response (photoaccumulation). Ntefidou *et al.*[11] reported that *Astasia longa*, a close relative of *E. gracilis*, and several mutant strains of *E. gracilis*, which have a cytologically altered or no PFB but have homologues

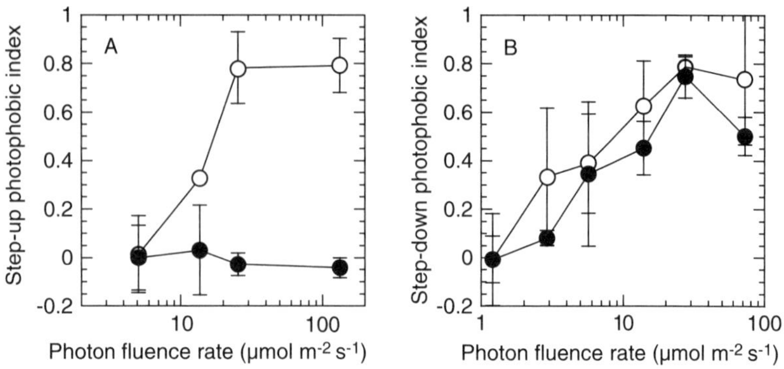

Figure 6. Fluence rate-response curves for the step-up (A) and the step-down (B) photophobic response of the RNAi-treated cells (open circles) and control cells (closed circles) at 450 nm. Average and standard deviation from three measurements are shown[9]

of PAC, lost their step-up photophobic responses after RNAi of PAC, demonstrating that at least in these strains, the PAC photoreceptor responsible for the step-up photophobic responses is not located in the PFB.

11.3.2. Phototaxis

Ntefidou *et al.*[11] reported that suppression of PAC expression by RNAi effectively suppresses both positive and negative phototaxis (Figure 7). This is convincing evidence that PAC or a PAC-related molecule is also involved in photoorientation of *Euglena*. However, the results should be interpreted carefully because the cellular mechanism of phototactic orientation has not been elucidated. It has long been hypothesized that the mechanism of phototaxis in *Euglena* is based on successive course corrections instigated through photophobic responses[40]: *i.e.* under unilateral illumination, the stigma (see Figure 2) periodically casts a shadow on the PFB due to rotation of the cell around its long axis, which causes successive photophobic responses resulting in course corrections toward or away from the light source. If this hypothesis holds, the RNAi-treated cells may well exhibit no phototaxis because the course correction mechanism does not work in the RNAi-treated cells that have lost their PFBs.

11.4. Photoactivation of PAC

11.4.1. Kinetic properties of PAC photoactivation

Although PAC appeared to be a photoreceptor for the step-up photophobic response, physiological evidence demonstrating that the photoactivation of PAC actually causes the step-up photophobic response is lacking. Recently, Yoshikawa *et al.*[41] reported the kinetic properties of *in vitro* activation of PAC by light, comparing them with those of the step-up photophobic response. They

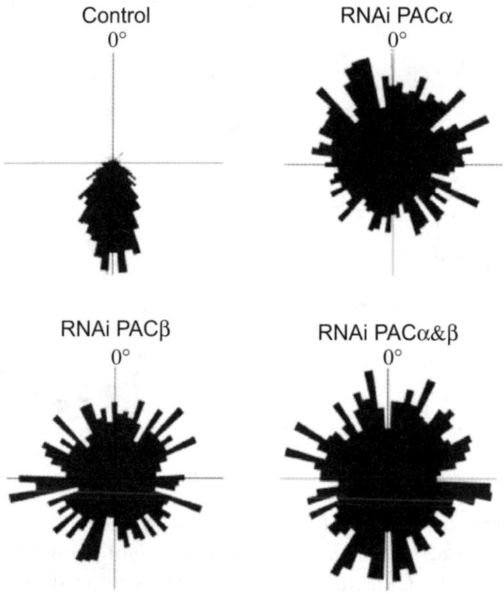

Figure 7. Tracks of Euglena cells at high irradiances summarized in circular histograms. Control cells swim with high precision away from the light source (800 W m^{-2}) whereas RNAi-treated cells show random swimming[11]

showed that activation of PAC is dependent both on photon fluence rate and the duration of irradiation and that reciprocity held well in the range of 2–50 μmol m^{-2} s^{-1} (total fluence of 1,200 μmol m^{-2}), suggesting that activation of PAC is entirely dependent on total photon fluence. They also examined the effects of intermittent irradiation on PAC activation and showed that intermittent irradiation using pulses of light and dark of equal length (0.1–180 s) caused activation of PAC to almost the same extent irrespective of the cycle periods. This means that elevation of PAC activity occurs only during the light period and that elevated PAC activity falls off within 0.1 s after the termination of irradiation. Such responsiveness is fast enough for PAC to mediate the step-up photophobic response that occurs with several subseconds latency.

Yoshikawa *et al.*[41] also reported wavelength dependency of PAC activation between 260–650 nm at equal quanta. The curve of wavelength dependency showed prominent peaks at 290, 390 and 450 nm, which agrees well both with the absorption spectrum of FAD and the action spectrum for the step-up photophobic response of *Euglena* (Figure 8). All the above properties of PAC activation confirm that PAC actually acts as a photoreceptor for the step-up photophobic response.

11.4.2. The change in intracellular cAMP levels upon blue light irradiation

Since *in vitro* activity of PAC was elevated by blue-light irradiation, it is easy to think that an increase in intracellular cAMP level evoked by photoactivation of

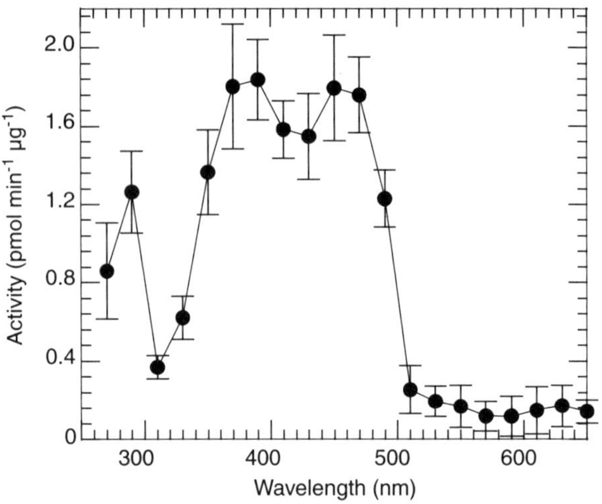

Figure 8. Wavelength dependency of PAC activation. The reaction mixture containing PAC (2 ng ml^{-1}) was irradiated with monochromatic light (from 270 to 650 nm, 10 μmol m^{-2} s^{-1}) from the Okazaki Large Spectrograph for 10 min, at 27°C[41]

PAC is a major cause of the step-up photophobic response. However, so far, there has been no experimental evidence that blue-light irradiation induces an increase in intracellular cAMP level of *Euglena*. Yoshikawa *et al.*[41] measured intracellular cAMP before and after onset of blue-light irradiation, which showed that the intracellular cAMP level remarkably increased within 1 s of the onset of irradiation and the increased cAMP level decreased within 10 s and gradually returned to the initial level even if irradiation continued. The time course of intracellular cAMP coincided well with the process of the step-up photophobic response (Figure 9). This strongly suggests that the increase in intracellular cAMP evoked by photoactivation of PAC is a key event in the step-up photophobic response.

11.4.3. Photoinduced spectral shift of F2 in PACα

The mechanism of photoactivation of PAC is a subject of great interest. However, it is difficult to obtain enough PAC from *Euglena* cells for photochemical and structural analyses because the efficiency of PAC purification is low; only several micrograms of PAC can be obtained from 30 L of *Euglena* culture. It is also difficult to obtain heterologously expressed PAC while keeping its activity intact because most of the expressed protein goes into the insoluble fraction, so-called inclusion bodies. Recently, Ito *et al.*[42] succeeded in obtaining the soluble recombinant flavin binding domain (F2) of PACα by heterologous expression in *E. coli* by fusing the domain to glutathion-S-transferase (GST). The recombinant F2 sample contained both FAD and FMN with trace amounts of riboflavin and showed a spectral red shift upon

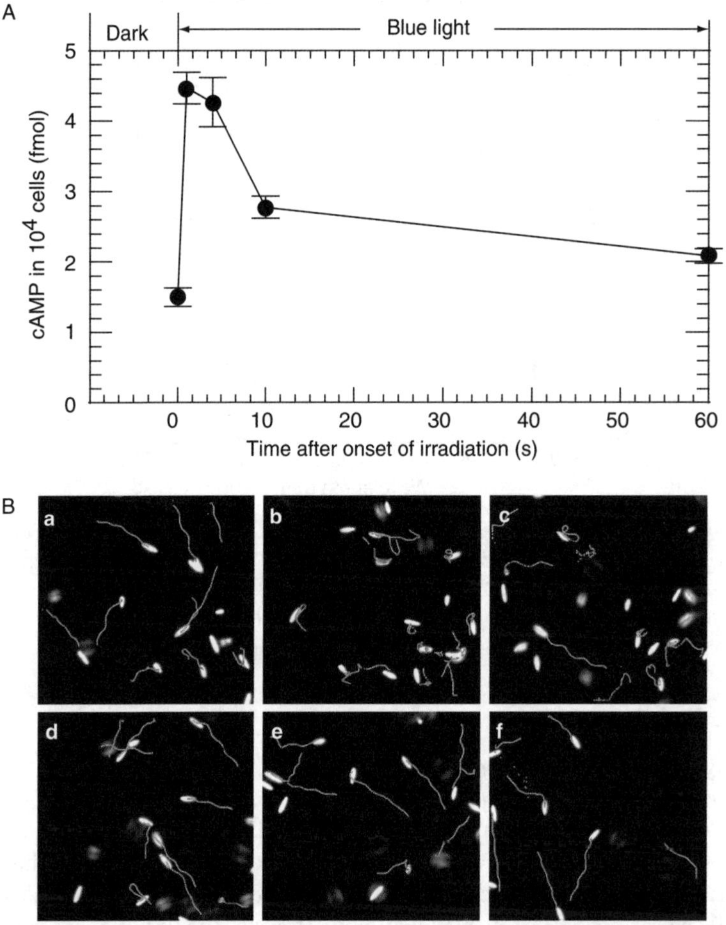

Figure 9. Relationships between intracellular cAMP level and the step-up photophobic response in Euglena. (A) The change in intracellular cAMP level upon blue light irradiation. Amounts of intracellular cAMP in Euglena cells were measured before and after onset of blue light irradiation (20 μmol m^{-2} s^{-1}). Bars indicate standard errors, n = 4–6.[41] (B) Tracks of Euglena cells before and after applying the step-up stimulus. The cells were observed using a phase contrast microscope equipped with a high-speed video camera under infrared illumination. Movement of each cell for 2 s was traced and plotted as green dots at 10 s before onset of irradiation (a), just after onset of irradiation (b), 4 s (c), 10 s (d), 30 s (e), and 60 s (f) after onset of irradiation

blue-light irradiation followed by recovery in darkness (Figure 10). Such photoinduced spectral shifts were first reported and well characterized for the BLUF domain of AppA.[12,43–45] The spectral shift is considered to be caused by alterations in π–π stacking and hydrogen bonding between FAD and the tyrosine residue (Tyr21). Recent reports on the crystal structures of BLUF domains of AppA[46] and a cyanobacterial protein Tll0078[47] indicate that the

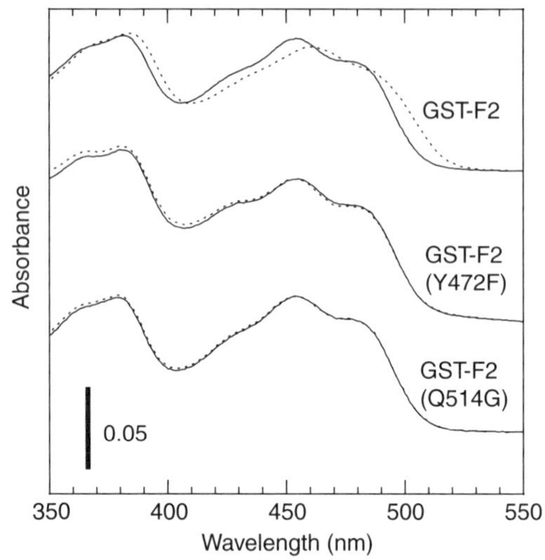

Figure 10. Absorption spectra of the recombinant F2 in PACα(GST-F2) and its mutants (GST-F2(Y472F) and GST-F2(Q514G)). The dark-adapted samples (solid line) were irradiated by a LED (465 nm, FWHM: 28 nm) at a photon fluence rate of 100 μmol m^{-2} s^{-1}. Spectra of light-irradiated samples (dotted line) were acquired at 60 s after onset of irradiation[42]

glutamine residue (Gln63 in AppA and Gln50 in Tll0078) play a crucial role in rearrangement of the hydrogen bond network to the flavin. Ito *et al.*[42] reported that when recombinant F2 proteins were mutated at the tyrosine (Tyr472) or glutamine (Gln514) residues, corresponding to Tyr21 and Gln63 in AppA, no photoinduced spectral shift was observed (Figure 10). Thus, the mechanism of photoactivation of the recombinant F2 in PACα seems essentially the same as that of the prokaryotic BLUF proteins, though the kinetic properties are slightly different from each other: *e.g.* the half-life was 34–44 s at 25°C for the recombinant PACα F2, whereas that reported for prokaryotic BLUF domains varies from *ca.* 3.5 s (Tll0078) to *ca.* 900 s (AppA). Han *et al.*[48] reported that a hybrid protein that consists of F1 from PACα and the C terminus of AppA is fully functional for light-dependent gene repression in *Rhodobacter sphaeroides*. This also implies that the primary process of photoactivation is common in PAC and AppA.

11.5. Future perspectives

PAC is a unique adenylyl cyclase in a sense that it can be activated by light without any other intervening signal transducers, such as G proteins. Since cAMP mediates various biological functions, PAC may serve as a cell biological tool for controlling cellular cAMP level by light: *i.e.* targeted heterologous expression of PAC in nerve and other cells would enable us to control cellular processes, such as axon guidance, synaptic plasticity, and cell differentiation. The first step to realize this idea would be functional expression of PAC in heterologous systems. In this sense, it is remarkable that Schröder-Lang *et al.*[49] succeeded in recording blue-light dependent conductance in *Xenopus* oocytes coexpressing PAC with cAMP-sensitive ion channels. We hope PAC, a product of pure biological study, will provide a useful tool for applied science and will contribute to human prosperity.

References

1. C. Darwin, *The Power of Movement in Plants*, John Murray, London, 1880.
2. J. Buder, Zur Kenntnis der phototaktischen Richtungsbewegungen, *Jahrb. Wiss. Bot.*, 1919, **58**, 105–220.
3. H. Senger (ed), Blue Light Responses: Phenomena and Occurrence in Plants and Microorganisms, Vols 1 and 2, CRC Press, Boca Raton, FL, 1987.
4. M. Watanabe, Action spectroscopy for photosensory processes, , in *CRC Handbook of Organic Photochemistry and Photobiology*, W. Horspool, F. Lenci (eds), CRC Press, Boca Raton, FL, 2003, 115-1 – 155-16.
5. M. Ahmad and A.R. Cashmore, *HY4* gene of *A. thaliana* encodes a protein with characteristics of a blue-light photoreceptor, *Nature*, 1993, **366**, 162–166.
6. C. Lin, D.E. Robertson, M. Ahmad, A.A. Raibekas, M.S. Jorns, P.L. Dutton and A.R. Cashmore, Association of flavin adenine dinucleotide with the *Arabidopsis* blue light receptor CRY1, *Science*, 1995, **269**, 968–970.
7. E. Huala, P.W. Oeller, E. Liscum, I.S. Han, E. Larsen and W.R. Briggs, *Arabidopsis* NPH1: a protein kinase with a putative redox-sensing domain, *Science*, 1997, **278**, 2120–2123.
8. J.M. Christie, P. Reymond, G.K. Powell, P. Bernasconi, A.A. Raibekas, E. Liscum and W.R. Briggs, *Arabidopsis* NPH1: a flavoprotein with the properties of a photoreceptor for phototropism, *Science*, 1998, **282**, 1698–1701.
9. M. Iseki, S. Matsunaga, A. Murakami, K. Ohno, K. Shiga, K. Yoshida, M. Sugai, T. Takahashi, T. Hori and M. Watanabe, A blue-light-activated adenylyl cyclase mediates photoavoidance in *Euglena gracilis*, *Nature*, 2002, **415**, 1047–1051.
10. M. Watanabe and M. Iseki, Discovery and characterization of photoactivated adenylyl cyclase (PAC), a novel blue-light receptor flavoprotein, from *Euglena gracilis*, , in *Handbook of Photosensory Receptors*, W.R. Briggs and J.L. Spudich (eds), Wiley-VCH, Weinheim, 2005, 447–460.
11. M. Ntefidou, M. Iseki, M. Watanabe, M. Lebert and D.-P. Häder, Photoactivated adenylyl cyclase controls phototaxis in the flagellate *Euglena gracilis*, *Plant Physiol.*, 2003, **133**, 1517–1521.
12. S. Masuda and C.E. Bauer, AppA is a blue light photoreceptor that antirepresses photosynthesis gene expression in *Rhodobacter sphaeroides*, *Cell*, 2002, **110**, 613–623.
13. S. Braatsch, M. Gomelsky, S. Kuphal and G. Klug, A single flavoprotein, AppA, integrates both redox and light signals in *Rhodobacter sphaeroides*, *Mol. Microbiol.*, 2002, **45**, 827–836.
14. A.C. Froehlich, Y. Liu, J.J. Loros and J.C. Dunlap, White Collar-1, a circadian blue light photoreceptor, binding to the frequency promoter, *Science*, 2002, **297**, 815–819.

15. Q. He, P. Cheng, Y. Yang, L. Wang, K.H. Gardner and Y. Liu, White collar-1, a DNA binding transcription factor and a light sensor, *Science*, 2002, **297**, 840–843.
16. D.-P. Häder, M. Lebert, (eds), *Comprehensive Series in Photosciences vol 1: Photomovement*, Elsevier, Amsterdam, 2001.
17. C. Barghigiani, G. Colombetti, B. Franchini and F. Lenci, Photobehavior of *Euglena gracilis*: action spectrum for the step-down photophobic response of individual cells, *Photochem. Photobiol.*, 1979, **29**, 1015–1019.
18. S. Matsunaga, T. Hori, T. Takahashi, M. Kubota, M. Watanabe, K. Okamoto, K. Masuda and M. Sugai, Discovery of signaling effect of UV-B/C light in the extended UV-A/blue-type action spectra for step-down and step-up photophobic responses in the unicellular flagellate alga *Euglena gracilis*, *Protoplasma*, 1998, **201**, 45–52.
19. P. Hegeman, W. Gärtner and R. Uhl, All-*trans* retinal constitutes the functional chromophore in *Chlamydomonas* rhodopsin, *Biophys. J.*, 1991, **60**, 1477–1489.
20. M.A. Lawson, D.N. Zacks, F. Derguini, K. Nakanishi and J.L. Spudich, Retinal analog restoration of photophobic responses in a blind *Chlamydomonas reinhardtii* mutant, *Biophys. J.*, 1991, **60**, 1490–1498.
21. T. Takahashi, K. Yoshihara, M. Watanabe, M. Kubota, R. Johnson, F. Derguini and K. Nakanishi, Photoisomerization of retinal at 13-ene is important for phototaxis of *Chlamydomonas reinhardtii*: simultaneous measurements of phototactic and photophobic responses, *Biochem. Biophys. Res. Comm.*, 1991, **178** 1273–1279.
22. J.S. Choi, Y.H. Chung, Y.J. Moon, C. Kim, M. Watanabe, P.S. Song, C.O. Joe, L. Bogorad and Y.M. Park, Photomovement of the gliding cyanobacterium *Synechocystis* sp. PCC6803, *Photochem. Photobiol.*, 1999, **70**, 95–102.
23. G. Nagel, D. Ollig, M. Fuhrmann, S. Kateriya, A.M. Musti, E. Bamberg and P. Hegemann, Channelrhodopsin-1: A light-gated proton channel in green algae, *Science*, 2002, **296**, 2395–2398.
24. O.A. Sineshchekov, K.-H. Jung and J.L. Spudich, Two rhodopsins mediate phototaxis to low- and high-intensity light in *Chlamydomonas reinhardtii*, *Proc. Natl. Acad. Sci. USA*, 2002, **99**, 8689–8694.
25. T. Suzuki, K. Yamasaki, S. Fujita, K. Oda, M. Iseki, K. Yoshida, M. Watanabe, H. Daiyasu, H. Toh, E. Asamizu, S. Tabata, K. Miura, H. Fukuzawa, S. Nakamura and T. Takahashi, Archaeal-type rhodopsins in *Chlamydomonas*: model structure and intracellular localization, *Biochem. Biophys. Res. Comm.*, 2003, **301**, 711–717.
26. M. Lebert, Phototaxis of *Euglena gracilis* – flavins and pterins, , in *Comprehensive Series in Photosciences 1: Photomovement*, D.-P. Häder and M. Lebert (eds), Elsevier, Amsterdam, 2001, 297–341.
27. P.A. Benedetti and F. Lenci, *In vivo* microspectrofluorometry of photoreceptor pigments in *Euglena gracilis*, *Photochem. Photobiol.*, 1977, **26**, 315–318.
28. B. Brodhun and D.-P. Häder, Photoreceptor proteins and pigments in the paraflagellar body of the flagellate *Euglena gracilis*, *Photochem. Photobiol.*, 1990, **52**, 865–871.
29. B. Brodhun and D.-P. Häder, A novel procedure to isolate the chromoproteins in the paraflagellar body of the flagellate *Euglena gracilis*, *J. Photochem. Photobiol. B: Biol.*, 1995, **28**, 39–45.
30. P.L. Walne and H.J. Arnott, The comparative ultrastructure and possible function of eyespots: *Euglena granulata* and *Chlamydomonas eugametos*, *Planta*, 1967, **77**, 325–353.
31. P.A. Kivic and M. Vesk, Structure and function in the euglenoid eyespot apparatus: the fine structure, and response to environmental changes, *Planta*, 1972, **105**, 1–14.

32. P.L. Walne, V. Passarelli, L. Barsanti and P. Gualtieri, Rhodopsin: a photopigment for phototaxis in *Euglena gracilis*, *Crit. Rev. Plant Sci.*, 1998, **17**, 559–574.
33. M. Gomelsky and S. Kaplan, *appA* a novel gene encoding a *trans*-acting factor involved in the regulation of photosynthesis gene expression in *Rhodobacter sphaeroides* 2.4.1, *J. Bacteriol.*, 1995, **177**, 4609–4618.
34. M. Gomelsky and S. Kaplan, AppA, a redox regulator of photosystem formation in *Rhodobacter sphaeroides* 2.4.1, is a flavoprotein. Identification of a novel FAD binding domain, *J. Biol. Chem.*, 1998, **52**, 35319–35325.
35. M. Gomelsky and G. Klug, BLUF: a novel FAD-binding domain involved in sensory transduction in microorganisms, *Trends Biochem. Sci.*, 2002, **27**, 497–500.
36. A. Danchin, Phylogeny of adenylyl cyclases, *Adv. Sec. Mess. Phosphoprot. Res.*, 1993, **27**, 109–162.
37. J.U. Linder and J.E. Schultz, The class III adenylyl cyclases: multi-purpose signaling modules, *Cell. Signal.*, 2003, **15**, 1081–1089.
38. A. Fire, S. Xu, M.K. Montgomery, S.A. Kostas, S.E. Driver and C.C. Mello, Potent and specific genetic interference by double-stranded RNA in *Caenorhabditis elegans.*, *Nature*, 1998, **391**, 806–811.
39. H. Ngô, C. Tschudi, K. Gull and E. Ullu, Double-stranded RNA induces mRNA degradation in *Trypanosoma brucei*, *Proc. Natl. Acad. Sci. USA*, 1998, **95**, 14687–14692.
40. A. Checcucci, Molecular sensory physiology of *Euglena*, *Naturwissenshaften*, 1976, **63**, 412–417.
41. S. Yoshikawa, T. Suzuki, M. Watanabe and M. Iseki, Kinetic analysis of the activation of photoactivated adenylyl cyclase (PAC), a blue-light receptor for photomovements of *Euglena*, *Photochem. Photobiol. Sci.*, 2005, **4**, 727–731.
42. S. Ito, A. Murakami, K. Sato, Y. Nishina, K. Shiga, T. Takahashi, S. Higashi, M. Iseki and M. Watanabe, Photocycle features of heterologously expressed and assembled eukaryotic flavin-binding BLUF domains of photoactivated adenylyl cyclase (PAC), a bluc-light receptor in *Euglena gracilis*, *Photochem. Photobiol. Sci.*, 2005, **4**, 762–769.
43. B.J. Kraft, S. Masuda, J. Kikuchi, V. Dragnea, G. Tollin, J.M. Zaleski and C.E. Bauer, Spectroscopic and mutational analysis of the blue-light photoreceptor AppA: a novel photocycle involving flavin stacking with an aromatic amino acid, *Biochemistry*, 2003, **42**, 6726–6734.
44. W. Laan, M.A. van der Horst, I.H. van Stokkum and K.J. Hellingwerf, Initial characterization of the primary photochemistry of AppA, a blue-light-using flavin adenine dinucleotide-domain containing transcriptional antirepressor protein from *Rhodobacter sphaeroides*: a key role for reversible intramolecular proton transfer from the flavin adenine dinucleotide chromophore to a conserved tyrosine?, *Photochem. Photobiol.*, 2003, **78**, 290–297.
45. S. Masuda and C.E. Bauer, The antirepressor AppA uses the novel flavin-binding BLUF domain as a blue-light-absorbing photoreceptor to control photosystem synthesis, , in *Handbook of Photosensory Receptors*, W.R. Briggs and J.L. Spudich (eds), Wiley-VCH Verlag GmbH & Co. KGaA, Weinheim, 2005, 433–445.
46. S. Anderson, V. Dragnea, S. Masuda, J. Ybe, K. Moffat and C. Bauer, Structure of a novel photoreceptor, the BLUF domain of AppA from *Rhodobacter sphaeroides*, *Biochemistry*, 2005, **44**, 7998–8005.
47. A. Kita, K. Okajima, Y. Morimoto, M. Ikeuchi and K. Miki, Structure of a cyanobacterial BLUF protein, Tll0078, containing a novel FAD-binding blue light sensor domain, *J. Mol. Biol.*, 2005, **349**, 1–9.

48. Y. Han, S. Braatsch, L. Osterloh and G. Klug, A eukaryotic BLUF domain mediates light-dependent gene expression in the purple bacterium *Rhodobacter sphaeroides* 2.4.1, *Proc. Natl. Acad. Sci. USA*, 2004, **101**, 12306–12311.
49. S. Schroder-Lang, D. Ollig, T. Schiereis, P. Hegemann and G. Nagel, Characterization of heterologously expressed photoactivated adenylyl cyclase, *Proceedings of the 11th International Conference on Retinal Proteins, Frauenchiemsee, Germany*, 2004, abstract P60.
50. T. Cavalier-Smith, Kingdom protozoa and its 18 phyla, *Microbiol. Rev.*, 1993, **57**, 953–994.
51. O.K. Okamoto and J.W. Hastings, Novel dinoflagellate clock-related genes identified through microarray analysis, *J. Phycol.*, 2003, **39**, 519–526.
52. M.X. Ruiz-Gonzalez and I. Marin, New insights into evolutionary history of type I rhodopsins, *J. Mol. Evol.*, 2004, **58**, 348–358.
53. H. Kawai, M. Kubota, T. Kondo and M. Watanabe, Action spectra for phototaxis in zoospores of the brown alga *Pseudochorda gracilis*, *Protoplasma*, 1991, **161**, 17–22.
54. H. Kawai, D.G. Müller, E. Fölster and D.-P. Häder, Phototactic response in the gametes of the brown alga, *Ectocarpus siliculosus*, *Planta*, 1990, **182**, 292–297.
55. S. Fujita, M. Iseki, S. Yoshikawa, Y. Makino, M. Watanabe, T. Motomura, H. Kawai and A. Murakami, Identification and characterization of a flagellar fluorescent protein of brown alga *Scytosiphon lomentaria* (Scytosiphonales, Phaeophyceae): a flavoprotein homologous to Old Yellow Enzyme, *Eur. J. Phycol.*, 2005, **40**, 159–167.
56. J. Saranak and K.W. Foster, Rhodopsin guides fungal phototaxis, *Nature*, 1997, **387**, 465–466.
57. S. Rajagopal, J.M. Key, E.B. Purcell, D.J. Boerema and K. Moffat, Purification and initial characterization of a putative blue light-regulated phosphodiesterase from *Escherichia coli*, *Photochem Photobiol.*, 2004, **80**, 542–547.
58. S. Masuda, K. Hasegawa, A. Ishii and T. Ono, Light-Induced structual changes in a putative blue-light receptor with a novel FAD binding fold sensor of blue-light using FAD (BLUF); Slr1694 of *Synechocystis* sp. PCC6803, *Biochemistry*, 2004, **43**, 5304–5313.
59. K. Okajima, S. Yoshihara, Y. Fukushima, X. Geng, M. Katayama, S.-I. Higashi, M. Watanabe, S. Sato, S. Tabata, Y. Shibata, S. Itoh and M. Ikeuchi, Biochemical and functional characterization of BLUF-type flavin-binding proteins of two species of cyanobacteria, *J. Biochem.*, 2005, **137**, 741–750.

Chapter 12

Mechanisms of Light Activation in Flavin-Binding Photoreceptors

JOHN T.M. KENNIS AND MAXIME T.A. ALEXANDRE

Department of Biophysics, Faculty of Sciences, Vrije Universiteit, De Boelelaan 1081, 1081HV, Amsterdam, The Netherlands

Abstract

Light, oxygen and voltage (LOV) and blue-light sensing using FAD (BLUF) domains constitute two classes of recently discovered photoreceptors that are found in plants, algae, fungi and bacteria. They are sensitive to blue light through a non-covalently bound flavin chromophore. LOV domains belong to the superfamily of PAS (PER/Arnt/Sim) sensor and regulatory domains, whereas BLUF domains make up a distinct family that shows no relationship to other sensor proteins. LOV domains undergo a photocycle involving light-driven covalent adduct formation between a conserved cysteine residue and the C(4a) atom of the flavin cofactor. In BLUF domains, light absorption initiates a photocycle that most likely involves a hydrogen-bond rearrangement between the flavin cofactor and the residues lining the flavin-binding pocket. In this chapter, an overview is given of the insights gained from biophysical studies into the functional properties of these novel photoreceptor proteins. The three-dimensional structures, steady-state and time-resolved spectroscopy and photocycles of LOV and BLUF domains are reviewed. Their photophysics and photochemistry are discussed in terms of proposed reaction mechanisms. Finally, the light-induced conformational changes that are required for the signaling function of these photoreceptor domains are considered.

12.1. Introduction

The past decade has witnessed the discovery and characterization of a myriad of novel photoreceptor proteins. Many of these photoreceptors are sensitive to blue light and belong to new classes of flavin-binding photoreceptor protein families: the light, oxygen or voltage (LOV) domains, the blue-light sensing using FAD (BLUF) domains and the cryptochromes. LOV and BLUF domains often form an integral part of a larger photoreceptor protein of modular composition where they constitute the light-sensitive domain, and relay the information to a signaling domain *via* dynamical changes in their conformation. Since the discovery of LOV and BLUF domains at the end of the 90s, our knowledge of these systems has progressed rapidly and in this chapter, an overview is given of the insights gained from functional and biophysical studies. The biophysical characterization of cryptochromes has significantly lagged that of LOV and BLUF proteins, and is not considered here. The physiological role of these novel flavin-binding blue-light receptors has been reviewed elsewhere.[1–9]

12.2. Light, Oxygen Or Voltage Domains

Light, Oxygen or Voltage (LOV)-domains[†] are flavin-binding photoreceptors that belong to a broad superfamily of proteins commonly known as PAS-domains.[10,11] PAS-domains consist of approximately 120 amino acids and are ubiquitous in nature, playing key roles in many sensory and regulatory processes, from light sensing in plants, algae and bacteria, oxygen sensing in bacteria[12,13] to potassium-channel regulation in the human heart.[14]

[†] The term "Light Oxygen Voltage (LOV) domain" was originally used to describe a subgroup of the large PAS domain superfamily characterized by elevated sequence homology. The term LOV domain has also been used in a more restricted sense to describe a subclass of FMN-based photoreceptors of the PAS domain family, which bind flavin and exhibit cysteinyl-adduct photochemistry.[35] Technically, the oxygen sensors NifL and AER, and the voltage sensitive N-terminal domain of the human potassium channel HERG are also members of the broader class of LOV domains. For clarity, we use the restricted LOV definition of Crosson *et al.* and do not refer to non-photoreceptor PAS domains as LOV domains.

Crystallographic and NMR studies have indicated a striking structural conservation among different PAS-domains, despite their widely varying sensor functions.[12–20] They can bind a range of cofactors to interact with a manifold of stimuli like hemes, flavins, *p*-coumaric acid, iron–sulfur centers or bind no cofactor at all. On the other hand, PAS-domains are able to communicate the information with a wide variety of output domains, such as histidine kinases, serine/threonine kinases, transcription factors or other PAS-domains.[10] These intriguing properties of PAS-domains, *i.e.*, their extremely versatile modes of input and output through a very conserved structural motif indicates that key aspects of biological sensing may proceed through a 'universal language,' and that with a proper choice of model system, important knowledge with a high degree of general significance can be obtained.

LOV domains have first been identified in the phototropin family of plant blue-light photoreceptors. The phototropins are serine/threonine kinases that undergo autophosphorylation in response to absorption of blue light,[21] and control several physiological responses including phototropism, light-mediated chloroplast movement and stomatal opening.[1,21–24] Two LOV domains, LOV1 and LOV2, each bind a flavin mononucleotide (FMN) chromophore and constitute the light-sensitive loci. Several phototropin homologues were readily discovered in plants, algae, fungi and bacteria and it was proposed that they act as photoreceptors in their respective organisms.[25–30,31–35].

A limited number of LOV domains have been characterized in some detail by biochemical, structural and spectroscopic methods, primarily *Avena sativa* (oat) phototropin (phot) 1 LOV2, *Adiantum capillus-veneris* (maidenhair fern) phy3 LOV2, *Chlamydomonas reinhardtii* phototropin (phot) LOV1 and LOV2, the LOV domain of the *Bacillus subtilis* YtvA protein and the PpSB2-LOV domain of *Pseudomonas putida.*

12.2.1. The Dark State of LOV Domains

Most LOV domains bind FMN as a chromophore, with the exception of the fungal White Collar 1 proteins, which bind flavin adenine dinucleotide FAD.[28,29] Chromophore exchange studies have demonstrated that *Avena sativa* phot1 LOV2 can exchange its FMN with FAD, riboflavin or 5′-malonyl-riboflavin while retaining photochemical activity.[36] At present, three structural models of LOV domains are available the X-ray structure of *Adiantum* phy3 LOV2,[17] the X-ray structure of *Chlamydomonas* phot LOV1,[19] and the NMR structure of *Avena sativa* phot1 LOV2.[18] The structure of the *Adiantum* LOV2-domain in its dark state is shown in Figure 1. It reveals the typical PAS-fold, consisting of a five-stranded antiparallel β-sheet, flanked by a helix-turn-helix motif, a single helical turn and a connector helix. The FMN chromophore is non-covalently held in place by polar interactions with the pyrimidine moiety and non-polar interactions with the dimethylbenzene ring, and further tethered to the apoprotein *via* a number of hydrogen bonds.[17,35,37] The crystal structure of *Chlamydomonas* phot LOV1 was recorded at 77 K and showed two conformations of the conserved cysteine, originating from a rotation around the Cα

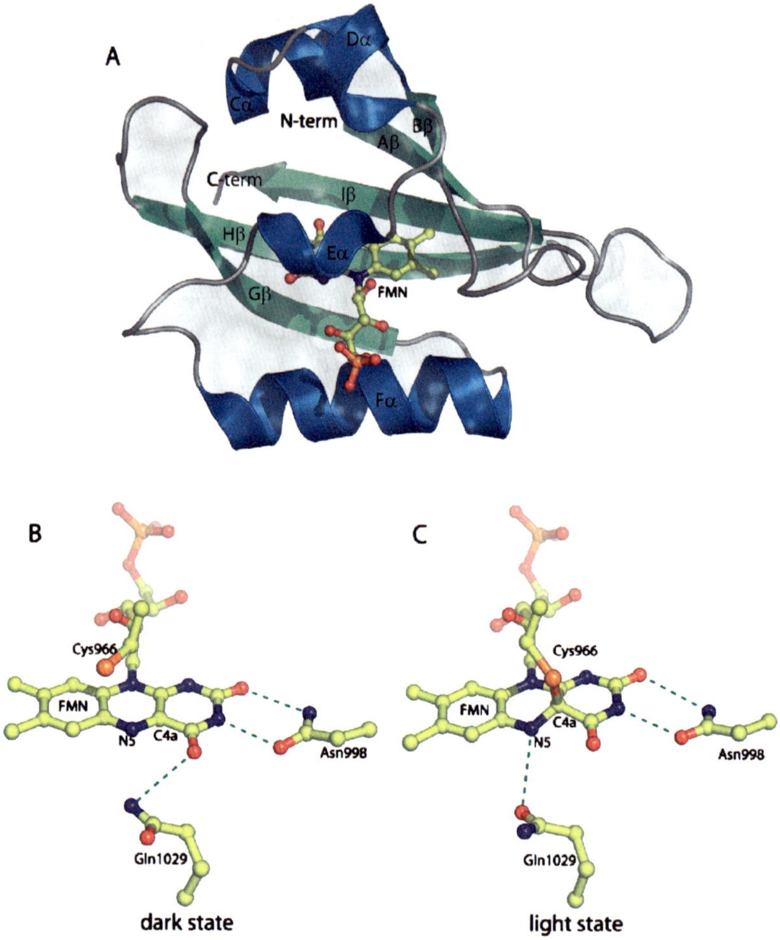

Figure 1. (A) X-ray structure of the *Adiantum* phy3 LOV2 domain; close-up of the crystal structure in the vicinity of the FMN chromophore in (B) the dark D447 state; and (C) the photoactivated S390 state

and Cβ bond and resulting in one conformer proximal to the C(4a) of flavin at 3.5 Å, and one distal at 4.4 Å from C(4a).[19]

The dark state of LOV domains is generally referred to as D447.[38] In the D447 state, the absorption spectrum of the FMN chromophore is dominated by three electronic transitions $S_0 \rightarrow S_1$ near 450 nm, $S_0 \rightarrow S_2$ near 360 nm and $S_0 \rightarrow S_3$ centerd at 270 nm, as shown in Figure 2 for the lower transitions. The one-electron transitions from HOMO to LUMO have π–π* character in the visible and *n*–π* in the UV at 270 nm.[39,40] The extinction coefficient of the FMN bound to LOV domains is similar to that of flavins in aqueous solution, and amounts to roughly 12,000 M^{-1} cm^{-1}.[41] The $S_0 \rightarrow S_1$ electronic transition exhibits three obvious vibronic structures $S_1(\nu = 0)$; $(\nu = 1)$; $(\nu = 2)$ located respectively near 475, 450 and 425 nm.[31,38,41,42] The vibronic structure in LOV

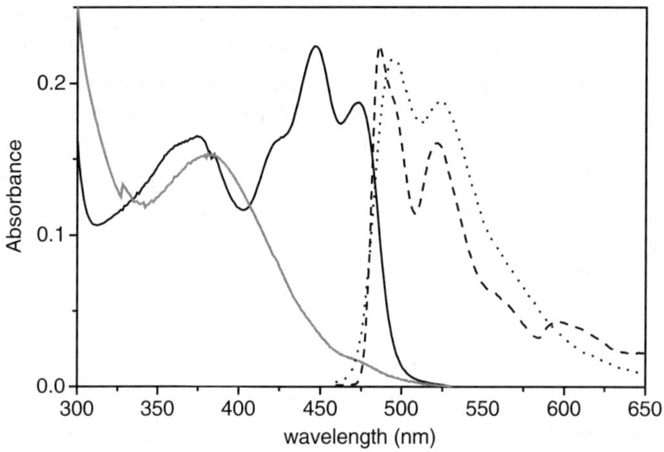

Figure 2. Absorption spectrum of the dark state D447 of *Avena sativa* phot1 LOV2 (solid line) and of its photoactivated S390 adduct state (gray line). The dotted and dashed lines indicate the fluorescence spectra at room temperature and 77 K, respectively.

domains is more pronounced than in most flavoproteins and is indicative of the tight binding of FMN through a network of van der Waals interactions with the dimethyl-benzene moiety and hydrogen bonding of the pyrimidine moiety to the well-ordered and rigid protein matrix. In aqueous solvent, the vibronic structure of FMN is smeared out because of solvent-induced inhomogeneous broadening or due to the loss of motional constraints imposed by the protein environment.[42,43] The vibrational mode spacing is also observed in the fluorescence spectrum (Figure 2, dotted line), with vibronic bands near 495 and 520 nm.[31,42,44,45] The Stokes shift due to dielectric relaxation of the surrounding protein matrix after excitation, is about 1100 cm^{-1}.[42]

12.2.2. The Photoactivated State of LOV Domains

Upon blue-light illumination, LOV domains undergo a photocycle that leads to the formation of a long-lived species with a prominent absorption band at 390 nm, referred to as S390 (Figure 2, gray line). The relatively long lifetime, from tens of seconds to minutes, or even hours, and its infinite lifetime below 250 K allow the steady-state recording of S390 under continuous illumination. The adduct state absorption spectrum is characterized by an overall blue-shift relative to D447: the S_1, S_2 and S_3 absorption bands shift from 450, 360 and 270 nm to 390, 305 and 240 nm in *Avena sativa* LOV2.[38] In Ytva-LOV and *Adiantum* phy3 LOV2, the photoproduct is slightly blue-shifted to 383 nm and 381 nm, respectively.[31,46] On the basis of the spectrum of a mutant form of mercuric reductase,[47] it was proposed that this species corresponds to a covalent FMN-C4a-thiol adduct.[41] This assignment was readily confirmed by a number of methods. An NMR study on isotope-labeled FMN bound to *Avena sativa*

LOV2 showed an upfield shift of C(4a) upon illumination, consistent with a sp^3 hybridization of C(4a), as expected for a FMN-C(4a)-thiol.[48] The X-ray structures of *Adiantum* phy3 LOV2 and *Chlamydomonas* phot LOV1 exhibited an obvious displacement of the sulfur atom of the conserved cysteine side chain toward and within covalent bonding distance of the FMN-C(4a) atom. Moreover, FMN-C(4a) adopted a tetrahedral geometry due to the adduct sp^3 hybridization state,[19,34] as shown in Figure 1C. Swartz *et al.* and Ataka *et al.* studied the adduct state relative to the dark state by use of FTIR spectroscopy.[49] Bleaches at 1580 and 1550 cm^{-1} were consistent with the FMN N5 = C(4a) stretch mode losing its double-bond character, and thus its coupling with the C(10a) = N1 in and out of phase stretch mode, respectively. The adduct state shows induced absorption bands at 1536 and 1516 cm^{-1}, which represent the stretching of C(10a) = N1 coupled with C(5a) = C(9a), the newly formed N5-H bend and the ring I stretch. Moreover, the C4 = O stretch frequency was upshifted by $\sim 10\ cm^{-1}$ because of uncoupling from the conjugated π-system of the isoalloxazine ring. Thus, the FTIR data were consistent with conversion of the C(4a) planar sp^2 to a tetrahedral sp^3 hybridization.

On the basis of fluorescence titrations on *Avena sativa* phot1 LOV2 and its cysteine mutants, Swartz *et al.* proposed that the sulfur group of the conserved cysteine existed as a thiolate ($-S^-$) rather than a thiol (–SH).[38] However, it was shown with FTIR spectroscopy that upon adduct formation, a S–H stretch band was bleached near 2570 cm^{-1}, which demonstrated that at least a significant fraction of the conserved cysteine exists as a thiol in the dark state.[50–52] Moreover, the S–H bleach of the *Chlamydomonas* LOV1 has a fine structure, which could be described by a superposition of two bleaching bands, centered at 2570 cm^{-1} and 2562 cm^{-1}. This was ascribed to the presence of two distinct cysteine conformations in the dark state, at room temperature. Sharp, single bleaches of the S–H band were seen for *Chlamydomonas* phot LOV2 and YtvA-LOV, indicating that in these LOV domains, the cysteine exists in a single conformation.[52]

At low temperature, photoadduct formation can still take place, although a significant non-reactive fraction remains in the D447 state at decreasing temperatures, and at 5 K no adduct is formed whatsoever.[53–55] This phenomenon may be related to freezing of essential protein motions necessary for adduct formation, possibly in combination with the population of different cysteine conformers. S390 exhibits little or no fluorescence;[31,42,56] Gauden *et al.* have reported a very weak fluorescence band for S390 in *Avena sativa* LOV2, with a broad maximum centered at 440 nm.[55]

12.2.3. The Photocycle of LOV Domains

12.2.3.1. Primary Reactions

Time-resolved fluorescence experiments have indicated a relatively long lifetime of the FMN singlet excited state: 2.4 ns in *A. sativa* phot1 LOV2,[44,55] 2.9 ns in *Chlamydomonas* LOV1[42] and 1.2 ns in *Chlamydomonas* LOV2.[57] In all cases, a single-exponential decay was observed. These values were shorter than those reported for FMN in aqueous solution, for which lifetimes of 2.9–4.7 ns were found, depending on pH.[42,55,58] Spectrally resolved streak camera experiments have shown that during the excited-state lifetime, no shifting of the fluorescence spectrum takes place.[55] In accordance with these observations, the fluorescence quantum yield in LOV domains is substantial, and ranges from 0.08 for *Chlamydomonas* LOV2,[57] 0.13 for *A. sativa* and *Adiantum* LOV2,[45] 0.17 for *Chlamydomonas* LOV1[42] and 0.22 for *B. subtilis* YtvA.[31] This is a quite remarkable fact given that the vast majority of flavoproteins and photoreceptor proteins are almost non-fluorescent.[59,60] Mutation of the cysteine in *Chlamydomonas* LOV1-C57S and *Avena sativa* LOV2-C39A lead to fluorescence quantum yields of 0.3 and 0.26, respectively.[42,44] For *Avena sativa* LOV2, an increase of the fluorescence quantum yield from 0.13 to 0.41 is observed on going from room temperature to 77 K.[55] The increase of fluorescence yield probably follows from internal properties of the protein-bound FMN, as the fluorescence emission of flavins in solution increases in a similar way upon lowering the temperature.[61]

Femtosecond to nanosecond transient absorption experiments performed on *Adiantum* phy3 LOV2 and *A. sativa* phot1 LOV2 were in line with the time-resolved fluorescence experiments.[44,45] Figure 3 shows the result of a femtosecond time-resolved transient absorption experiment on *Avena sativa* phot1 LOV2, with excitation at 400 nm. The time-resolved data were fitted with a sequential model, whereby the species follow one another, *i.e.*, 1 → 2 → 3 →. Three components with time constants of 1.2 ps, 2.2 ns and a non-decaying component were required for an adequate description of the data. The three absorption difference spectra shown in Figure 3 represent the spectral evolution of the LOV2 domain after excitation: the first spectrum (dotted line), which has a lifetime of 1.2 ps, is formed upon excitation of the LOV2 domain with a femtosecond flash at 400 nm. It shows ground-state bleach near 450 nm and a stimulated emission band near 540 nm. It represents a vibrationally 'hot' singlet excited state of FMN. The second difference spectrum (solid line), which is formed in 1.2 ps and has a lifetime of 2.2 ns corresponds to an increase and blueshift of the stimulated emission signal at 530 nm, and can be attributed to the vibrationally relaxed singlet excited state of FMN. It shows bleaching of the D447 ground state at 425, 447 and 475 nm, an excited-state absorption at 500 nm, stimulated emission at 530 nm and excited-state absorption at wavelengths longer than 610 nm. The third difference spectrum (gray line) follows from the FMN singlet excited state in 2.2 ns. It shows ground-state bleach near 450 nm, and a broad excited-state absorption that ranges from 500 to 700 nm, with a maximum near 660 nm. It is very similar to the transient spectra of *Avena sativa* phot1 LOV2 and *Chlamydomonas* LOV1 observed on the tens of nano- to micro-second timescales (see below), and is assigned to the triplet state of FMN.[38,43,45] The results shown in Figure 3 agree well with those reported earlier for *Avena sativa* phot1 and *Adiantum* phy3 LOV2.[44,45] Further evidence that the primary photoproduct in LOV2 indeed corresponds to a FMN triplet state comes from femto–nanosecond infrared spectroscopy, where a downshift of C=N stretches from 1550 cm^{-1} to 1492 and 1437 cm^{-1}, and a C=O stretch

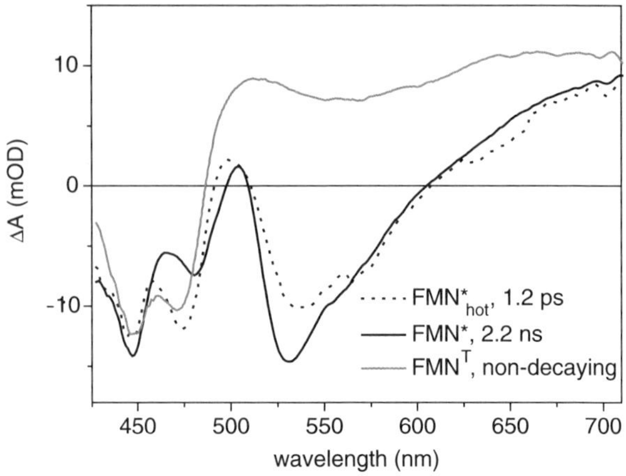

Figure 3. Results from a femto- to nano-second transient absorption study on *Avena sativa* phot1 LOV2. The dotted line represents the absorption difference spectrum of vibrationally 'hot' singlet excited FMN and has a lifetime of 1.2 ps. The solid line denotes vibrationally relaxed singlet excited FMN, and has a lifetime of 2.2 ns. The gray line denotes the FMN triplet excited state, which is formed directly from the FMN singlet excited state and does not decay on the time scale of the experiment, which was 5 ns.

downshift from 1684 to 1659 cm^{-1} was observed after a few nanoseconds.[62] These frequency shifts corresponded well with those observed for the triplet state observed and calculated in a flavin model compound.[63]

Apart from different singlet excited-state lifetimes, the FMN spectral evolution and triplet yield in LOV2 are very similar to those observed for FMN in aqueous buffer. Thus, the LOV primary photochemistry is particularly simple and can be described in terms of two electronic states, the FMN singlet excited and the FMN triplet excited state. No additional time constants or spectral evolution were observed, which indicates that the FMN triplet states are formed directly from the FMN singlet excited state *via* intersystem crossing (ISC). This excludes a mechanism for triplet formation *via* a light-induced charge-separated state (as a result of electron transfer form tryptophan or tyrosine residues to FMN, for example), which recombines on the nanosecond timescale to form the triplet state. Such radical-pair mechanism is for instance in effect in photosynthetic reaction centers, where under the circumstance that secondary electron transfer is blocked, it can result in (bacterio)chlorophyll triplet yields of up to unity.[64]

A remarkable aspect revealed by the ultrafast experiments was that despite a two-fold shortening of the singlet excited-state lifetime of FMN in *Avena sativa* phot1 and *Adiantum* phy3 LOV2 ($\sim$2 ns) as compared to FMN in solution (4.7 ns), the triplet yield was essentially the same and estimated at 60%. This implied that the rate of ISC in LOV domains is enhanced as compared to FMN in aqueous solution. The ISC rate was estimated to increase from $(7.8\ ns)^{-1}$

in water to $(3.3\ ns)^{-1}$ in LOV2. A similar ISC-rate enhancement was observed in *Chlamydomonas* LOV1.[42] It was proposed that the vicinity of the sulfur nucleus in the reactive cysteine to the FMN chromophore would be responsible for this rate increase.[42,45] In accordance with this interpretation, the singlet excited-state lifetime of *Chlamydomonas* LOV1 and *Avena sativa* phot1 LOV2 mutants lacking the reactive cysteine was increased to 4.6 ns and 4.3 ns, respectively.[42,44] Radiative and internal conversion rates of $(16\ ns)^{-1}$ and $(7.4\ ns)^{-1}$ were estimated for LOV2, which correspond to FMN triplet, fluorescence and non-radiative yields of 0.60, 0.13 and 0.27, respectively.[45] A somewhat higher estimate of the FMN triplet yield in *Avena sativa* phot1 LOV2 of 0.86 has been given by Schüttrigkeit *et al.*[44] By photoacoustic methods, a triplet yield of 0.6 was reported for *Chlamydomonas* LOV1 and YtvA LOV, and 0.46 for the *Pseudomonas putida* LOV domain.[31–32,65] Islam *et al.* reported a triplet yield of 0.35 for *Chlamydomonas* LOV1 and 0.25 for its C57S mutant and free FMN in solution.[66] However, these numbers appear to be underestimated, possibly because of short-pulse excitation in the FMN S_2 band.[65]

12.2.3.2. Nano-/Micro-Second Processes

Swartz *et al.* and Kottke *et al.* have studied the photocycles of *Avena sativa* phot1 LOV2 and *Chlamydomonas* LOV1 by means of nanosecond flash photolysis spectroscopy. Consistent with the ultrafast spectroscopic data, they found that the FMN triplet state was formed within their time resolution of about 30 ns. In LOV2, the FMN triplet state, denoted as L660, evolved into the adduct state S390 with a time constant of 2 μs. The ground-state bleach signal near 447 nm was diminished by 50% with this evolution, which was interpreted as a deactivation of 50% of the FMN triplet states to the ground state. The adduct state then decayed with a single time constant of 63 s to the dark-adapted form.[38] Thus, the adduct state follows directly from the FMN triplet state, without an apparent intermediate. Subsequent work showed that adduct formation rate in LOV2 is essentially independent of pH. Remarkably, by H/D exchange, the adduct formation rate slowed down by nearly a factor of 5, indicating that it was limited by proton transfer events.[67] In *Chlamydomonas* LOV1, two distinct lifetimes of 800 ns and 4 μs were observed for the FMN triplet state, both of which led to formation of the covalent adduct in a stoichiometric fashion, *i.e.*, contrary to *Avena sativa* LOV2, no significant deactivation of the FMN triplet took place in *Chlamydomonas* LOV1. The two lifetimes were assigned to two distinct populations of FMN triplets, presumably resulting from the presence of two different cysteine rotamers previously identified in the LOV1 crystal structure.[19] Hence, efficient adduct formation proceeds from either of the two conformations, possibly *via* a fast interconversion from the distal to the proximal conformer. In YtvA, the triplet state decays mono-exponentially into the adduct form in 1.4 μs.[31] *Chlamydomonas* LOV2 displays a faster triplet decay of 500 ns, in line with its very high efficiency of adduct formation of 0.9.[68] In cysteine-less mutants, where the triplet mainly decays to the ground state, the lifetimes are 70 μs for *Avena sativa* LOV2 and 27 μs for *Chlamydomonas* LOV1.[38,43]

Losi and co-workers have determined the light-induced energy and volume changes in *Bacillus subtilis* YtvA-LOV and of *Chlamydomonas* LOV1 by photoacoustic methods.[31,65,69] The energy of the L660 triplet species was estimated at 200 kJ mol^{-1}, or about 16800 cm^{-1}. In excellent agreement with this observation, the *Avena sativa* LOV2 domain shows a FMN phosphorescence band at low temperature around 600 nm ($\sim$16700 cm^{-1}), depicted as the dashed line in Figure 2.[55] The energy content of the adduct state was rather high and estimated to 113–180 kJ mol^{-1}, *i.e.*, 45–70% of the initial photon energy, for YtvA and *Chlamydomonas* LOV1, respectively, pointing to a strained photoreceptor conformation. Adduct state formation was accompanied by a volume contraction of about -9 ml mol^{-1} and -17 ml mol^{-1} for these two LOV domains, which suggested a loss of entropy and conformational flexibility

Relatively few studies have directly addressed the quantum yield of adduct formation by quantitative methods. Kasahara *et al.* have assessed the quantum yield for LOV1 and LOV2 of *Avena sativa*, *Arabidopsis*, rice and *Chlamydomonas* phototropin by time-dependent saturation of the fluorescence after illumination, and found that for the plant LOV2 domains and for *Chlamydomonas* LOV1 and LOV2, the adduct yield varied between 0.27 and 0.35. Salomon *et al.* estimated a slightly higher yield for *Avena sativa* phot1 LOV2 of 0.44 (ref. 56). Strikingly, the plant phototropin LOV1 domains exhibited a very low adduct yield that was 5 to 10 times below that of LOV2[41,56] The origin of this extensive loss process and its significance for the physiological role in plant LOV1 remain obscure. Utilizing photoacoustic methods, Losi and co-workers found adduct yields of 0.49 in YtvA-LOV and 0.42 in PpSB2-LOV. For *Chlamydomonas* phot LOV1 an adduct yield of 0.6 was reported, at odds with the numbers found by Kasahara *et al.*[31–32,56,65]

In the cysteine-less mutants, which are incapable of forming a covalent adduct, the neutral one-electron reduced form of FMN, $FMNH^{\cdot}$ is readily formed upon illumination with strong white or blue light.[70] Most likely, the FMN triplet reacts with redox-active amino acids like tyrosine and tryptophan to result in the neutral semiquinone. The neutral radical reacts very slowly with oxygen, indicating that the flavin radical is stabilized by the protein matrix and well shielded from the solvent. In wild-type LOV, the triplet is very likely protected from electron transfer processes with intrinsic aromatic residues, located at about 12 Å,[17,19] by reacting with the conserved cysteine instead.[70]

12.2.3.3. Adduct Cleavage and Dark-State Recovery

The thermal recovery of the dark state proceeds rather slowly in LOV domains, and varies between a few seconds for *Arabidopsis* phot2 LOV2, to tens of seconds or minutes in most plant and algal phototropin LOV domains, to several hours for the *Bacillus subtilis* YtvA LOV protein and *Adiantum* phy3 LOV1.[31,38,41,43,71] The *Arabidopsis* FKF1 and ZTL clock proteins never recover after illumination.[25] Phot2 LOV2 consistently has shorter lifetimes than phot1 LOV2, which may be related to phot2's function of measuring higher fluences than phot1.[1] Interestingly, the full-length phototropin photoreceptor, as well as LOV1 + LOV2 tandem proteins exhibit dark recovery times that are significantly longer than the isolated domains, indicating that LOV1 and LOV2 interact to retard the dark regeneration to the ground state.[56] Iwata *et al.* have shown that the Q1029L mutant of *Adiantum* phy3 LOV2 has an almost 10-fold slower recovery rate than the wild type. Considering that the Q1029L mutant showed decreased structural changes upon illumination as compared to the wild type, these authors proposed that the recovery rate was related to the amplitude of the protein conformational changes that follow adduct formation.[71]

A base-catalyzed mechanism for adduct cleavage was proposed, by which the proton at the N5 site of FMN would be abstracted by a basic group somewhere in the protein.[19,38,43] The adduct decay slowed down by a factor of 3 upon H/D exchange, which suggests that the basic group could be in contact with the chromophore *via* a network of hydrogen bonds. The pronounced pH dependence of adduct decay in LOV1, with a pK between 5 and 6 supported the base-catalyzed hypothesis, and histidine residues near the surface or the phosphate group of the FMN itself were thought to act as the active base. The latter idea has become doubtful since it was observed that LOV2 reconstituted with riboflavin actually showed an increased dark recovery rate as compared to native FMN-binding protein.[36] A small pH dependence with a pK similar to that in LOV1 was found in *Avena sativa* phot1 LOV2.[67]

Near-UV light is capable of breaking the covalent adduct, as shown by femtosecond transient absorption spectroscopy on the photoaccumulated adduct state on *Adiantum* phy3 LOV2. Upon 400 nm excitation, four kinetic components were required to describe the ensuing dynamics, with time constants of 500 fs, 9 ps, 100 ps and a non-decaying component. The non-decaying component (which corresponded to the long-lived ‘photoproduct’ that is formed upon photolysis of the adduct state) agreed well with the dark-minus-light (D447-S390) difference spectrum of LOV2, demonstrating that in 100 ps, the dark ground state is regenerated from the photoexcited adduct state. The spectral evolution was interpreted with a very short excited-state lifetime of the adduct of 500 fs, which agrees with the previous observation that S390 is essentially non-fluorescent.[31,55,56] The spectra of the 9 and 100 ps components resembled those associated with a charge-transfer complex between an oxidized flavin and a cysteine thiolate anion. A reaction mechanism was invoked in which light-driven bond rupture by electron transfer from flavin to cysteine in 500 fs, after which the resulting charge-transfer complex relaxes in multiple steps of 9 and 100 ps to the dark conformation D447. The quantum yield of the reaction was estimated between 0.2 and 0.25. A similar light-driven back reaction was proposed for the *Chlamydomonas* LOV1 and LOV2 domains, albeit at a lower yield.[43,68,72] These observations implied that the LOV2 domain acts as reversible photochromic switch, a property shared with other classes of photoreceptor proteins such as phytochromes and xanthopsins.[60] The physiological significance of the light-driven adduct cleavage remains, however, uncertain.

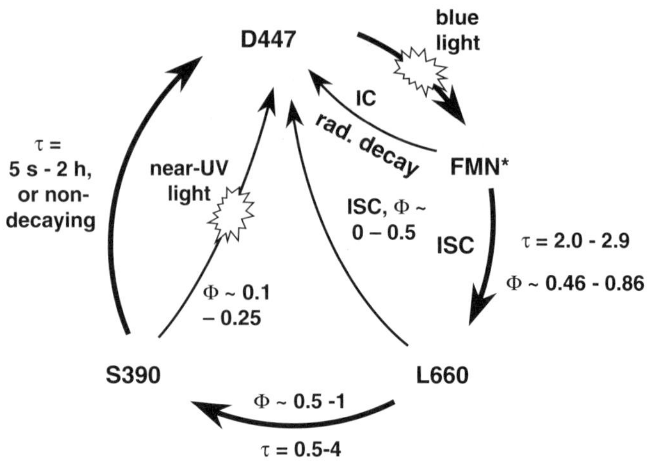

Figure 4. Photocycle scheme of LOV domains, based on available data for plant phototropin LOV1 and LOV2, *Chlamydomonas* phototropin LOV1 and LOV2, *Bacillus subtilis* YtvA, and *Pseudomonas putida* PpSB2-LOV. τ denotes boundaries for measured time constants, Φ boundaries of reported reaction yields in the various LOV domains. See text for details.

The photocycle scheme depicted in Figure 4 summarizes the results of the kinetic experiments and quantum yield determinations. The molecular species and the boundaries for the lifetimes and quantum yields by which they interconvert are indicated, as observed for the various LOV domains. The *Chlamydomonas* phot LOV2 domain appears to fall outside this scheme, as there is evidence for an adduct formation pathway that bypasses the FMN triplet state.[57] Given that the photochemistry of only a small fraction of the known LOV domains has been investigated, we anticipate that more of such increased complexities may need to be added to the LOV photocycle scheme.

It is important to note that the photocycle scheme only involves the LOV dynamics that are observed by visible/UV absorption and fluorescence spectroscopy. Such spectroscopies only probe the physicochemical state of the FMN chromophore, and changes in the conformation of the protein, even in the immediate vicinity of the FMN, will as a rule not be sensed by the UV/visible absorption spectrum. Therefore, the photocycle scheme is far from complete in the sense that it does not include the structural–functional conformational dynamics that underlies the signaling function of LOV domains.

12.2.4. Adduct Formation in LOV Domains: Mechanistic Considerations

In contrast to the consensus that appears to exist on the (spectroscopically distinguishable) intermediates in the LOV photocycle, the mechanism by which covalent adduct formation occurs in the LOV domains is a matter of considerable debate. Broadly speaking, two reaction mechanisms for adduct formation have been put forward: (1) an ionic mechanism and (2) a radical-pair mechanism. The ionic mechanism was first put forward by Crosson and Moffat and Swartz *et al.*, and was based on the three-dimensional structure of LOV2, results from time-resolved spectroscopy and early quantum-chemical calculations by Song, which indicated that upon promotion of flavin to the triplet state, electron density migrates from the N1 to the N5 atom of the flavin chromophore, thereby increasing the electronegativity of the N5 site and decreasing that of C(4a).[17,38,73] According to the ionic model, which is schematically depicted in Figure 5 (upper panel), the sharply increased basicity of N5 in the FMN triplet state triggers its protonation. The proton would either be donated by the conserved cysteine itself[17,19,45] or by another nearby group[38,74] (the former is shown in the Figure 5). This event would change the double bond of N5 = C(4a) to a single bond, leaving a very reactive carbo-cation at the C(4a) position. This site would have a sp^3 hybridization, which would decrease the distance to the cysteine, which could be important in the progress of the reaction. Subsequently, a nucleophilic attack by the cysteine thiolate on the C(4a) carbo-cation would occur, leading to formation of the covalent FMN-C4a-thiol adduct. These events could occur sequentially[45] or in a concerted fashion.[17,19] In any of these cases, the sharp increase in the pK_a of N5 upon triplet formation is the switch that drives the photoreaction by proton abstraction of a nearby donor. Evidence for the ionic mechanism was presented by Kennis *et al.* These authors found that the triplet spectra of free FMN in solution were pH dependent; at neutral pH, the triplet spectrum of free FMN recorded after several nanoseconds exhibited two distinct absorption bands near 650 and 710 nm. At pH 2, only one band was observed at 660 nm, which had previously been assigned to protonation of the FMN triplet at the N5 site.[75] The triplet spectra of LOV2-bound FMN of *Adiantum* phy3 and *Avena* phot1 could be well approximated by a superposition of these two species, which was interpreted as a partial protonation of the flavin triplet at N5, taking place on a nanosecond timescale.[45] Further evidence for the ionic model was provided by quantum-chemical calculations based on the *Chlamydomonas* LOV1 structure, which showed that upon excitation of the FMN triplet, the N5 atom of FMN could abstract a proton from proximal cysteine conformers.[19]

Kay *et al.* discussed the unlikelihood of an ionic mechanism for adduct formation by noting that in such case, the covalent adduct has to be formed in its triplet state, which would be energetically highly unfavorable. A direct formation of the adduct singlet ground state from the FMN triplet state would be a spin-forbidden process, and, if preceded by a triplet-to-singlet transition to the FMN ground state, they argued that the FMN would deprotonate rapidly before the adduct could be formed. Instead, by observing $FMNH^{\cdot}$ neutral semiquinone radicals in cysteine mutants of *Avena sativa* phot1 LOV2, these authors proposed that adduct formation proceeds *via* a radical-pair mechanism,[70] schematically represented in Figure 5 (lower panel). Upon promotion of the FMN chromophore to the triplet state, a hydrogen would be transferred from the conserved cysteine to the N5 of the flavin, resulting in an

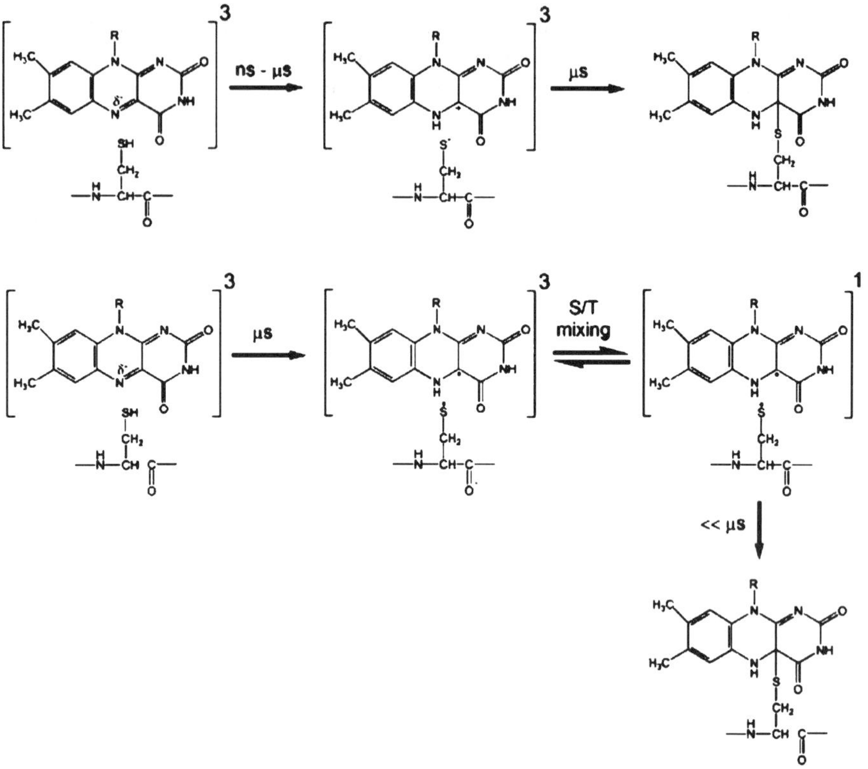

Figure 5. Proposed reaction mechanisms for light-driven covalent flavin-C(4a)-cysteinyl adduct formation in LOV domains; (upper) the ionic mechanism; (lower) the radical-pair mechanism. See text for details.

FMNH$^{\cdot}$–H_2C–S$^{\cdot}$ radical pair. In the neutral flavin semiquinone radical, the unpaired electron density resides at the C(4a) atom. Such a radical pair would be formed in a triplet configuration, however, the close proximity of the (heavy-atom) sulfur radical to the isoalloxazine ring causes a strong spin-orbit coupling, inducing a rapid triplet–singlet inter-conversion. Once the radical pair obtains an appreciable singlet character, radical-pair recombination between the H_2C–S$^{\cdot}$ and unpaired electron at C(4a) may take place to result in the FMN-C4a-thiol adduct. In the context of the proposed mechanism, the trigger for adduct formation could either be an electron transfer from the cysteine to the flavin, followed by proton transfer, or a concerted mechanism, whereby a net hydrogen transfer from the thiol to the flavin takes place.[53,70,76,77] Quantum chemical calculations favor the latter possibility.[39,78] However, Schleicher *et al.* reported that at 77 K the absorption of the covalent adduct was red-shifted to 405 nm, which they assigned to the zwitterionic (SH$^{\cdot +}$---C(4a)$^{\cdot}$–N5^{-}) species, which would be formed by initial electron transfer from the cysteine to the flavin.

It is difficult to obtain direct spectroscopic evidence for the proposed radical-pair mechanism since little flavin radical character has been observed in the photocycle of LOV domains.[38,43–45] However, by their nature, such radicals would be a short-lived intermediate between the flavin triplet and covalent adduct states, and thus easily escape detection. Quantum-chemical calculations on the *Chlamydomonas* LOV1 domain have preferred the radical-pair mechanism over the ionic mechanism for adduct formation because of the more favorable energetics of the transient states involved.[39,78] It should be noted, however, that the energy content of the covalent adduct state, and thereby that of the radical-pair intermediate, was appreciably underestimated as compared to the calorimetric experiments by Losi *et al.*[65,78]

Further evidence for the radical-pair mechanism has come from transient EPR studies, where no protonation of the FMN triplet was detected for LOV1 and LOV2, or free FMN at 77 K.[53,79] At low temperature, the FMN triplet state in the LOV domain can be photoaccumulated, and FMN triplet FTIR spectra that were recorded in this way showed no deprotonation of the conserved cysteine, but rather the formation of a hydrogen bond of the cysteine to FMN, as evidenced by a pronounced band-shift of the S–H stretch rather than a ground-state bleach.[80] Note, however, that in these low-temperature experiments, the triplet species under study involved a non-reactive fraction of LOV domains that did not form the adduct state. In a *Chlamydomonas* LOV1 mutant where the conserved cysteine was replaced by a methionine, a covalent adduct formation to N5 was detected even though the –SH group with its readily detachable proton was replaced by the less reactive $-S-CH_3$.[76] Also, analogies with other flavoproteins or free flavin in solution have supported the radical-pair model. In a mutant of old yellow enzyme and free FMN in the presence of indole, the C(4a)-peroxyflavin and a C(4a)-indole adduct proceeds from the neutral semiquinone radical *via* initial electron transfer, followed by proton transfer and radical pair recombination.[63,81]

Interestingly enough, time-resolved Raman and laser-flash photolysis studies of free FMN have shown that at low pH, the protonated triplet state can be reduced to form the neutral semiquinone radical.[75] Thus, the rate-limiting step in the radical-pair mechanism may actually be a proton-transfer reaction, as it necessarily is in the ionic model. The reported FMN triplet-state protonation[45] and the 5-fold slowdown of the adduct formation upon deuteration[67] are therefore consistent with both proposed mechanisms. Obviously, the question which reaction mechanism governs adduct formation in LOV domains is far from settled. More detailed studies that aim to address the transient local structures on the nano- to micro-second timescales of the LOV domain and its site-directed mutants will be key to resolve this issue.

12.2.5. Light-Induced Conformational Changes in LOV Domains

Surprisingly, the crystal structures of *Adiantum* phy3 LOV2 and *Chlamydomonas* LOV1 showed little overall structural changes in the adduct state.[19,34]

The changes were mainly confined to the immediate environment of the flavin chromophore, as shown in Figures 1B and C. The FMN was distorted by adduct formation and tilted by $\sim 8^\circ$. Moreover, a conserved glutamine changed orientation and had become a hydrogen-bond acceptor of the N5 proton of FMN, rather than a donor to the C4=O carbonyl group. In LOV2, a conserved volume was identified that stretched from the FMN to a conserved salt bridge at the molecular surface. The residues constituting the conserved volume showed a small but significant concerted movement toward the surface upon photoactivation. It was proposed that a light-induced modulation of the salt bridge could influence the binding affinity to an interaction partner, either by an actual conformational change at the surface, or entropically by a change in the molecular flexibility.[35] In a site-directed mutant of YtvA where the conserved salt bridge was removed, only small differences in α-helical and β-sheet and turn content were seen as compared to wild type, in the dark as well as in the light, utilizing CD spectroscopy. These results were interpreted in that the salt bridge stabilizes locally the protein structure and participates in the regulation of the photocycle but has negligible effects on the overall structure.[82]

In contrast to the results from X-ray crystallography, FTIR studies indicated significant blue light-driven structural changes in LOV2 domains. Iwata *et al.* demonstrated that *Adiantum* phy3 LOV2 exhibits structural changes that involve a H-bond loosening of helices and a tightening of the β-sheet, accompanied by a small turns motion, suggesting that the tertiary structure of LOV2 is modified upon blue-light activation. Below 150 K, the helix and sheet motions were frozen.[54] Interestingly, mutation of the conserved glutamine that connects the FMN chromophore and PAS β-sheet (Figure 1B and C) resulted in an FTIR difference spectrum resembling those recorded at low temperature, pointing at an inhibition of most of the protein motions.[46]

Adiantum phy3 LOV1 showed only small changes, which were nearly temperature independent. Thus, in contrast to LOV2, the phy3 LOV1 surface remains mostly unaltered upon adduct formation.[71] These FTIR results were consistent with those reported on *Chlamydomonas* LOV1, where almost all differential signals were assigned to FMN modes and none could definitely be assigned to the protein.[51] In agreement with these findings, no light-dependent conformational changes were found in *Arabidopsis* phot1 and phot2 LOV1 domains during small-angle X-ray scattering (SAXS) experiments.[83] It was shown that in *Arabidopsis* phot1 and phot2, inactivation of the LOV2 domain abolishes phot1 autophosphorylation and phototropin-activated responses, including phototropic curvature and chloroplast movement. Photoactivation of LOV1 alone is not sufficient to induce phototropism or chloroplast avoidance movement.[84] If protein structural changes are larger in LOV2 than in LOV1 as suggested, they are correlated with the functional differences between LOV2 and LOV1. Indeed, Salomon *et al.* have shown that *Avena sativa* phot 1 LOV1, either functional or inactivated (LOV1 and LOV1-C39A, respectively), is found as a homodimer in solution, whereas phot1 LOV2 is monomeric.[85] Thus, LOV1 seems to play a structural and perhaps regulatory role in phototropin.

Sequence alignment of LOV1 and LOV2 domains from all phototropins identified to date reveals the presence of a very conserved segment associated only with LOV2. It is located outside the LOV2 domain and involved in the linker region between LOV2 and the kinase domain. Using NMR spectroscopy, Harper *et al.* have shown that the 40 amino acids at the *C*-terminal of *Avena sativa* phot1 LOV2 form a helical secondary structure, termed Jα.[18] This helix is amphipatic and binds to the solvent-exposed β-sheet surface of LOV2, burying hydrophobic surfaces on both sides. Upon illumination of the LOV2 domain, the LOV2/Jα interaction is eliminated, Jα loses its secondary structure and partially unfolds. These observations are in agreement with a far-UV CD study on the same LOV2 domain, which suggested a reversible loss of α-helicity during the photocycle,[67] and the increased dynamics of helices noted by Iwata in phy3-LOV2.[54] Moreover, SAXS experiments on *Arabidopsis* phot1 and phot2 LOV2 domains, expressed with and without the *N*-terminal linker polypeptides, were consistent with light-induced structural changes in the linker region outside the PAS core.[83]

Harper *et al.* demonstrated the constitutive kinase activation of *Avena sativa* phot1 expressed in insect cells, by introducing point mutations that disrupt the LOV2/Jα interaction. The mutations were located along three consecutive turns of the Jα helix surface which are in contact with the LOV2 β-sheets, thus establishing that unbinding and unfolding of the Jα helix is the propagating structural change outside the LOV2 domain, which regulates kinase signaling in phototropin 1.[86]

Thus, from FMN-cysteinyl adduct formation a tertiary structure change follows, which functions as a "switch" and somehow causes the Jα helix to dissociate from the PAS β-sheet. As mentioned earlier, a FTIR study by Nozaki *et al.* of a conserved glutamine point mutant of the *Adiantum* phy3 LOV2 domain has shown cancellation of helix and sheet light-induced conformational changes compared to wild type (ref. 46). In the *Adiantum* phy3 LOV2 crystal structure, this glutamine 1029 side chain switches hydrogen bond with the FMN chromophore, by rotation, from C(4)=O donor to N(5)–H acceptor in the dark and adduct state, respectively, as shown in Figure 1B and C. Moreover the FMN ring moiety is sandwiched between a helix, which contains the reactive cysteine (Cys966) and antiparallel β-sheets that contain this glutamine. In this view, it is likely that this H-bond switch leads to β-sheet motion and subsequent Jα unbinding and unfolding by loss of van der Waals complementarities. In *Avena sativa* LOV2, the conserved glutamine is located on the Iβ strand, where an isoleucine residue from the Jα-helix docks. In the aforementioned work by Harper *et al.*, it was shown that point mutation of this isoleucine (and adjacent alanine and isoleucine) were able to disrupt the interaction between the Jα helix and the LOV2 core. Hence, this hydrogen-bond switching glutamine residue may form a link between adduct formation and the larger conformational changes in the LOV core, which lead to Jα-helix dissociation.[46,86]

12.3. Blue-Light Sensing Using FAD Domains

BLUF domains are a distinct family of flavin-binding photoreceptors that show no significant relationship to other sensor proteins in sequence or structure. The BLUF domain was first discovered as the *N*-terminal part of the flavoprotein AppA from *Rhodobacter (Rb.) sphaeroides*[87] AppA is a two-component protein that can sense blue light at its *N*-terminal domain, and oxygen tension at its cysteine-rich *C*-terminal domain, and was shown to control photosynthesis gene expression in response to high-intensity blue light irradiation and variation of the oxygen tension.[87–89,6–7] Other BLUF domains were characterized in the light-activated photoactivated adenylyl cyclase (PAC) of *Euglena gracilis*,[4,90] where it regulates a step-up photophobic response, and in the YcgF protein of *E. coli*, which has a phosphodiesterase as *C*-terminal enzymatic domain and has an as of yet unknown function.[88,91] Single BLUF domains, which have been characterized include Slr1694 from the cyanobacterium *Synechocystis PCC6803*,[92] Tll0078 from *Thermosynechococcus elongatus BP1*[93] and BlrB from *Rhodobacter sphaeroides*. Slr1694 and Tll0078 have roles in negative phototaxis,[92] while BlrB is of unknown function.[94] At present, our knowledge of the functional and biophysical properties of BLUF domains is not nearly as detailed as the picture we have for the LOV domains, but is rapidly progressing.

12.3.1. The Dark and Photoactivated States of BLUF Domains

BLUF domains bind a single molecule of oxidized FAD, resulting in a $S_0 \rightarrow S_1$ absorption band near 450 nm and a $S_0 \rightarrow S_2$ absorption near 360 nm. The absorption spectrum of the AppA BLUF domain (hereafter referred to as AppA) is shown in Figure 6 (solid line) and exhibits vibronic fine structure of both electronic transitions, with features at 420, 445 and 475 nm ($S_0 \rightarrow S_1$), and 363 and 377 nm ($S_0 \rightarrow S_2$).[87,95,96] Other BLUF domains show similar absorption features, but with an overall slight blue-shift for the cyanobacterial BLUF domains[93] or a substantial red-shift for the *E. coli* YcgF protein.[91] Upon illumination, a long-lived signaling state is formed, which shows a ˜10–15 nm red-shifted absorption spectrum, shown in Figure 6 as the dashed line. The light-minus dark difference spectrum is shown as the gray line. This red-shifted state has a lifetime of several seconds in the cyanobacterial BLUF domains to 1800 s in AppA.[94–98] Deuteration and dehydration slow down the dark recovery in Slr1694 and AppA by approximately a factor of 4 and 20, respectively.[99,100] Notably, addition of a large excess of imidazole to AppA speeds up the dark recovery time by two orders of magnitude.[100]

In AppA, heterologous expression in *E. coli* leads to the incorporation of a mixture of FAD, FMN and riboflavin. Reconstitution of AppA with any one of these flavins resulted in a fully photochemically active receptor, and raised questions whether *in vivo*, BLUF domains exclusively bind FAD or may bind other flavins as well.[101]

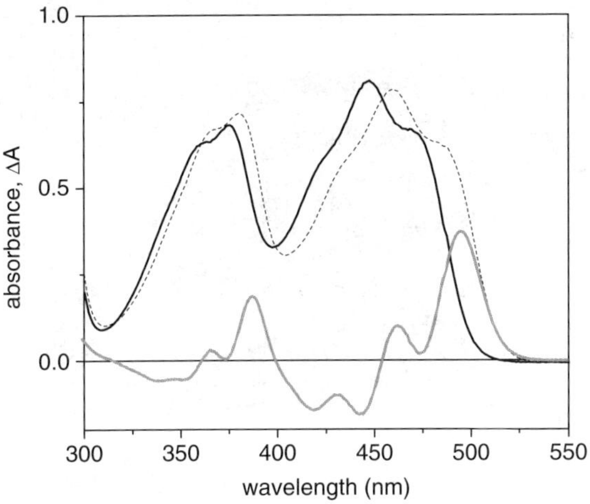

Figure 6. Absorption spectrum of the AppA BLUF domain in the dark state (solid line) and photoactivated state (dashed line). The gray line denotes the light-minus-dark difference spectrum.

Three BLUF domain X-ray structures are available: *Rb. sphaeroides* AppA determined at 2.3 Å resolution,[102] *Thermosynechococcus* Tll0078 at 2 Å resolution[93] and *Rb. sphaeroides* BlrB at 1.9 Å resolution.[94] The structures showed similar features, and indicated an $\alpha + \beta$ sandwich, with two α-helices located on one side of an antiparallel five-stranded β-sheet, as shown in Figure 7A for AppA.[102] The BLUF domain fold was identified as ferredoxin-like. The flavin cofactor was non-covalently bound and sandwiched between the two α-helices, with the plane of its isoalloxazine ring oriented perpendicular to the β-sheet. The BlrB and Tll0078 proteins bind one and two additional α-helices at the *C*-terminal, respectively, which run perpendicular to the strands of the β-sheet, and were thought to play a role in oligomeric assembly.

Figure 7B and C show a close-up of the flavin-binding pocket of AppA. As in the LOV domains, the flavin dimethylbenzene ring is surrounded by hydrophobic residues, whereas the pyrimidine ring is involved in an extensive network of hydrogen bonds. A conserved glutamine (Gln63) appears to be of chief importance, since its side chain is capable of making two different sets of hydrogen bonds, depending on its orientation. Unfortunately, the resolution of the AppA structure was not sufficient to determine the orientation of Gln63. Anderson *et al.* proposed that the orientation shown in Figure 7B corresponded to the dark state of AppA, with the amino group of Gln63-donating hydrogen bonds to the conserved tyrosine and N5 of the flavin, and the carbonyl moiety of Gln63 accepting a hydrogen bond from the conserved tryptophan. Because this conformation involves one additional hydrogen bond, it would be preferred over that shown in Figure 7C, with the amino moiety of Gln63 donating a hydrogen bond to the C4=O carbonyl of flavin, and the Gln63 carbonyl group accepting a hydrogen bond from the conserved tyrosine.

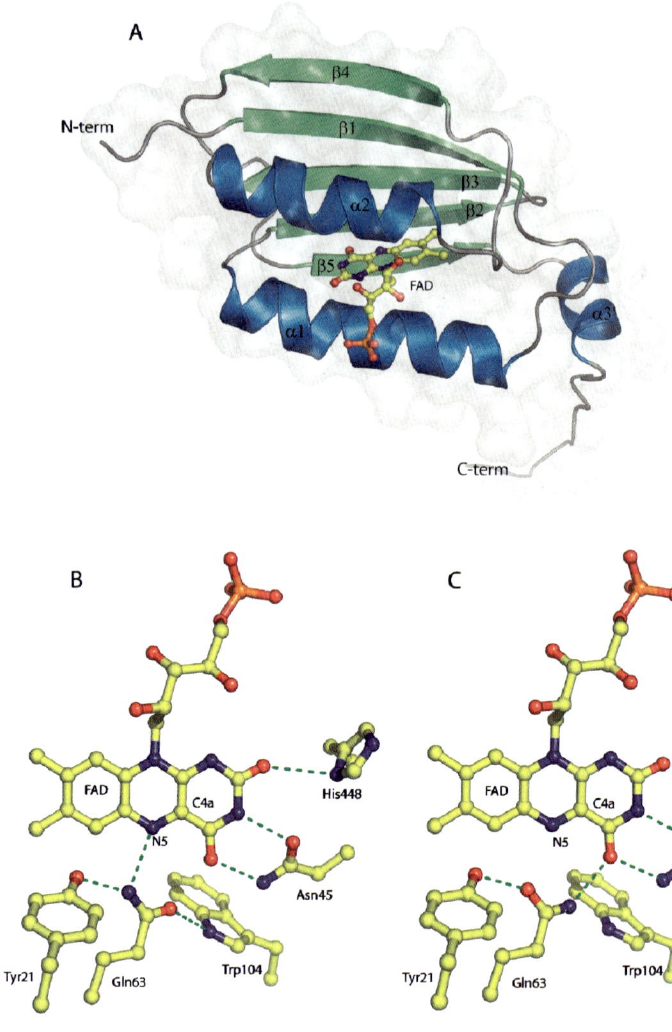

Figure 7. (A) X-ray structure of the *Rhodobacter sphaeroides* AppA BLUF domain. Panel (B) and (C) represent a close-up of the crystal structure in the vicinity of the FAD chromophore in two proposed orientations of the conserved glutamine.

For the Tll0078 and BlrB BLUF domain structures, another viewpoint was taken. The dark conformation of the conserved glutamine was oriented similar to the proposed photoactivated conformation of AppA. Here, the amino moiety of the Gln would donate hydrogen bonds to the N5 and C4═O of the flavin, and accept a hydrogen bond from the conserved tyrosine.[93,94]

The origin of the red-shift in the absorption spectrum upon illumination was originally proposed to result from altered π–π stacking interaction between tyrosine and the flavin isoalloxazine ring, along with a hydrogen bond rearrangement between N5 and the tyrosine.[95] However, with the X-ray structures at hand, the distance between the flavin and tyrosine is too large for this stacking to occur. Laan *et al.* proposed that in the signaling state, the FAD cofactor would deprotonate at the N3 position, thereby protonating a tyrosinate residue.[96] Again, the X-ray structure indicates that flavin and tyrosine are oriented unfavorably for such processes to occur.

Masuda and co-workers have performed comprehensive UV–visible absorption and FTIR studies on *Synechocystis* Slr1694[97,103] and the AppA BLUF domain.[104] Illumination with blue-light shifts the C4=O stretch vibration of the flavin from 1713 to 1693 cm^{-1}, and the C2=O from 1670 to 1653 cm^{-1}, indicating that hydrogen bonds to the flavin carbonyls are formed or strengthened in the light state. Moreover, the C=N stretch bands at 1510 cm^{-1} (corresponding to coupled C4a=N5 and N1=C10a in- and out-of phase stretches) upshifts to 1533 and 1544 cm^{-1}, which indicates that hydrogen bonds, possibly to the N5 of the flavin, may exist in the dark state and break upon formation of the signaling state. Changes in the Amide I region near 1639–1608 cm^{-1} signaled structural changes in the protein, most likely β-sheets and turns, and to a less extent the α-helices.[97]

Upon dehydration of the sample, or by lowering the temperature to −35°C, the red-shift of the UV–visible absorption spectrum still occurred upon photoactivation, but most of the changes in Amide I, C2=O and C=N stretch bands disappeared from the FTIR spectrum. This indicated that under these circumstances, only the hydrogen-bond formation to C4=O of the flavin took place.[103] Accordingly, the authors proposed that the hydrogen-bond formation to C4=O established the primary event of the Slr1694 photocycle, which would trigger the further hydrogen-bond rearrangements and conformational changes in the protein. Importantly, the experiment demonstrated that the ~10 nm shift of the UV–visible absorption spectrum in BLUF domains after illumination is almost entirely defined by hydrogen bonding to the flavin C4=O, and only weakly dependent on hydrogen bonding to C2=O and N5, and to the structural changes in the protein. In fact, it was demonstrated that during the dark relaxation in Slr1694, decay of the C4=O frequency shift occurred simultaneously with that of the red-shifted UV–visible absorption, but that the shifts of C2=O, C=N, Amide I and other protein modes lagged those of the UV–visible absorption by a factor of 1.5 to 3.[99]

FTIR spectroscopy on the AppA BLUF domain showed overall similarities with those reported for Slr1694, but also revealed marked differences.[104] By reconstitution of unlabeled FAD in isotope-labeled apoprotein, it was established that only one feature in the FTIR difference spectrum, namely the C4=O stretch downshift from 1707 to 1684 cm^{-1} could be assigned to the FAD chromophore. The remainder of the vibrational bands, primarily showing amplitude in the region from 1610 to 1709 cm^{-1} could be assigned to changes in the Amide I vibrations of the protein. Most notably, the large up-shift of the C=N stretches at 1510 cm^{-1}, which was prominent in Slr1694 and signaled the light-induced loss of a hydrogen bond to the N5 of the flavin, was absent in the AppA FTIR difference spectrum. The application of Raman spectroscopy to the AppA BLUF domain corroborated many of the features observed by FTIR. In addition, it was suggested that the formation of the signaling state was accompanied by structural modifications of the N10-ribityl moiety.[105]

12.3.2. The BLUF Photocycle

Time-resolved spectroscopy has been applied to BLUF domains to characterize their photocycles and to determine any intermediate on the reaction pathway toward the red-shifted signaling state. So far, only AppA and, to a less extent, the *Thermosynechococcus* Tll0078 BLUF domain have been investigated in any detail. Contrary to the situation in LOV domains, a highly multi-exponential excited-state decay of FAD was reported by use of time-resolved fluorescence methods, with lifetimes of 25, 150, 670 ps (dominant) and 3.8 ns in AppA,[106] and 120 ps (dominant), 710 ps and 4.6 ns in the *Thermosynechococcus* BLUF domain.[98]

Femtosecond time-resolved transient spectroscopy on AppA revealed a multi-exponential decay of excited FAD with time constants similar to those found in fluorescence, although one less component was found, namely 90 ps, 590 ps and 2.7 ns. Concomitant with the decay of excited FAD, the rise of a species with a narrow absorption-difference band near 495 nm was detected, which was nearly identical to the long-living red-shifted signaling state of AppA (gray line in Figure 6). Moreover, a broad-induced absorption in the red (600–700 nm) was detected.[106] To account for these observations, the time-resolved data on AppA were described by a kinetic model that comprised three molecular states: the FAD singlet excited-state FAD*, the FAD triplet state FAD^T, and the red-shifted AppA signaling state $AppA_{RED}$, represented by the photocycle scheme depicted in Figure 8. Three fractions of FAD* were assumed to represent the multi-exponential fluorescence decay, with lifetimes of 90 ps, 590 ps and 5 ns, respectively. The presence of multiple lifetimes for FAD* was related to heterogeneity in the ground state conformation of AppA. The former two evolved into the red-shifted signaling state $AppA_{RED}$, and all three fractions exhibited an ISC rate of $(5\ ns)^{-1}$ to the triplet state FAD^T. The resulting difference spectra of the species involved are shown in Figure 8 (lower panel). The difference spectra of the FAD^* (solid line) and FAD^T (dashed line) excited states resemble those reported earlier,[38,43,45] including the FMN singlet and triplet spectra shown in Figure 3 for LOV domains. The difference spectrum for the $AppA_{RED}$ state (gray line) is virtually identical to that observed under steady-state illumination (Figure 6, gray line). It was accordingly concluded that the red-shifted signaling state of the AppA BLUF domain is formed on the ultrafast timescale, directly from the FAD singlet excited state, without any apparent intermediate with time constants of 90 and 590 ps, and remains stable over 12 decades of time. In parallel with the signaling state, the FAD triplet state is formed from the FAD singlet excited state at 9% efficiency. Consistent with the observations from femtosecond spectroscopy, nanosecond flash photolysis measurements

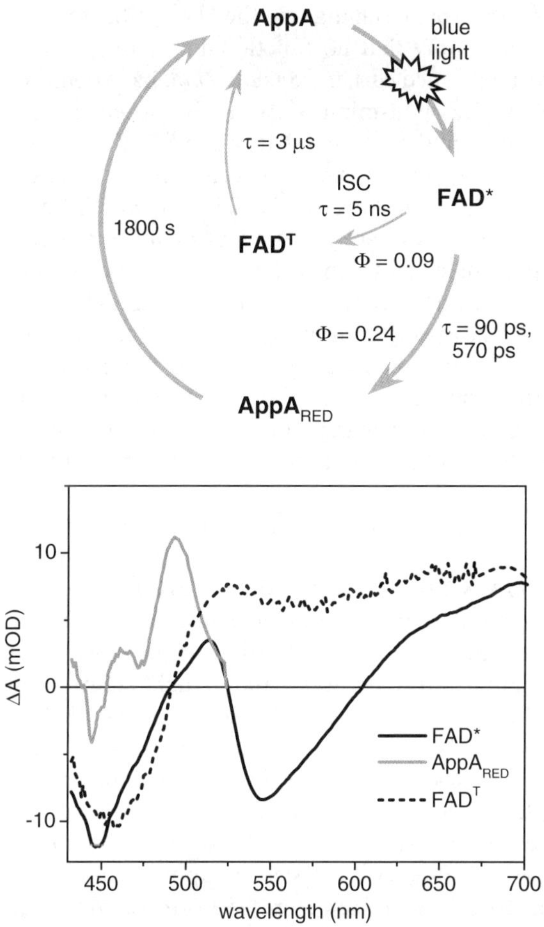

Figure 8. (upper) Photocycle scheme for the AppA BLUF domain; (lower) absorption difference spectra of the FAD*, $AppA_{RED}$ and FAD^{T} intermediates, as determined from femto- and nano-second transient absorption spectroscopy. See text for details.

indicated that formation of the signaling state was complete within the time resolution of 10 ns, and showed no further changes up to 10 ms. The nanosecond flash photolysis experiments also showed that the formation of FAD^{T} represents a side reaction, and is not coupled to the physiologically relevant reaction pathway. FAD^{T} decays to the molecular ground state in 3 μs. The quantum yield of the signaling-state formation was determined at 0.24 by comparing the absorbance changes of AppA with a reference compound. Figure 8 summarizes the results in the photocycle scheme for AppA.[106,107]

Little kinetic information is available for other BLUF domains. Nanosecond flash photolysis on the *Thermosynechococcus* Tll0078 BLUF domain revealed that the signaling state was formed within the time resolution of ~10 ns, as in AppA, with a quantum yield of 0.29.[98]

The absence of absorbance changes in the UV–visible region from 1 ns to 10^3 s in AppA does not imply that no functional structural changes occur in this time window. As mentioned earlier, Masuda *et al.* have demonstrated that most of the features in the light-minus-dark FTIR spectrum of AppA may be assigned to changes in the protein backbone.[104] Moreover, the UV–visible absorption is sensitive only to the hydrogen-bonding strength to the C4=O moiety of the isoalloxazine ring.[99,103] Well-studied photoreceptor proteins, such as rhodopsins and photoactive yellow protein demonstrate that the global changes in the protein structure in response to the chromophore photoactivation take place on the microsecond to millisecond timescales.[60] Further studies, using time-resolved X-ray diffraction, step-scan and rapid-scan FTIR, transient grating or photoacoustic spectroscopy would be required to assess the relation between the initial changes around the C4=O moiety of the flavin and the ensuing functional structural changes on and around the chromophore, and the conformational changes that finally result in the signaling state of the protein.

12.3.3. The BLUF Photoactivation Mechanism

Based on the AppA X-ray structure and available FTIR data, Anderson *et al.* have proposed a photoactivation model that involves a hydrogen-bond rearrangement accompanying a ~180° rotation of the conserved glutamine, corresponding to Figures 7 B (dark state) and C (light state). In this view, photon absorption would lead to the breaking of hydrogen bonds from the glutamine amino moiety to the N5 of flavin and to the tyrosine, and formation of a hydrogen bond to the C4=O of the flavin. Concomitantly, the carbonyl moiety of the glutamine would break the hydrogen bond from the conserved tryptophan, and accept one from tyrosine. These events would be consistent with the ~10 nm red-shift of the visible spectrum upon illumination,[105] and with the FTIR results on BLUF domains described in the previous section[97,103,104]. In support of this hypothesis, mutation of the conserved tyrosine in AppA (Y21) and Slr1694 (Y8) and the glutamine (Q50) in *Thermosynechococcus* Tll0078 results in a complete loss of photocycling activity[93,95,96,99] Anderson *et al.* suggested that the conserved glutamine functions as an anchor tying the conserved tryptophan to the flavin. Disruption of the hydrogen bond between glutamine and the tryptophan would allow increased motion on the β5 strand, which might be important in creating the global changes that alter the interaction of full-length AppA with its interaction partner.

With the *Rhodobacter sphaeroides* BlrB structure at hand, Jung *et al.* noted that rotation of the conserved glutamine would be unlikely because of its strong interaction with the protein backbone. Instead, they proposed an alternative reaction mechanism which would imply a light-induced proton transfer from arginine 32 to the C2=O of the flavin.[94] It should be noticed, however, that mutation of the corresponding residue in Tll0078, asparagine 31, to alanine did not abolish the photocycle activity of this protein.[93] Kita *et al.* suggested that a slight shifting of the conserved glutamine toward C4=O would accompany signaling-state formation in Tll0078, strengthening their hydrogen bond.[93]

As mentioned above, no transient intermediate between the FAD singlet excited state and 10 nm red-shifted signaling state was detected in the AppA BLUF domain, suggesting that a putative hydrogen-bond rearrangement or any other proposed reaction would be initiated directly from the singlet excited state of FAD. Based on the pronounced kinetic isotope effect on dark relaxation, Masuda proposed a photocycle model where the N5 of the flavin would act as a short-lived proton acceptor on a light-driven proton transfer pathway from one nearby amino acid side group to another, leading to hydrogen-bond formation to the C4═O or C2═O moieties of the flavin. In their view, the increased basicity of N5 upon promotion of the flavin to the singlet excited state would be the trigger for this reaction.[99]

The application of femtosecond transient absorption spectroscopy on *Synechocystis* Slr1694 and *E. coli* YcgF revealed a more complex dynamical pattern than in AppA. The FAD excited-state lifetime was significantly shorter, and the red-shifted species was formed on the ~100 ps timescale *via* a neutral FAD radical intermediate (Gauden *et al. Proc. Natl. Acad. Sci. (USA)* in press and Key *et al.* manuscript in preparation). Importantly, protonation of the FAD neutral radical takes place at the N5 position, which suggests that the BLUF domain photochemistry mainly occurs in the vicinity of the conserved tyrosine, glutamine and tryptophan residues. Assuming that the basic photochemistry is conserved among BLUF domains, the neutral FAD semiquinone radical probably forms an integral part of the AppA photoreaction as well, but may not be detectable due to a rate-limiting initial electron or proton transfer rate.

Surprisingly, mutation of the conserved tryptophan (W104) in AppA to a phenylalanine results in an increased quantum efficiency of signaling-state formation, from 0.24 in the wild type to 0.37 in the W104F mutant. Moreover, the FAD singlet excited-state lifetime in the W104F mutant was significantly longer than in wild-type AppA. It was proposed that in wild-type AppA, the FAD singlet excited state is partially deactivated *via* a hydrogen, electron, or a coupled electron–proton transfer from Trp104 to FAD, and subsequent rapid recombination to the ground state. In the W104F mutant, the contribution of this deactivation pathway would be significantly reduced due to the lower electron/proton transfer capacities of the phenylalanine, resulting in a longer excited-state life-time and increased quantum yield of signaling-state formation.[100]

In summary, despite the recent growth in BLUF-domain-related investigations, the photochemistry that constitutes their light activation remains unclear. Significant gaps remain in our knowledge, which need to be filled in by further biophysical and biochemical experimentation, and by theoretical work.

12.4. Outlook

In recent years, the newly emerging field of flavin-based blue-light photoreceptors has attracted significant attention, and by the use of advanced biophysical, biochemical and theoretical methods, our understanding of how these proteins function at the molecular level has progressed at an impressive pace. Still, many challenges lie ahead. So far, our view of signal propagation through and outside light-sensitive domains mainly derives from steady-state light-minus-dark experiments, whereby putative 'snapshots' of intermediate states are taken by modifying the system or the conditions, such as lowering the temperature of introducing mutations. 'True' dynamical–structural methods, such as time-resolved X-ray crystallography and time-resolved vibrational spectroscopy could provide for a complete picture of the course of events that form the basis of LOV- and BLUF-mediated signaling. All LOV and BLUF domains interact with signaling domains to communicate the perceived photic information to the organism, yet very little is known of the dynamic–structural basis of these processes. To address these issues, the biophysical and biochemical studies described in this chapter may be extended to full-length photoreceptor proteins, interacting with signaling partners further downstream. Such studies will provide us with a genuine molecular understanding of light-mediated signaling.

Acknowledgments

We thank our co-workers Magdalena Gauden, Wouter Laan, Sergey Yeremenko, Jason Key, Ivo van Stokkum, Klaas Hellingwerf, Rienk van Grondelle, Sean Crosson, Keith Moffat, Marloes Groot and Peter Hegemann for enabling some of the work described in this chapter and stimulating discussions. We gratefully acknowledge Winslow Briggs, John Christie, Kevin Gardner, Trevor Swartz, Roberto Bogomolni, Robert Bittl, Stefan Weber and Christopher Kay for enlightening discussions and continuing interest in our work. We thank Jason Key, Sergey Yeremenko and Anjali Pandit for help with preparing the figures, and Jason Key for critically reading the manuscript. J.T.M.K. was supported by the Earth and Life Sciences council of the Netherlands Foundation for Scientific Research (NWO-ALW) *via* a VIDI grant. M.T.A.A. was supported by NWO-ALW through the 'Molecule to Cell' programme.

References

1. J. M. Christie and W. R. Briggs, Blue-light sensing and signaling by the phototropins, , in: *Handbook of Photosensory Receptors*, W.R. Briggs and J.L. Spudich (eds), Wiley-VCH Verlag GmbH & Co., Weinheim, 2005, 277–304.
2. R. B. Celaya and E. Liscum, Phototropins and associated signaling: providing the power of movement in higher plants, *Photochem. Photobiol.*, 2005, **81**, 73–80.
3. A. Batschauer, Plant cryptochromes: their genes, biochemistry, and physiological roles, , in: *Handbook of Photosensory Receptors*, W.R. Briggs and J.L. Spudich (eds), Wiley-VCH Verlag GmbH & Co., Weinheim, 2005, 211–236.

4. M. Watanabe and M. Iseki, Discovery, characterization, and prospect of photo-activated adenylyl cyclase (PAC), a novel blue-light receptor flavoprotein from *Euglena gracilis*, , in: *Handbook of Photosensory Receptors*, W.R. Briggs and J.L. Spudich (eds), Wiley-VCH Verlag GmbH & Co., 2005, 447–460.
5. A. R. Cashmore, Cryptochromes: enabling plants and animals to determine circadian time, *Cell*, 2003, **114**, 537–543.
6. S. Masuda and C. E. Bauer, The antirepressor AppA uses the novel flavin-binding BLUF domain as a blue-light-absorbing photoreceptor to control photosystem synthesis, , in: *Handbook of Photosensory Receptors*, W.R. Briggs and J.L. Spudich (eds), Wiley-VCH Verlag GmbH & Co., Weinheim, 2005, 433–447.
7. S. Braatsch and G. Klug, Blue light perception in bacteria, *Photosynth. Res.*, 2004, **79**, 45–57.
8. M. Gomelsky and G. Klug, BLUF: a novel FAD-binding domain involved in sensory transduction in microorganisms, *Trends Biochem. Sci.*, 2002, **27**, 497–500.
9. J. C. Dunlap and J. J. Loros, Neurospora photoreceptors, , in: *Handbook of Photosensory Receptors*, W.R. Briggs and J.L. Spudich (eds), Wiley-VCH Verlag GmbH & Co., Weinheim, 2005, 371–389.
10. B.L. Taylor, I.B. Zhulin, PAS domains: internal sensors of oxygen, redox potential, and light, *Microbiol. Mol. Biol. Rev.*, 1999, **63**, 479–506.
11. W. R. Briggs and E. Huala, Blue-light photoreceptors in higher plants, *Annu. Rev. Cell Dev. Biol.*, 1999, **15**, 33–62.
12. W. Gong, B. Hao, S. S. Mansy, G. Gonzalez, M. A. Gilles-Gonzalez and M. K. Chan, Structure of a biological oxygen sensor: a new mechanism for heme-driven signal transduction, *Proc. Natl. Acad. Sci. USA*, 1998, **95**, 15177–15182.
13. H. Miyatake, M. Mukai, S.-Y. Park, S. Adachi, K. N. Tamura, H. K. Nakamura, T. Tsuchiya, T. Iizuka and Y. Shiro, *J. Mol. Biol.*, 2000, **301**, 415–431.
14. J. H. Morais Cabral, A. Lee, S. L. Cohen, B. T. Chait, M. Li and R. Mackinnon, Crystal structure and functional analysis of the HERG potassium channel *N*-terminus: a eukaryotic PAS domain, *Cell*, 1998, **95**, 649–655.
15. G. E. Borgstahl, D. R. Williams and E. D. Getzoff, 1.4 A structure of photoactive yellow protein, a cytosolic photoreceptor: unusual fold, active site, and chromophore, *Biochemistry*, 1995, **34**, 6278–6287.
16. P. Dux, G. Rubinstenn, G. W. Vuister, R. Boelens, F. A. Mulder, K. Hard, W. D. Hoff, A. R. Kroon, W. Crielaard, K. J. Hellingwerf and R. Kaptein, Solution structure and backbone dynamics of the photoactive yellow protein, *Biochemistry*, 1998, **37**, 12689–12699.
17. S. Crosson and K. Moffat, Structure of a flavin-binding plant photoreceptor domain: insights into light-mediated signal transduction, *Proc. Natl. Acad. Sci. USA*, 2001, **98**, 2995–3000.
18. S. M. Harper, L. C. Neil and K. H. Gardner, Structural basis of a phototropin light switch, *Science*, 2003, **301**, 1541–1544.
19. R. Fedorov, I. Schlichting, E. Hartmann, T. Domratcheva, M. Fuhrmann and P. Hegemann, Crystal structures and molecular mechanism of a light-induced signaling switch: the Phot-LOV1 domain from *Chlamydomonas reinhardtii*, *Biophys. J.*, 2003, **84**, 2474–2482.
20. C. A. Amezcua, S. M. Harper, J. Rutter and K. H. Gardner, Structure and interactions of PAS kinase *N*-terminal PAS domain: model for intramolecular kinase regulation, *Structure*, 2002, **10**, 1349–1361.

21. J. M. Christie, P. Reymond, G. K. Powell, P. Bernasconi, A. A. Raibekas, E. Liscum and W. R. Briggs, *Arabidopsis* NPH1: a flavoprotein with the properties of a photoreceptor for phototropism, *Science*, 1998, **282**, 1698–1701.
22. J. M. Christie, M. Salomon, K. Nozue, M. Wada and W. R. Briggs, LOV (light, oxygen, or voltage) domains of the blue-light photoreceptor phototropin (nph1): binding sites for the chromophore flavin mononucleotide, *Proc. Natl. Acad. Sci. USA*, 1999, **96**, 8779–8783.
23. T. Kagawa, T. Sakai, N. Suetsugu, K. Oikawa, S. Ishiguro, T. Kato, S. Tabata, K. Okada and M. Wada, *Arabidopsis* NPL1: a phototropin homolog controlling the chloroplast high-light avoidance response, *Science*, 2001, **291**, 2138–2141.
24. T. Kinoshita, M. Doi, N. Suetsugu, T. Kagawa, M. Wada and K. Shimazaki, phot1 and phot2 mediate blue light regulation of stomatal opening, *Nature*, 2001, **414**, 656–660.
25. T. Imaizumi, H. G. Tran, T. E. Swartz, W. R. Briggs and S. A. Kay, FKF1 is essential for photoperiodic-specific light signalling in *Arabidopsis*, *Nature*, 2003, **426**, 302–306.
26. K. Y. Huang and C. F. Beck, Phototropin is the blue-light receptor that controls multiple steps in the sexual life cycle of the green alga *Chlamydomonas reinhardtii*, *Proc. Natl. Acad. Sci. USA*, 2003, **100**, 6269–6274.
27. P. Ballario, C. Talora, D. Galli, H. Linden and G. Macino, Roles in dimerization and blue light photoresponse of the PAS and LOV domains of Neurospora crassa white collar proteins, *Mol. Microbiol.*, 1998, **29**, 719–729.
28. R. Ambra, B. Grimaldi, S. Zamboni, P. Filetici, G. Macino and P. Ballario, Photomorphogenesis in the hypogeous fungus Tuber borchii: isolation and characterization of Tbwc-1, the homologue of the blue-light photoreceptor of Neurospora crassa, *Fungal Genet. Biol.*, 2004, **41**, 688–697.
29. Q. Y. He, P. Cheng, Y. H. Yang, L. X. Wang, K. H. Gardner and Y. Liu, White collar-1, a DNA binding transcription factor and a light sensor, *Science*, 2002, **297**, 840–843.
30. A. Idnurm and J. Heitman, Light controls growth and development *via* a conserved pathway in the fungal kingdom, *PLOS Biol.*, 2005, **3**, 615–626.
31. A. Losi, E. Polverini, B. Quest and W. Gartner, First evidence for phototropin-related blue-light receptors in prokaryotes, *Biophys. J.*, 2002, **82**, 2627–2634.
32. U. Krauss, A. Losi, W. Gartner, K. E. Jaeger and T. Eggert, Initial characterization of a blue-light sensing, phototropin-related protein from *Pseudomonas putida*: a paradigm for an extended LOV construct, *Phys. Chem. Chem. Phys.*, 2005, **7**, 2804–2811.
33. A. Losi, The bacterial counterparts of plant phototropins, *Photochem. Photobiol. Sci.*, 2004, **3**, 566–574.
34. S. Crosson and K. Moffat, Photoexcited structure of a plant photoreceptor domain reveals a light-driven molecular switch, *Plant Cell*, 2002, **14**, 1067–1075.
35. S. Crosson, S. Rajagopal and K. Moffat, The LOV domain family: photoresponsive signaling modules coupled to diverse output domains, *Biochemistry*, 2003, **42**, 2–10.
36. H. Durr, M. Salomon and W. Rudiger, Chromophore exchange in the LOV2 domain of the plant photoreceptor phototropin1 from oat, *Biochemistry*, 2005, **44**, 3050–3055.
37. S. Crosson, LOV-domain structure, dynamics, and diversity, , in: *Handbook of Photosensory Receptors*, W.R. Briggs and J.L. Spudich (eds), Wiley-VCH Verlag GmbH & Co., Weinheim, 2005, 323–336.

38. T. E. Swartz, S. B. Corchnoy, J. M. Christie, J. W. Lewis, I. Szundi, W. R. Briggs and R. A. Bogomolni, The photocycle of a flavin-binding domain of the blue light photoreceptor phototropin, *J. Biol. Chem.*, 2001, **276**, 36493–36500.
39. C. Neiss and P. Saalfrank, *Ab Initio* quantum chemical investigation of the first steps of the photocycle of phototropin: a model study, *Photochem. Photobiol.*, 2003, **77**, 101–109.
40. C. Neiss, P. Saalfrank, M. Parac and S. Grimme, Quantum chemical calculation of excited states of flavin-related molecules, *J. Phys. Chem. A*, 2003, **107**, 140–147.
41. M. Salomon, J. M. Christie, E. Knieb, U. Lempert and W. R. Briggs, Photochemical and mutational analysis of the FMN-binding domains of the plant blue light receptor, phototropin, *Biochemistry*, 2000, **39**, 9401–9410.
42. W. Holzer, A. Penzkofer, M. Fuhrmann and P. Hegemann, Spectroscopic characterization of flavin mononucleotide bound to the LOV1 domain of Phot1 from *Chlamydomonas reinhardtii*, *Photochem. Photobiol.*, 2002, **75**, 479–487.
43. T. Kottke, J. Heberle, D. Hehn, B. Dick and P. Hegemann, Phot-LOV1: photocycle of a blue-light receptor domain from the green alga *Chlamydomonas reinhardtii*, *Biophys. J.*, 2003, **84**, 1192–1201.
44. T. A. Schuttrigkeit, C. K. Kompa, M. Salomon, W. Rudiger and M. E. Michel-Beyerle, Primary photophysics of the FMN binding LOV2 domain of the plant blue light receptor phototropin of *Avena sativa*, *Chem. Phys.*, 2003, **294**, 501–508.
45. J. T. M. Kennis, S. Crosson, M. Gauden, I. H. M. van Stokkum, K. Moffat and R. van Grondelle, Primary reactions of the LOV2 domain of phototropin, a plant blue-light photoreceptor, *Biochemistry*, 2003, **42**, 3385–3392.
46. D. Nozaki, T. Iwata, T. Ishikawa, T. Todo, S. Tokutomi and H. Kandori, Role of Gln1029 in the photoactivation processes of the LOV2 domain in *Adiantum* phytochrome3, *Biochemistry*, 2004, **43**, 8373–8379.
47. S. M. Miller, V. Massey, D. Ballou, C. H. Williams, M. D. Distefano, M. J. Moore and C. T. Walsh, Use of a site-directed triple mutant to trap intermediates – demonstration that the flavin-C(4a)-thiol adduct and reduced flavin are kinetically competent intermediates in mercuric ion reductase, *Biochemistry*, 1990, **29**, 2831–2841.
48. M. Salomon, W. Eisenreich, H. Durr, E. Schleicher, E. Knieb, V. Massey, W. Rudiger, F. Muller, A. Bacher and G. Richter, An optomechanical transducer in the blue light receptor phototropin from *Avena sativa*, *Proc. Natl. Acad. Sci. USA*, 2001, **98**, 12357–12361.
49. T. E. Swartz, P. J. Wenzel, S. B. Corchnoy, W. R. Briggs and R. A. Bogomolni, Vibration spectroscopy reveals light-induced chromophore and protein structural changes in the LOV2 domain of the plant blue-light receptor phototropin 1, *Biochemistry*, 2002, **41**, 7183–7189.
50. T. Iwata, S. Tokutomi and H. Kandori, Photoreaction of the cysteine S–H group in the LOV2 domain of adiantum phytochrome3, *J. Am. Chem. Soc.*, 2002, **124**, 11840–11841.
51. K. Ataka, P. Hegemann and J. Heberle, Vibrational spectroscopy of an algal Phot-LOV1 domain probes the molecular changes associated with blue-light reception, *Biophys. J.*, 2003, **84**, 466–474.
52. T. Bednarz, A. Losi, W. Gartner, P. Hegemann and J. Heberle, Functional variations among LOV domains as revealed by FT-IR difference spectroscopy, *Photochem. Photobiol. Sci.*, 2004, **3**, 575–579.
53. E. Schleicher, R. M. Kowalczyk, C. W. M. Kay, P. Hegemann, A. Bacher, M. Fischer, R. Bittl, G. Richter and S. Weber, On the reaction mechanism of

adduct formation in LOV domains of the plant blue-light receptor phototropin, *J. Am. Chem. Soc.*, 2004, **126**, 11067–11076.

54. T. Iwata, D. Nozaki, S. Tokutomi, T. Kagawa, M. Wada and H. Kandori, Light-induced structural changes in the LOV2 domain of *Adiantum* phytochrome3 studied by low-temperature FTIR and UV–visible spectroscopy, *Biochemistry*, 2003, **42**, 8183–8191.
55. M. Gauden, S. Crosson, I. H. M. van Stokkum, R. van Grondelle, K. Moffat and J. T. M. Kennis, Low-temperature and time-resolved spectroscopic characterization of the LOV2 domain of *Avena sativa* phototropin 1, , in: *Femtosecond Laser Applications in Biology*, S. Avrillier and J. M. Tualle (eds), Proceedings of SPIE, SPIE Bellingham, 2004, 97–104.
56. M. Kasahara, T. E. Swartz, M. A. Olney, A. Onodera, N. Mochizuki, H. Fukuzawa, E. Asamizu, S. Tabata, H. Kanegae, M. Takano, J. M. Christie, A. Nagatani and W. R. Briggs, Photochemical properties of the flavin mononucleotide-binding domains of the phototropins from *Arabidopsis*, rice, and *Chlamydomonas reinhardtii*, *Plant Physiol.*, 2002, **129**, 762–773.
57. W. Holzer, A. Penzkofer, T. Susdorf, M. Alvarez, S. D. M. Islam and P. Hegemann, Absorption and emission spectroscopic characterisation of the LOV2-domain of phot from *Chlamydomonas reinhardtii* fused to a maltose binding protein, *Chem. Phys.*, 2004, **302**, 105–118.
58. A. Visser, A. vanHoek, N. V. Visser, Y. Lee and S. Ghisla, Time-resolved fluorescence study of the dissociation of FMN from the yellow fluorescence protein from *Vibrio fischeri*, *Photochem. Photobiol.*, 1997, **65**, 570–575.
59. P. A. W. Van den Berg and A. J. W. G. Visser, Tracking molecular dynamics of flavoproteins with time-resolved fluorescence spectroscopy, , in: *New Trends in Fluorescence Spectroscopy: Applications to Chemical and Life Sciences*, B. Valeur and J.-C. Brochon (eds), Springer, Berlin, 2001, 457–485.
60. M. A. van der Horst and K. J. Hellingwerf, Photoreceptor proteins, "star actors of modern times": a review of the functional dynamics in the structure of representative members of six different photoreceptor families, *Acc. Chem. Res.*, 2004, **37**, 13–20.
61. M. Sun, T. A. Moore and P. S. Song, Molecular luminescence studies of flavins. I. The excited states of flavins, *J. Am. Chem. Soc.*, 1972, **94**, 1730–1740.
62. M. T. A. Alexandre, L. J. G. W. van Wilderen, R. van Grondelle, K. J. Hellingwerf, M. L. Groot and J. T. M. Kennis, Primary reactions of the LOV2 domain of phototropin studied with ultrafast mid-infrared spectroscopy, , in: *Photosynthesis: Fundamental Aspects to Global Perspectives*, A. van der Est and D. Bruce (eds), The International Society of Photosynthesis, Lawrence, Kansas, 2005, 1035–1037.
63. C. B. Martin, M. L. Tsao, C. M. Hadad and M. S. Platz, The reaction of triplet flavin with indole. A study of the cascade of reactive intermediates using density functional theory and time resolved infrared spectroscopy, *J. Am. Chem. Soc.*, 2002, **124**, 7226–7234.
64. A. J. Hoff and J. Deisenhofer, Photophysics of photosynthesis. Structure and spectroscopy of reaction centers of purple bacteria, *Phys. Rep.*, 1997, **287**, 2–247.
65. A. Losi, T. Kottke and P. Hegemann, Recording of blue light-induced energy and volume changes within the wild-type and mutated Phot-LOV1 domain from *Chlamydomonas reinhardtii*, *Biophys. J.*, 2004, **86**, 1051–1060.
66. S. D. M. Islam, A. Penzkofer and P. Hegemann, Quantum yield of triplet formation of riboflavin in aqueous solution and of flavin mononucleotide bound

to the LOV1 domain of photl from *Chlamydomonas reinhardtii*, *Chem. Phys.*, 2003, **291**, 97–114.

67. S. B. Corchnoy, T. E. Swartz, J. W. Lewis, I. Szundi, W. R. Briggs and R. A. Bogomolni, Intramolecular proton transfers and structural changes during the photocycle of the LOV2 domain of phototropin 1, *J. Biol. Chem.*, 2003, **278**, 724–731.
68. H. M. Guo, T. Kottke, P. Hegemann and B. Dick, The phot LOV2 domain and its interaction with LOV1, *Biophys. J.*, 2005, **89**, 402–412.
69. A. Losi, B. Quest and W. Gartner, Listening to the blue: the time-resolved thermodynamics of the bacterial blue-light receptor YtvA and its isolated LOV domain, *Photochem. Photobiol. Sci.*, 2003, **2**, 759–766.
70. C. W. M. Kay, E. Schleicher, A. Kuppig, H. Hofner, W. Rudiger, M. Schleicher, M. Fischer, A. Bacher, S. Weber and G. Richter, Blue light perception in plants – detection and characterization of a light-induced neutral flavin radical in a C450A mutant of phototropin, *J. Biol. Chem.*, 2003, **278**, 10973–10982.
71. T. Iwata, D. Nozaki, S. Tokutomi and H. Kandori, Comparative investigation of the LOV1 and LOV2 domains in *Adiantum* phytochrome3, *Biochemistry*, 2005, **44**, 7427–7434.
72. W. Holzer, A. Penzkofer and P. Hegemann, Absorption and emission spectroscopic characterisation of the LOV2-His domain of phot from *Chlamydomonas reinhardtii*, *Chem. Phys.*, 2005, **308**, 79–91.
73. P. S. Song, On basicity of excited state of flavins, *Photochem. Photobiol.*, 1968, **7**, 311–313.
74. T. E. Swartz and R. A. Bogomolni, LOV-domain photochemistry, , in: *Handbook of Photosensory Receptors*, W.R. Briggs and J.L. Spudich (eds), Wiley-VCH Verlag GmbH & Co., Weinheim, 2005, 305–323.
75. M. Sakai and H. Takahashi, One-electron photoreduction of flavin mononucleotide: time-resolved resonance Raman and absorption study, *J. Mol. Struct.*, 1996, **379**, 9–18.
76. T. Kottke, B. Dick, R. Fedorov, I. Schlichting, R. Deutzmann and P. Hegemann, Irreversible photoreduction of flavin in a mutated Phot-LOV1 domain, *Biochemistry*, 2003, **42**, 9854–9862.
77. R. Bittl, C. W. M. Kay, S. Weber and P. Hegemann, Characterization of a flavin radical product in a C57M mutant of a LOV1 domain by electron paramagnetic resonance, *Biochemistry*, 2003, **42**, 8506–8512.
78. M. Dittrich, P. L. Freddolino and K. Schulten, When light falls in LOV: a quantum mechanical/molecular mechanical study of photoexcitation in Phot-LOV1 of *Chlamydomonas reinhardtii*, *J. Phys. Chem. B*, 2005, **109**, 13006–13013.
79. R. M. Kowalczyk, E. Schleicher, R. Bittl and S. Weber, The photoinduced triplet of flavins and its protonation states, *J. Am. Chem. Soc.*, 2004, **126**, 11393–11399.
80. Y. Sato, T. Iwata, S. Tokutomi and H. Kandori, Reactive cysteine is protonated in the triplet excited state of the LOV2 domain in *Adiantum* phytochrome3, *J. Am. Chem. Soc.*, 2005, **127**, 1088–1089.
81. B. J. Brown, J.-W. Hyun, S. Duvvuri, A. Karplus and V. Massey, The role of glutamine 114 in old yellow enzyme, *J. Biol. Chem.*, 2002, **277**, 2138–2145.
82. A. Losi, E. Ghiraldelli, S. Jansen and W. Gartner, Mutational effects on protein structural changes and interdomain interactions in the blue light sensing LOV protein YtvA, *Photochem. Photobiol.*, 2005, **81**, 1145–1152.
83. M. Nakasako, T. Iwata, D. Matsuoka and S. Tokutomi, Light-induced structural changes of LOV domain-containing polypeptides from *Arabidopsis* phototropin 1

and 2 studied by small-angle X-ray scattering, *Biochemistry*, 2004, **43**, 14881–14890.

84. J. M. Christie, T. E. Swartz, R. A. Bogomolni and W. R. Briggs, Phototropin LOV domains exhibit distinct roles in regulating photoreceptor function, *Plant J.*, 2002, **32**, 205–219.
85. M. Salomon, U. Lempert and W. Rudiger, Dimerization of the plant photoreceptor phototropin is probably mediated by the LOV1 domain, *FEBS Lett.*, 2004, **572**, 8–10.
86. S. M. Harper, J. M. Christie and K. H. Gardner, Disruption of the LOV-J alpha helix interaction activates phototropin kinase activity, *Biochemistry*, 2004, **43**, 16184–16192.
87. S. Masuda and C.E. Bauer, AppA is a blue light photoreceptor that antirepresses photosynthesis gene expression in *Rhodobacter sphaeroides*, *Cell*, 2002, **110**, 613–623.
88. M. Gomelsky and S. Kaplan, AppA, a redox regulator of photosystem formation in *Rhodobacter sphaeroides* 2.4.1, is a Flavoprotein. Identification of a novel FAD binding domain, *J. Biol. Chem.*, 1998, **273**, 35319–35325.
89. M. Gomelsky and S. Kaplan, AppA, a novel gene encoding a trans-acting factor involved in the regulation of photosynthesis gene-expression in *Rhodobacter sphaeroides* 2.4.1, *J. Bacteriol.*, 1995, **177**, 4609–4618.
90. M. Iseki, S. Matsunaga, A. Murakami, K. Ohno, K. Shiga, K. Yoshida, M. Sugai, T. Takahashi, T. Hori and M. Watanabe, A blue-light-activated adenylyl cyclase mediates photoavoidance in *Euglena gracilis*, *Nature*, 2002, **415**, 1047–1051.
91. S. Rajagopal, J. M. Key, E. B. Purcell, D. J. Boerema and K. Moffat, Purification and initial characterization of a putative blue light-regulated phosphodiesterase from *Escherichia coli*, *Photochem. Photobiol.*, 2004, **80**, 542–547.
92. K. Okajima, S. Yoshihara, Y. Fukushima, X. X. Geng, M. Katayama, S. Higashi, M. Watanabe, S. Sato, S. Tabata, Y. Shibata, S. Itoh and M. Ikeuchi, Biochemical and functional characterization of BLUF-type flavin-binding proteins of two species of cyanobacteria, *J. Biochem.*, 2005, **137**, 741–750.
93. A. Kita, K. Okajima, Y. Morimoto, M. Ikeuchi and K. Miki, Structure of a cyanobacterial BLUF protein, Tll0078, containing a novel FAD-binding blue light sensor domain, *J. Mol. Biol.*, 2005, **349**, 1–9.
94. A. Jung, T. Domratcheva, M. Tarutina, Q. Wu, W. Ko, R. Shoeman, M. Gomelsky, K.H. Gardner and I. Schlichting, Structure of a bacterial BLUF photoreceptor: insights into blue-light mediated signal transduction, *Proc. Natl. Acad. Sci. USA*, 2005, **102**, 12350–12355.
95. B. J. Kraft, S. Masuda, J. Kikuchi, V. Dragnea, G. Tollin, J. M. Zaleski and C. E. Bauer, Spectroscopic and mutational analysis of the blue-light photoreceptor AppA: a novel photocycle involving flavin stacking with an aromatic amino acid, *Biochemistry*, 2003, **42**, 6726–6734.
96. W. Laan, M. A. van der Horst, I. H. van Stokkum and K. J. Hellingwerf, Initial characterization of the primary photochemistry of AppA, a blue-light-using flavin adenine dinucleotide-domain containing transcriptional antirepressor protein from *Rhodobacter sphaeroides*: a key role for reversible intramolecular proton transfer from the flavin adenine dinucleotide chromophore to a conserved tyrosine?, *Photochem. Photobiol.*, 2003, **78**, 290–297.
97. S. Masuda, K. Hasegawa, A. Ishii and T. Ono, Light-induced structural changes in a putative blue-light receptor with a novel FAD binding fold sensor of blue-light using FAD (BLUF); Slr1694 of *Synechocystis* sp. PCC6803, *Biochemistry*, 2004, **43**, 5304–5313.

98. Y. Fukushima, K. Okajima, Y. Shibata, M. Ikeuchi and S. Itoh, Primary intermediate in the photocycle of a blue-light sensory blue-light FAD-protein, Tll0078, of *Thermosynechococcus* elongatus BP-1, *Biochemistry*, 2005, **44**, 5149–5158.
99. K. Hasegawa, S. Masuda and T. A. Ono, Spectroscopic analysis of the dark relaxation process of a photocycle in a sensor of blue light using FAD (BLUF) protein Slr1694 of the cyanobacterium *Synechocystis* sp PCC6803, *Plant Cell Physiol.*, 2005, **46**, 136–146.
100. W. Laan, M. Gauden, S. Yeremenko, R. van Grondelle, J.T.M. Kennis and K.J. Hellingwerf, On the mechanism of activation of the BLUF domain of AppA, *Biochemistry*, 2006, **45**, 51–60.
101. W. Laan, T. Bednarz, J. Heberle and K. J. Hellingwerf, Chromophore composition of a heterologously expressed BLUF-domain, *Photochem. Photobiol. Sci.*, 2004, **3**, 1011–1016.
102. S. Anderson, V. Dragnea, S. Masuda, J. Ybe, K. Moffat and C. Bauer, Structure of a novel photoreceptor, the BLUF domain of AppA from *Rhodobacter sphaeroides*, *Biochemistry*, 2006, **44**, 7998–8005.
103. K. Hasegawa, S. Masuda and T. Ono, Structural intermediate in the photocycle of a BLUF (sensor of blue light using FAD) protein Slr1694 in a cyanobacterium *Synechocystis* sp PCC6803, *Biochemistry*, 2004, **43**, 14979–14986.
104. S. Masuda, K. Hasegawa and T. Ono, Light-induced structural changes of apoprotein and chromophore in the sensor of blue light using FAD (BLUF) domain of AppA for a signaling state, *Biochemistry*, 2005, **44**, 1215–1224.
105. M. Unno, R. Sano, S. Masuda, T. A. Ono and S. Yamauchi, Light-induced structural changes in the active site of the BLUF domain in AppA by Raman spectroscopy, *J. Phys. Chem. B*, 2005, **109**, 12620–12626.
106. M. Gauden, S. Yeremenko, W. Laan, I. H. M. van Stokkum, J. A. Ihalainen, R. van Grondelle, K. J. Hellingwerf and J. T. M. Kennis, Photocycle of the flavin-binding photoreceptor AppA, a bacterial transcriptional antirepressor of photosynthesis genes, *Biochemistry*, 2005, **44**, 3653–3662.
107. P. Zirak, A. Penzkofer, T. Schiereis, P. Hegemann, A. Jung and I. Schlichting, Absorption and fluorescence spectroscopic characterization of BLUF domain of AppA from *Rhodobacter sphaeroides, Chem. Phys.*, 2005, **315**, 142–154.

Subject Index